Psychobiology of Stress

NATO ASI Series

Advanced Science Institutes Series

A Series presenting the results of activities sponsored by the NATO Science Committee, which aims at the dissemination of advanced scientific and technological knowledge, with a view to strengthening links between scientific communities.

The Series is published by an international board of publishers in conjunction with the NATO Scientific Affairs Division

A Life Sciences	Plenum Publishing Corporation
B Physics	London and New York
C Mathematical	Kluwer Academic Publishers
and Physical Sciences	Dordrecht, Boston and London
D Behavioural and Social Sciences	
E Applied Sciences	
F Computer and Systems Sciences	Springer-Verlag
G Ecological Sciences	Berlin, Heidelberg, New York, London,
H Cell Biology	Paris and Tokyo

Series D: Behavioural and Social Sciences - Vol. 54

Psychobiology of Stress

edited by

Stefan Puglisi-Allegra
Institute of Psychobiology and Psychopharmacology,
National Research Council of Italy (C.N.R.),
Rome, Italy

and

Alberto Oliverio
Department of Genetics and Molecular Biology,
University of Rome,
Rome, Italy

Kluwer Academic Publishers

Dordrecht / Boston / London

Published in cooperation with NATO Scientific Affairs Division

Proceedings of the NATO Advanced Research Workshop on
Psychobiology of Stress
Sorrento, Italy
August 28–September 2, 1988

Library of Congress Cataloging in Publication Data

Psychobiology of stress / [edited by] Stefano Puglisi-Allegra, Alberto
  Oliverio.
      p.   cm. -- (NATO ASI series. Series D, Behavioural and social
  sciences ; vol. 54)
    ISBN 0-7923-0682-1 (alk. paper)
    1. Stress (Physiology) 2. Stress (Psychology) 3. Psychobiology.
  4. Psychoneuroendocrinology.   I. Puglisi-Allegra, Stefano, 1946-
  II. Oliverio, Alberto. III. Series: NATO ASI series. Series D,
  Behavioural and social sciences ; no. 54.
    [DNLM: 1. Stress--physiopathology. 2. Stress, Psychological-
  -physiopathology.   QT 162.S8 P974]
  QP82.2.S8P78   1990
  612--dc20
  DNLM/DLC
  for Library of Congress   90-4142

ISBN 0-7923-0682-1

Published by Kluwer Academic Publishers, WITHDRAWN
P.O. Box 17, 3300 AA Dordrecht, The Netherlands.

Kluwer Academic Publishers incorporates the publishing programmes of
D. Reidel, Martinus Nijhoff, Dr W. Junk and MTP Press.

Sold and distributed in the U.S.A. and Canada
by Kluwer Academic Publishers,
101 Philip Drive, Norwell, MA 02061, U.S.A.

In all other countries, sold and distributed
by Kluwer Academic Publishers Group,
P.O. Box 322, 3300 AH Dordrecht, The Netherlands.

Printed on acid-free paper

Table of Contents

vi

PREFACE

From a historical point of view the first studies on the response of the organism to stressful situations in general, and on the psychobiology of stress in particular, are probably those of Cannon and de la Paz, the physiologists who showed in 1911 that the adrenal medulla and the sympathetic system are involved in emergency situations. Cannon noted that the venous blood of cats frightened by barking dogs contained adrenaline, a response of the organism which was prevented by adrenalectomy or by section of the splanchnic nerve innervating the adrenal medulla. Cannon suggested that the adrenal medulla was acting in concert with the sympathetic nervous system, so that both systems were activated during stress. The role of the sympathetic system in response to stressful events was later emphasized by the experiments carried out by Maickel et al. (1967) and by Mason (1968): these authors clearly showed that stressors activate the sympathetic system causing it to release adrenaline and noradrenaline.

This line of research may be contrasted with that carried out by Hans Selye, centered on the role of the adrenal cortex in the stress response. Selye's findings and theories originated the so-called hypothalamic – pituitary – adrenal cortex (HPA) model of stress: in short, during stress adrenocorticotropic hormone is released from cells of the anterior pituitary and elicits secretion of glucocorticoids from the adrenal cortex.

The adaptive value of sympathetic activation and corticosteroid secretion is evident in both Cannon's and Selye's conceptions. Adrenaline causes an activation of the glucose supply by accelerating the degradation of glycogen in the liver and by diverting the blood from the viscera and muscles; in addition to that, lipolytic processes are also enhanced. Also glucocorticoids stimulate glycogenolysis and lipolysis, increase plasma glucose, stimulate the conversion of aminoacids to glucose, thus increasing the metabolic supply in order to cope with the stressful situation.

Though Cannon's and Selye's findings indicated that the organism might cope with stressful events through a number of peripheral reactions, most stress theories did not take into sufficient account the role of the brain and the wide range of psychobiological events related to stress. As a matter of fact stress responses are not limited to a concerted reaction of sympathetic-adrenal or HPA systems: they also implicate many other neurochemical (monoaminergic) and neuroendocrinological (peptidergic) reactions at the brain level which result in a number of behavioral patterns. An integrated picture of central and peripheral modifications produced by stressors emerges today from the studies centered on brain peptidergic and aminergic neurons. For example, if we consider the HPA system we may see that the noradrenergic, dopaminergic, serotonergic and cholinergic inputs from the hypothalamus alter the release of corticotropin releasing factor (CRF), the peptide stimulating the release of ACTH from the pituitary.

If it is therefore meaningless to consider the HPA and sympathetic-adrenal systems as two separate aspects of stress reactions it is also evident that a number of peptides modulate stress responses at the central and peripheral level by acting on neural, endocrine and immune systems, thus defining a group of cells and reaction patterns whose function may be to integrate information through a psychoneuroendocrinoimmune network: within this network a psychic event results in different neuroendocrinoimmune modifications leading in turn to behavioural changes which are important manifestations of the organism's adjustments or maladjustments objective or subjective stressors. This book is mostly centered on the psychobiology of stress: though it does not cover the wide array of laboratory and clinical approaches to this expanding research field it presents a number of main research lines centered on models and theories of stress, on its neurochemical correlates, on different genetic and ontogenetic factors which modulate stress reactions in animals and man.

References

Cannon, W.B. and la Paz, D. 'Emotional stimulation of adrenal secretion', *American Journal of Physiology*, 28, 64-70 (1911)

Maickel, R.P., Matussek, N., Stern, D.N. and Brodie, B.B. 'The sypathetic nervous system as a homeostatic mechanism', *Journal of Pharmacology and Experimental Therapeutic*, 157: 103-110 (1967)

Mason, J.W. 'A review of psychoendocrine research on the sympathetic-adrenal medullary systema', *Psychosomatic Medicine*, 30: 631-653 (1968)

Selye, H. Stress in Health and Disease, Butterworths, Boston (1976)

Stefano Puglisi-Allegra
Alberto Oliverio

Acknowledgements

The authors wish to thank the NATO Science Committee.

STRESS: ETHOLOGICAL IMPLICATIONS

P.R. Wiepkema
Agricultural University
Marijkeweg 40, 6709 PG Wageningen
The Netherlands

Ethologists claim to study animal behaviour as biologists. Therefore they focus on qualitative (form) and quantitative (frequency, intensity, sequence) aspects of behaviour as it occurs under natural conditions. Central ethological questions are, among others, 1) the causation and adaptive value of specific behaviour patterns and 2) the way in which individual behaviour develops and results from the interplay between genetic disposition and varying environmental agents or conditions. Ethologists were, and some still are, reluctant to incorporate psychological (mental) factors in their analyses and models of animal behaviour. This reservedness has to do with the fear that such factors could reintroduce the anthropomorphic and often empty explanations of animal behaviour as happened in the first quarter of this century. Just this former attitude in the study of animal behaviour was one of the main reasons to look for another approach: ethology (cf. Tinbergen, 1951). However, with the arrival of cognitive ethology, in which animals are conceived as well – adapted information – processing unities (Roitblat, 1987), opinions and concepts appear to change essentially (Marler and Terrace, 1984).

It is not my intention to review ethology: not even in a nutshell. What I am going to do is to pick up some ethological topics that seem to be relevant for the present discussion on stress and at least helped me to understand the behaviour of (farm) animals kept under chronic stress conditions. The selection of the topics is certainly tentative, since until recently ethologists rarely related their findings and models to stress. Most, and even recent textbooks of ethology do not, or only superficially, mention the concept of stress. An interesting exception is the recent book of Huntingford and Turner (1987) on animal conflicts. Presumably the often quite artificial context of many studies on animal stress hindered ethologists in appreciating stress as a reference to quite significant and natural phenomena. Why could bridging the gap between ethology and stress research be profitable?

Goals and conflicts

The first ethological topic deals with behaviour systems (programs) and goals (needs). Detailed ethological analysis has revealed the existence of species-specific systems, each serving a specified goal (Toates, 1986). For instance, all members of a given species practice a similar age-dependent feeding program, a preening program, a reproductive program, etc. Each program appears to be a non-random composition and sequence of basic units

1

S. Puglisi-Allegra and A. Oliverio (eds.), Psychobiology of Stress, 1–13.
© 1990 *Kluwer Academic Publishers. Printed in the Netherlands.*

(behavioural elements) like sniffing, gnawing, scratching, chasing, etc. These programs are species-specific and clearly adapted to the ecological niche of that species. Most of these programs are characterized by two successive phases: 1) a flexible introductory phase (appetitive behaviour) with large inter- and intraindividual differences and 2) a consummatory phase of which the final elements are very uniform in all species members (for instance, grasping food, threatening conspecific, resting during sleep, etc.). Although programs serve different goals, they may share behavioural elements. Especially elements of the introductory phase appear to be multipurpose.

This state of affairs not only implies the existence of a substantial number of separate, although interwoven, goals (needs or interests) per individual, but also that the fulfillment or realization of these goals rests on separate and only partially overlapping behaviour programs. Such a design makes behavioural conflicts likely if two or more goals have to be realized simultaneously and the two (or more) programs involved are incompatible. For instance, it is difficult if not impossible to perform simultaneously feeding and resting or preening and copulating or attacking and escaping. This type of conflict has been investigated extensively by ethologists (cf. Hinde, 1970); it is worthwhile to mention some results, since they may represent ethological counterparts of stress.

Therefore the second ethological topic implies conflict behaviour, which occurs when two conflicting tendencies are active and mutually exclusive with respect to their behavioural expressions. For instance, in social interactions, like fighting and courting, ambivalent behaviour may occur, when the tendencies to approach and to avoid are about equally strong. This conflict behaviour is often characterized by the simultaneous expression of the two conflicting tendencies, which may result in a circling movement and typical body postures (Kruyt, 1964). Under similar conditions a second and most intriguing type of conflict behaviour may occur. For instance, Rowell (1961) showed in a beautiful experiment that chaffinches, approaching a well-known feeding site but deterred by the presence of a predator model at that same site, started to perform preening just there, where the tendencies to approach and to avoid the feeding site were equally high. These short-lasting and out – of – context behaviours, which have been called displacement (Tinbergen, 1951) or interruptive behaviours are often performed in an exaggerated way.

A most intriguing point is that a given conflict often is associated with a specific displacement. For instance, Kruyt (1964) found that in fighting cocks (experiencing episodes of conflicting tendencies to approach and to avoid each other) the future winner performs a lot of groundpecking during the fight, while the future loser may preen. Although several mechanisms have been suggested to explain the causal mechanism underlying displacement behaviour (Hinde, 1970), until now there is no adequate answer to the question why specific conflicts have been associated with specific displacement behaviour (but see Andrew (1956) about the role of the autonomic nervous system during a conflict). Later on we will see that also during chronic stress such specific relationships appear to exist.

A third category of conflict behaviour may occur when, for instance, the organism is ready to attack a given object or conspecific, but at the last moment seems to change its mind and attacks a nearby substitute: redirected behaviour (Tinbergen, 1951). Displacement and redirected behaviour may also occur, when the performance of a given program is thwarted and/or does not lead to the expected goal-situation.

In ethology the story of conflicts has been described and explained in terms of conflicting tendencies; drives or motivations, giving rise to sometimes new and ritualized behaviour, which in the course of evolution gained great communicative value (Baerends, 1975).

However, since the increased appreciation of cognitive processes in animal behaviour it is promising to propose a somewhat different look at what also may happen during these behavioural conflicts. If cognition is characterized by collecting, storing, evaluating and implementing information by the organismen, then it will be clear that such a processing of information is optimal only if the organism deals with unequivocal information. Only under such conditions can clear-cut decisions be made. In this respect, conflict situations appear to represent a state of reduced reliability of information about relevant environmental changes or events. From this point of view it is not too difficult to consider conflict states as described in ethology as temporary phases of uncertainty, during which the animal is less sure about how to deal with the actual situation. In brief, conflict behaviours as mentioned above could quite well be considered as reflecting temporary states of reduced certainty in the organism. In this context it is interesting to refer to Feekes (1972) work on conflicting behaviour in fighting cocks. On the basis of her data she concluded that groundpecking of the winning cock could have a reassuring significance for its performer.

Cognition

It is clarifying to see the process of adaptation of living organisms to a specific ecological niche as a gradual gain of information in the course of evolution and during individual life. Natural selection produced genetic information about what is more or less constant and equal for all conspecifics; this has been crystallized in a species-specific morphology and behaviour. Although this adaptation may be described as a key-lock relationship, there is something extremely peculiar in this relationship. Within a species there is much individual variation – the keys are not constant – , and the same holds for many environmental characteristics. Since the separate keys do not fit equally well the available locks, we may expect all sorts of frictions between individuals and variable aspects of their niche. These frictions play on the individual level, and in Vertebrates they may be basic to stress.

On the individual level learning is a second source of information gain, by which adaptation to more detailed and variable, but predictable, parts of the environment become possible. Per individual this enhances the capability to distinguish information from error.

Vertebrates, the type of living organisms relevant for the present discussion, gain individual information in early life (pre- and postnatal) and when adult. It is relevant to distinguish these age-related types of information gain. The former takes place while the brain is still differentiating and information gained in this period may deeply influence neural maturation; this may lead to irreversible bias of the brain and resulting behavioural capabilities and strategies (Blass, 1986).

In contrast, adult learning has a much more reversible and flexible character: things can be learned and unlearned. Although early-life experiences significantly determine the individual's adaptability in later life, I will not discuss this point now (cf. Wiepkema, 1987).

The main point for the present discussion appears to be the capability of individual Vertebrates to inform themselves about spatial and temporal (= causal) relationships in their environment. The first type of information gain has been investigated in many, and quite diverse, studies on exploratory behaviour (cf. Archer and Birke, 1983). The indisputable conclusion from this research is that Vertebrates when placed in a new environment give priority to learning spatial relationships and by doing so at least develop a sort of cognitive map. As a result experienced animals quite precisely know their way and where interesting places are located, even when such places are used rarely. Ongoing exploratory behaviour keeps the animal informed about possible and recent spatial changes.

The second type of learning is the classical one in which animals show a perfect capability to associate successive events, when such events are causally related and significant for the individuals involved. These successive events may be an operant (E_1) and a relevant reward (E_2), or an external event (E_1) preceding an interesting environmental change (E_2). Of course, higher order combinations of this basic pattern are possible and exist. Animals can only learn a relationship between E_1 and E_2 if:

$$P_{E_2/E_1} \neq P_{E_2/no\text{-}E_1} \text{ (Dickinson, 1980)}$$

Both types of associating successive events have been described and analysed under the headings of operant learning and classical conditioning respectively.

The significant point in the foregoing is that operant learning enhances the individuals capability to control (and by this also to predict) environmental changes, whereas classical conditioning enhances the capability to predict those environmental changes that cannot be controlled. The latter changes include, for instance, the arrival of a predator, change in weather conditions, behaviour of the dominant in a group from the viewpoint of the subordinate etc. In present-day farms, where feeding has been automatised, animals can well predict the arrival of food, but as a rule cannot control the same event. It could well be that for some individuals of a given species predicting environmental (including social) changes is a more successful coping strategy than to control such changes, while for other individuals the reverse holds true. Such a differentiation may be present in the so-called passive and active copers respectively, as described by Bohus et al. (1987).

Finally, the occurrence of adequate anticipatory actions (like enzymatic or hormonal changes, specific movements or postures) after, for instance, perceiving a signal, strongly suggests the existence of expectancies that prepare future behaviour.

Two aspects of animal learning seem to contribute to the severeness of stressful situations. The first has to do with species-specific constraints in learning. Although animals are able to associate successive events readily, they cannot combine any E_1-E_2 sequence. For instance, in golden hamsters Shettleworth (1975) demonstrated that not every element of their behavioural repertoire could be used equally well as an operant to obtain food. In a similar way, not every preceding event can be interpreted as a signal for a successive and relevant environmental change (Domjan and Wilson, 1972). Obviously, in the course of evolution and, maybe, during maturation of the brain, the organism has become predisposed or becomes predisposed for certain E_1-E_2 associations to the exclusion of many others.

These species-specific constraints in associative capabilities may also be reflected in the phenomenon of autoshaping, when the organism transforms an initial and neutral operant (E_1) gradually to one that is coloured by the type of reward (E_2) it brings about. For instance, pressing a bar to obtain water may change gradually into pressing and licking this bar anticipating the uptake of water. As we will see later, this type of predisposition may significantly contribute to the performance of a specific category of disturbed behaviours in modern husbandry systems and zoos: the so called injurious behaviours.

The second aspect refers to the fact that biological processes always change over time. As a rule in the first phase of learning predictability and/or controllability (= P/C) is low; later on P/C may reach its maximum value of 1. It is not without significance that the early phase of learning has been associated with conflict behaviours and/or emotional expressions (cf. Simonov, 1986) and neuro-endocrine activities, that differ from those in the final phase.

While for good reasons the first phase could be called an emotional one, the final phase may obtain a very routine-like character. In fact each behaviour program or parts of it may obtain this routine character when P/C has become high and of a proven reliability. When discussing learning events we should never mix up both phases, since inevitably this will lead to great confusion and contradictory statements.

Taking together the foregoing remarks, it seems plausible to consider Vertebrates as characterized by a high capability to gain, to store and to apply species-specific information in an individual manner. This cognitive capability implies expectancies about (nearby) future events and is only limited by species-specific predispositions.

Conflicts and stress

Since 1980 I have been involved in the biological aspects of so-called animal welfare questions in present-day husbandry systems. These systems appear to overtax many farm animals as indicated by the frequent occurrence of abnormal behaviour (Wiepkema et al. 1983), of reproductive problems (cf. Barnett et al., 1987) and of relatively high morbidity and mortality (cf. Smidt, 1983). All these negative features are economically compensated for by low labour costs and the large number of animals kept per farm. What is going on in our modern stalls?

About 7 years ago I became aware of present-day ideas on stress and especially a number of chapters in the book "Coping and Health" (edited by Levine and Ursin, 1980) were eye-openers. Most striking was the idea that stress could be conceived as a state of the organism resulting from a significant decrease in P/C of relevant environmental changes. This low P/C appeared to me identical to the low certainty level of animals during a conflict as mentioned before. Moreover, the whole P/C concept so nicely fits present day thoughts on learning, that the use of this concept as a core idea in questions on animal welfare appears to me a most promising one.

I started with the following definitions of stress 1) acute stress: state of the organism when suddenly P/C of relevant environmental changes has been reduced.

2) chronic stress: state of the organism when during a long period P/C of relevant environmental changes is extremely low.

Two comments have to be added. First, environmental changes include all changes in the external and internal environment (events), and the performance of specific behaviours (operants). Second, the description of chronic stress does not imply a continuous state of stress. Presumably chronic stress is a discontinuous process depending on when (for instance, time of the day) the environmental changes involved normally occur.

Basically stress appears to reflect a cognitive defect characterized by losing grip on relevant environmental changes temporally or permanently .

I will not deal with acute stress and its temporary character in any detail. My only remark is that because of its strong similarity with ethological conflicts, this type of stress will also be characterized by the occurrence of conflict behaviour like redirected, displacement and agonistic behaviour. I want to focus on behavioural, and some physiological, changes associated with chronic stress. Good examples hereof are available from farm animals kept in modern husbandry systems.

For reasons of economy, technological possibilities and of development in automation, many farms have undergone an enormous scale enlargement; they may house hundreds (cattle) to thousands (pigs) or tens of thousands (poultry) of animals per farm. The highly uniform environment per individual has been reduced in size and structure drastically, while taking care of these animals is mainly a technological affair (cf. Pond, 1983).

My first example stems from sows kept for reproduction in the so-called intensive husbandry systems. Most of these sows, when pregnant, are housed in separate pens during a period of 3-4 months. As a rule, the same sows have been tethered by means of a chain fixed to the floor and, for instance, a neckcollar. By this treatment the sows are very restricted in their control of physical and social aspects of their environment; without doubt a chronic stress period of about 3-4 months has to be assumed. In the course of some weeks most of these sows start to develop stereotypies, which are characterized by their relatively constant form, repeated performance and apparent uselessness. Moreover, quality and quantity of these stereotypies are typical of individual sows.

Cronin (1985) made a detailed analysis of the development of these stereotypies in 9 sows from the first day of tethering on. Although each individual had its own time course, they all went through 3 comparable stages preceding the performance of true stereotypies. Immediately after tethering, most animals tried to break out: first stage with an average duration of 45 minutes. The second stage (average duration 1 day) was characterized by rest and/or exhaustion. In the following period (average duration 14 days) the sows again started to break out, but now in a much less intensive way. Moreover, each sow followed an individual way and first signs of repeating behaviour sequences were observed. In the fourth stage a restricted number of behavioural elements appeared to be selected and real stereotypies were performed for many hours a day. Performance of the same stereotypies could be maintained during many successive months. These stereotypies only occurred during day-time and in the hours between the two daily feeding periods. These stereotypies did not occur during the night hours indicating the discontinuous character of the chronic stress state involved. The form of these stereotypies (c. Cronin and Wiepkema, 1984) is different from those stereotyped behaviours performed just before daily feeding time (Rushen, 1985).

Since these stereotypies appeared to be very sensitive to naloxone (Cronin et al., 1987), it was concluded that in some way the performance of stereotypies had been associated with beta-endorphines (fig. 1).

It is quite well possible that performing stereotypies is reinforcing, because of an associated beta-endorphine release. If so, stereotypies might be less useless than thought hitherto.

It is interesting that Cronin (1985) also found some evidence that older-established stereotypies (being in use during a long period of time by the same individual) were less sensitive to naloxone then more recent ones. This may imply that gradually stereotypies loose their original support (of beta-endorphines), and become part of the routine repertoire in a way similar to what Panksepp et al. (1980) suggested with respect to the development of normal social behaviour. It is tempting to assume that by performing stereotypies for many hours a day, the pregnant sows changed the relationship with their actual environment drastically. In fact the animal may have shaped a new environment, in which many of the original disturbing factors have been negated and a new balance has been established.

Recently we found that only on the day of tethering could large increases in plasma level of ACTH, beta-endorphine and of cortisol be observed. After the first day of tethering these parameters were normal again or even subnormal. Moreover, there were large individual differences in behaviour and endocrine changes. Some animals reacted weakly, others quite

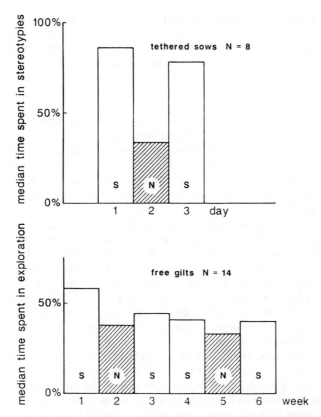

Fig. 1. Performance of stereotypies by tethered sows (upper graph) or of exploratory behaviour in freely moving young sows during a period of about 2 hours following an injection of 5 ml saline (S) or of 5 ml naloxone solution (N) (dose 1 mg/kg pig). Upper graph: injections were given once a day on three successive days; lower graph: once a week one injection was given during 6 successive weeks. The effect of naloxone was always significant and by far the strongest for stereotypies. Unpublished data of H. van Praag and K. van Reenen.

intensively. It is possible that the two types of coping, as described by Bohus and his colleagues (this volume), also exist in pigs.

Although the foregoing data support the idea that performing stereotypies may have great coping value under conditions of chronic stress, one comment has to be added. The stereotypies performed by the organism are not arbitrary: they appear to be related to the actual conflict. For instance, Keiper (1969) showed that in caged canaries one stereotypy (spot-picking) was related with the way the animals were fed, whereas route-tracing (another stereotypy) was caused by the (small) size of the cage.

A similar phenomenon may be present in the development of stereotypies of separately housed (crated) veal calves. Such calves are fed twice a day with a milk substitute during a period of about 20 weeks; then the calves are slaughtered. During the first 10 weeks of their life these calves develop a stereotypy – biting/gnawing specific parts of their crate – which declines in the second half of their life. During this latter period the same animals may develop a second stereotypy, called tongue-playing (Wiepkema et al., 1987).

The different time-course and quality of both stereotypies suggest the involvement of two successive conflict periods in the developing calf. It is plausible that these young mammals go through successive developmental stages, during which they pass critical periods needed for specific interactions with their environment and determining normal maturation. Our crated calves could not move around, could not perform and develop social play and, finally, they could not take in roughage (grass, hay) needed for a normal development of rumination (behaviourally and physiologically).

With respect to the performance of tongue-playing we found something quite peculiar. This tongue-playing, presumably a ritualized form of grasping grass or hay (Sambraus, 1985), does not develop in each individual. The same holds for the presence of abomasal ulcers (in the pyloric region), when the calves are about 20 weeks old (and slaughtered). The intriguing point is that the presence of abomasal ulcers was predicted by a low performance level of tongue-playing. Abomasal ulcers show no relationship with individual performance level of gnawing/biting.

Assuming a chronic stress condition in these crated veal calves – and this seems very likely – the data suggest that in developing animals chronic stress may consist of a sequence of different and only partially overlapping periods of specific conflicts, each associated with its own and specific behavioural and/or physiological deviations. This reminds one of the specific relationships between the type of displacement behaviour (groundpecking or preening) found in fighting cocks as indicated before. In the calves the specific link between tongue-playing and abomasal ulcers may be that both can be conceived as aberrations in the complete food intake and food processing program (which in the case of the veal calves also has to deal with an overfeeding regime).

Besides these stereotypies, another type of disturbed behaviour may occur in animals that chronically experience a significant environmental shortage: injurious behaviours. This category includes all activities by which an organism may damage itself or conspecifics. In modern husbandry examples of injurious behaviour are not uncommon.

Most of these injurious behaviours occur in group-housed farm animals, that cannot or need not perform normal foraging behaviour. For instance, grower pigs, kept on a concrete floor and fed twice a day high quality and ready-for-use pellets, redirect their natural but now needless foraging behaviour, like rooting, on penates. This rooting may change into tailbiting by which the pigs may damage each other severely (Van Putten, 1969; Ruiterkamp, 1985; Schouten, 1986). A comparable example is known from laying hens kept on a wire floor on which they cannot forage in a natural way (scratching, ground-pecking). Such hens tend to start featherpecking each other; this leads to severe damage (Blokhuis and Arkes, 1984). A third example is found in grouphoused veal calves fed with milk that has to be drunk (not sucked!) from open buckets. Such calves often start to suck each other and by this may damage each other (De Wilt, 1987).

The common factor in all these examples seems to be a long-term superfluousness of species-specific operants normally used to obtain food. A quite comparable phenomenon has been described by Breland and Breland (1961) in their sensational article on the "misbehaviour" of organisms. Concerning the remarks made earlier on constraints in learning and autoshaping, it is plausible that the farm animals just mentioned farm animals experience a chronic problem (stress), since their natural operants appear to fail with respect to the control of food intake. Why should animals under such long-term conditions, during which specific natural operants are really superfluous, redirect this behaviour in such a risky and damaging way? We urgently need to understand much better the biological significance of constraints in learning – the existence of specific predispositions – much better in order to explain the coping significance of this category of disturbed behaviour.

In order not to simplify too much, I have to add that in farm animals other factors also contribute to the occurrence of injurious behaviours. Climatic conditions and differences in natural disposition of the individuals involved are also responsible. In sum, however, it appears justified to state that phases of chronic stress can be recognized by the occurrence of two types of disturbed behaviours: 1) stereotypies and 2) injurious behaviours. There seems to be a specific relationship between the type of conflict involved and the sort of disturbed behaviour performed.

Social support

If these behavioural and physiological aberrations are mainly due to a reduced P/C of environmental changes, one could expect that social support of conspecifics may significantly influence quantity and quality of these aberrations. In the case of domestic animals this social support may also stem from their human caretakers. Strong evidence in favour of the beneficial effect of social support amoung conspecifics stems from the research of von Holst (1986). He demonstrates that in tree shrews (a primitive mammalian) some combinations of adults (pairs) are highly fatal, while other combinations fit within some seconds after the first encounter, and have strong stress reducing effects on both members (amicable behaviour, reduction of blood pressure). The class of "calming down relationship" has been known for long in motheroffspring relationships (Mason, 1978; Mendoza et al., 1980; Kaplan and Cubiccioti, 1980). Presumably this type of relationship will be of utmost relevance in stabilizing social organizations of adult vertebrates.

The role of social support stemming from human caretakers has convincingly been demonstrated by Nerem et al. (1980). They reported a significant reduction of atherosclerosis in rabbits handled in a consistent and friendly way, as compared to rabbits that were dealt with according to normal laboratory practice. In animal husbandry, Hemsworth and coworkers (cf. Hemsworth and Barnett, 1987) showed that handling pigs in a friendly and consistent way significantly enhanced growth rate and reproductive performance, two parameters that in modern farm practice are relatively low as a result of chronic stress conditions. The data of Seabrook (1984) tell a similar story. Farmers that actively and consistently interact with their cows, realize a higher milk yield per cow per year than farmers that do not. All these data strongly suggest that good social contacts may imply a strong stress reducing factor, which in animal husbandry results in better (re)production per animal.

Recently Kastelein and I found that in sealions stereotypies, like circular swimming, are significantly reduced in periods when the animals receive much attention from their trainer and caretaker, as compared to periods when this attention is practically absent (Kastelein and Wiepkema, 1988). In fact, in our present investigations on the development of stereotypies in tethered sows, this human-animal relationship presents a serious, methodological problem. In order to integrate behavioural and physiological data derived from the same individual sow, we have implanted a permanent heart-catheter in our tethered sows. However, this treatment forced us to interact much more frequently and intensively with these operated sows than with the unoperated ones. It is our strong impression that these intensive contacts have much to do with the fact that it is just these operated (and tethered) sows that do not rapidly develop clear stereotypies.

The area of social relationships between conspecifics, or man and his animals, is a rather new and most significant one when we are interested in the question of how social organisms cope with stressful conditions. Conspecifics may support or disturb each other as a result of individual characteristics and mutual relationships that often are overlooked in our investigations.

Emotional expressions

During conflict situations vertebrates often perform behaviours that reflect activity changes of the autonomic nervous system; usually they are called emotional expressions. To this category of activities belong changes in blood circulation through the skin and other organs, changes in heart beat frequency and bloodpressure, piloerection, pupil diameter or saliva secretion, face expressions, ear- and tail movements, utterances like yelping, screaming, peeping, laughing and others (cf. Darwin, 1872).

Some of these activity changes may reflect anticipatory actions and/or may have great communicative value. In a number of cases these two aspects appear to be not the most dominating ones. Two examples may illustrate this latter remark.

As part of his investigations on metabolism of race and riding horses, Bruin (Res. and Advis. Institute for cattle, sheep and horse husbandry, Lelystad, the Netherlands) frequently excercised these horses on a treadmill. He found a perfect and expected relationship between heart rate and running speed of these horses. The interesting finding is that, in some horses, heart rate during running with a speed of say 7 m per sec. could be lowered from 160 to 150 beats per min by standing next to the horse, tapping him in a friendly manner on the neck and saying some reassuring words. Such a change cannot be explained in metabolic terms. This finding indicates that, in at least some horses, heart rate during fast running is not determined only by metabolic requirements. A small but significant part of total heart rate might be due to some uncertainty of the horse about its running situation. This emotional component of heart rate appears to be reduced by reassuring behaviour or social support given by the trainer.

A second example stems from the work of my colleague Schenk (Schenk et al., 1985) who investigated the conditions that determine the occurrence of "gakelen", a specific call of laying hens (fig. 2). He found that hens produce such a call when, for example, they learn by experience to expect food at a given time of the day and at a well-defined location, but then on one occasion they do not receive this food. Under these conditions "gakelen" is more intensive when the hens are food deprived (fig. 2). The same call is also produced when a laying hen is ready to enter and to use her laying nest, but finds it closed: under these conditions "gakelen" is very intensive. Finally , "gakelen" is performed when hens go to a place where they usually dustbath, say a tray with sand, but find the tray inaccessible. In this case "gakelen" is not very intensive, which corresponds with the finding that under ad lib. conditions hens do not dustbath each day.

These data can be summarized by stating that hens perform "gakelen", when they expect some goal situation, but experience an unforeseen change and/or a blocked goal situation. The call's intensity appears to reflect the strength of the expectation and/or the significance of the goal. I am inclined to describe the hen's situation as one of a reduced P/C (certainty) of relevant environmental changes. "Gakelen" may express some form of negative emotion (disappointment) associated with a reduced P/C. Although "gakelen" may have a

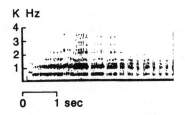

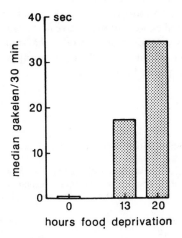

Fig. 2. The amount of gakelen in 11 laying hens at 08.00 AM after 0 (control), 13 or 20 hours of food deprivation. Upper part: sonagram of one gakel performance. In the control situation food was available, in the other two situations the hens expected food, but did not receive it. Unpublished data of M. Groot.

communicative value (until now we have no good evidence for this idea) it may basically be an emotional expression.

Both examples accommodate the view that, in vertebrates, emotional expressions are associated with changes in P/C. Events that decrease P/C presumably facilitate the performance of negative emotional expressions (and underlying emotions), whereas events that increase P/C and bring nearby a wished goal situation presumably are accompanied with positive emotional expressions (and underlying emotions). Negative and positive simply refer to the measurable tendency of the organism to avoid or to repeat the conditions involved.

The relationship between emotions and certainty-changes has been elaborated extensively by Simonov (1986). After discussing much Russian research on emotional processes (behavioural and physiological), Simonov states, that emotion is a reflection by the brain of vertebrates of any need and the probability to satisfy that need as evaluated on genetic and individual experience. This view nicely corresponds with the one put forward here, in which emotions are assumed to be closely connected with each change in the reliability of available information.

If all this is true, it is likely that emotional expressions are probable as long as the outcome of behaviour programs in use has not yet been established. This aspect is most strikingly demonstrated in learning processes, which often start with a quite emotional beginning phase. Later on, when the program has become a routine, emotional expressions may be virtually absent. Should for this reason the tendency to behave emotionally decrease with age?

From the foregoing, the significant point appears to me that in many cases vertebrates do not solve conflicts in an automatonlike manner, but in an emotional way. Why should this be so? Presumably emotions have great impact on the way individuals shape their flexible behaviour according to actual needs and individually assessed possibilities to satisfy those needs (cf. Livesey, 1986). Conflict situations (states of uncertainty) that cannot be solved automatically may therefore be associated with emotions.

It appears to be very promising to collect and analyze in a systematic way data concerning under which conditions emotional expressions (positive and negative ones) are most likely to occur. Such data are neccessary to develop a model that not only describes emotional behaviour as part of conflict states, but also explains it causally and functionally. This may elucidate the significance of emotions during stress.

References

Andrew, R.J., 1956. Some remarks on behaviour in conflict situations, with special reference to Emberiza sp. Br.*J. Animal. Behav.* 4, 41-45.

Archer, J. and L. Birke (Eds), 1983. Exploration in animals and humans. van Nostrand Reinhold, U.K.

Baerends, G.P., 1975. An evaluation of the conflict hypothesis as an explanatory principle for the evolution of displays. In: Function and evolution in behaviour (Eds. G. Baerends, C. Beer and A. Manning). Clarendon Press, Oxford, pp. 187-227.

Blass, E.M. (ed). 1986. Handbook of Neurobiology 8. Developmental psychobiology and developmental neurobiology. Plenum Press, New York.

Barnett, J.L., P.H. Hemsworth, C.G. Winfield and K.A. Fahy. 1987. The effects of pregnancy and parity number on behavioural and physiological responses related to the welfare status of individual and group-housed pigs. *Appl. Anim. Behav. Sci.* 17, 229-243.

Blokhuis, H.J. and J.G. Arkes, 1984. Some observations on the development of feather-pecking in poultry. *Appl. Anim. Behav. Sci.* 12, 145-157.

Bohus, B., R.F. Benus, D.S. Fokkema, J.M. Koolhaas, C. Nyakas, G.A. van Oortmerssen, A.J.A. Prins, A.J.H. de Ruiter, A.J.W. Scheurink and A.B. Steffens. 1987. Neuroendocrine states and behavioral and physiological stress responses. In: "Neuropeptides and brain function". (Eds. E.R. de Kloet, V.M. Wiegant and D. de Wied). Progr. Brain Research. 72, 57-70.

Breland, K. and M. Breland, 1961. The misbehavior of organisms. Am. Psychologist 16, 681 – 684.

Cronin, G.M. 1985. The development and significance of abnormal stereotyped beahviours in tethered sows. Ph.D. Thesis, Agric. Univer. Wageningen, the Netherlands.

Cronin, G.M. and P.R. Wiepkema, 1984. An analysis of stereotyped behaviour in tethered sows. *Ann. Rech. Vét.* 15, 263-270.

Cronin, G.M., P.R. Wiepkema and J.M. van Ree, 1987, Endorphins implicated in stereotypies of tethered sows. *Experientia,* 42, 198-199.

Darwin, C., 1872. The expression of emotions in man and animals. Murray, London.

Dickinson, A. 1980. Contemporary animal learning theory. Cambridge Univ. Press., Cambridge.

Domjan, M. and N.E. Wilson, 1972. Specificity in aversion learning in the rat. Psychon. *Science,* 26, 143-145.

Feekes, F., 1972., "Irrelevant" ground pecking in agonistic situations in Burmese Red Junglefowl (Gallus gallus spadiceus). *Behaviour,* 43, 186-326.

Hemsworth, P.H. and J.L. Barnett, 1987. Human-animal interactions. The Vet. Clin. North America, 3, 339-356.

Hinde, R.A., 1970. Animal behaviour. McGraw Hill, New York.

Holst, D. von, 1986. Vegetative and somatic components of tree shrews'behavior. *J. auton. nerv. system. Suppl.* 657-670.

Huntingford, F. and A. Turner, 1987. Animal conflict. Chapman and Hall, London.

Kaplan, J.N. and D.D. Cubicciotti 111, 1980. Early perceptual experience and the development of social preference in squirrel monkeys. In: "Maternal influences and early behavior". (Eds. R.W. Dell and W.P. Smotherman). MTP Press, Lancaster, pp. 253-270.

Kastelein, R.A. and P.R. Wiepkema, 1988. the significance of training for the behaviour of Stellar sealions (Eumetopias jubata) in human care. *Aquatic Mammals*, 14, 39-41.

Keiper, R., 1969, Causal factors of stereotypies in caged birds. *Anim. Behav.* 17, 114-119.

Kruyt, J.P. 1964. Ontogeny of social behaviour in Burmese Red Junglefowl. Behaviour, Suppl. XII, 1-201.

Livesey, P.J.,1986. Learning and emotion. A biological synthesis. Vol. 1. Evolutionary processes. Hillsdale: Lawrence Erlbaum.

Levine, S. and H. Ursin (eds). 1980, Coping and health, Plenum Press, New York.

Marler, P. and H.S. Terrace (eds), 1984. The biology of learning Dahlem Workshop Reports. Springer, Berlin.

Mason, W.A. 1978. Social experience and primate cognitive development. In: "The development of behaviour (Eds. G.M. Burghardt and M. Bekoff). Garland, New York, pp. 233-251.

Mendoza, S.P., C.L. Coe, W.P. Smotherman, J. Kaplan and S. Levine. 1980. Functional consequences of attachment: a comparison of two species. In: "Maternal influences and early behavior". (Eds. R.W. Dell and W.P. Smotherman). MTP Press, Lancaster, pp. 235-252.

Nerem, R.M., M.J. Levesque and J.F. Cornhill, 1980, Social environment as a factor in diet induced atherosclerosis. *Science*, 208, 1475-1476.

Panksepp, J., B.H. Herman, T. Vilberg, P. Bishop and F.G. De Eskinazi. 1980, Endogenous opioids and social behaviour. *Neuroscience and Biobehavioral Rev.* 4, 473-487.

Pond, W.G., 1983. Modern pork production. *Scientific American*, 248, 5, 78-87.

Putten, G. van, 1969. An investigation into tail-biting among fattening pigs. Br. Vet. J. 125, 511-517.

Roitblat, H.L., 1987, Introduction to comparative cognition. Freeman, New York.

Rowell, C.H.F., 1961. Displacement grooming in the chaffinch. *Anim. Behav.* 9, 38-63.

Ruiterkamp, W.A. 1985. The behaviour of grower pigs in relation to housing systems. *Neth. J. Agric. Sci.* 35, 67-70.

Rushen, J.P., 1985. Stereotypies, aggression and the feeding schedules of tethered sows. *Appl. Anim. Behav. Sci.* 14, 137-147.

Sambraus, H.H. 1985. Mouthbased anomalous syndromes. In: World Animal Science, 5, Ethology of farm animals (Ed A.F. Fraser). Elsevier, Amsterdam, pp. 391-422.

Schenk, P.M., F.M. Meijser und H.J.G.A.M. Limpens, 1985, Gakeln als Indikator für Frustration bei Legehennen. KTBL-Schrift, Darmstadt, 299, 65-81.

Schouten, W.G.P. 1986. Rearing conditions and behaviour in pigs. Ph.D. Thesis. Agricultural University Wageningen, the Netherlands.

Seabrook, M.f., 1984. The psychological interaction between the stockman and his animals and its influence on performance of pigs and dairy cows. *Vet. Rec.* 115, 84-87.

Shettleworth, S.J., 1975. Reinforcement and the organization of behaviour in golden hamsters: hunger, environment and food reinforcement. *J. Exp. Psychol: Animal Behaviour Processes.* 1, 56-87.

Simonov, G.P.V., 1986. The emotional brain. Plenum Press, New York.

Smidt, D. (ed), 1983. Indicators relevant to farm animal welfare. Martinus Nijhoff, Boston.

Tinbergen, N., 1952. The study of instinct. Clarendon Press, Oxford.

Toates, F., 1986, Motivational systems. Cambridge Univ. Press, Cambridge.

Wiepkema, P.R., D.M. Broom, I.J.H. Duncan and G. van Putten, 1983, Abnormal behaviour in farm animals, CEC-Report, pp. 1-16, Brussel.

Wiepkema, P.R., 1987, Developmental aspects of motivated behavior in domestic animals. *J. animal Science.* 65, 1220-1227.

Wiepkema, P.R., K.K. and Hellemond, P. Roessingh and H. Romberg, 1987, Behaviour and abomasal damage in individual veal calves. *Appl. Anim. Behav. Sci.* 18, 257-268.

Wilt, J.G. de, 1987, Development and prevention of preputial sucking in veal calves. *Neth. J. Agric. Sci.* 35, 78-80.

SHORT AND LONG TERM PHYSIOLOGICAL AND NEUROCHEMICAL ADAPTATIONS TO SOCIAL CONFLICT

Klaus A. Miczek[1], Michael L. Thompson[2] and Walter Tornatzky[1]
Departments of Phychology[1] and Pharmacology[2]
Tufts University
Medford and Boston, Massachusetts

Social conflict engenders behavioral and physiological phonemena that are relevant to the concept of stress. Attack and threat by an opponent leads to profound behavioral, physiological and neurochemical reactions, some adaptive, others maladaptive, by animals that exhibit defensive and submissive responses (Miczek et al., 1989 in press). Here, we focus on the short-and long-term submissive reactions at the level of behavior, endocrine and cardiovascular physiology as well as certain neurochemical systems in mice and rats that confront an aggressive opponent.

Submissive behavior in mice and rats

Short-term reactions

Ethological analyses have yielded detailed accounts of the species-specific, interactive patterns of agonistic behavior as well as their ontogeny and functions (e.g:, Banks 1962, Ginsburg and Allee 1942, Scott and Frederickson 1951, Brain 1981). When confronting each other in agonistic situations, rodents display characteristic acts, movements, and postures, and transmit specific visual, olfactory and auditory signals (e.g. Grant and Mackintosh 1963). Agonistic behavior patterns are organized in sequences of well-defined behavioral elements that follow each other with high probability. Moreover, like the bursts of action potentials in neurons, agonistic behavior is episodic with bursts of high-intensity interactions alternating with periods of behavioral quiescence (e.g. Miczek et al. 1989). The potential and capacity to engage in submissive as well as aggressive behavior is present in each individual. Different types of agonistic behavior such as defensive-submissive responses and aggressive behavior are not opposite poles on a continuum, but rather separate modes of behavior.

The salient elements of agonistic behavior in mice and rats share many similarities, although the social organizations of *Mus musculus* and *Rattus norvegicus* differ markedly. Mice that confront a resident male or a lactating female are threatened, attacked and pursued vigorously (e.g., Brain 1981, Noirot et al. 1975, Haney et al. in press). The immediate reaction is evasion and flight, and more rarely, retaliatory attacks. Usually, the attacked mouse assumes a defensive upright postures with extended forelimbs, oriented toward the threatening opponent. If flight is prevented, a severely attacked mouse displays defeat or

S. Puglisi-Allegra and A. Oliverio (eds.), Psychobiology of Stress, 15–30.
© 1990 *Kluwer Academic Publishers. Printed in the Netherlands.*

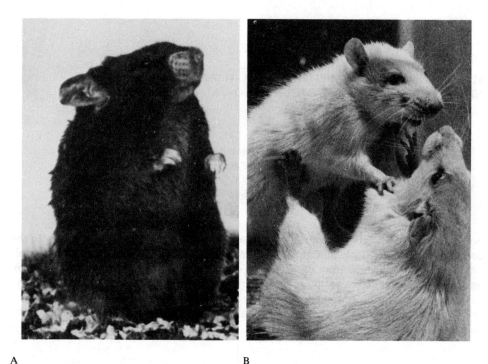

A   B

Figure 1. A. The defeated mouse in characteristic posture. (From Miczek, Thompson, and Shuster (1982), Science, 215:1520-1522). *B.* A submissive-supine posture by a rat in response to an aggressive posture by an opponent (Miczek 1973, unpublished).

submissive behavior in the form on an upright posture, the toward the approaching attacker (Fig. 1A, Ginsburg and Allee 1942, Miczek et al. 1982). These immediate behavioral reactions may terminate an ongoing burst of attacks or even prevent them altogether (e.g., Eibl-Eibesfeldt 1075, Blanchard and Blanchard 1988).

In rats, threats and attacks prompt rapid evasive movements and rarely provoke retaliatory attacks: most frequently, a 'defensive upright posture' is exhibited in reaction to sideways threats, a 'submissive supine posture' in reaction to a full aggressive posture, and a 'crouch' while being groomed in the neck region or while being investigated in the anogenital region (Fig. 1B). When bitten or being threatened, rats squeal; but more frequent and louder are the ultrasounds in the 20 to 30 kHz range after a rat has been attacked and while it is threatened to be attacked (Sales and Sewell 1972). This repertoire of acts and postures in reaction to attack may be associated with profound alterations in opioid-mediated pain responses (e.g., Miczek and Fier, unpublished observations).

Long-term consequences of submission

Aside from the behavioral responses (i.e. defensive upright postures, etc.) that are characteristic of submission in the short-run, direct conflict situation, other, more pervasive behavioral alterations that appear even in the absence of direct conflict are associated with long-term subordinate status. In some cases these may have adaptive significance, enabling the individual to coexist within the social group. For example, animals seem to learn when to act submissive. Mice that have previously experienced defeat surrender much more readily in subsequent aggressive encounters, especially if matched against a previously victorious opponent (Ginsburg and Allee, 1942; Scott and Marston, 1953; Nock and Leshner, 1976, Siegfried et al. 1982, 1984, Kulling et al. 1987). Subordinate individuals also modulate their activity when in areas that are marked with the urine of dominant males, thereby decreasing their likelihood of a confrontation with a dominant male (cotton rats: Summerlin and Wolfe 1971; mice: Boshop and Chevins, 1987). Marking behavior is also altered by subordinate males. Whereas dominant male mice distribute urine in small amounts throughout a territory, subordinates often deposit their urine in large pools at the edge of the test arena (Desjardins et al. 1973; Bishop and Chevins, 1987).

Access to resources such as food, prime habitat and reproductive females is also determined by social rank. Subordinate males are typically forced to live in less desirable habitats, which in the case of the hispid cotton rat means occupying patchy areas devoid of shrub cover to protect against predators (Spencer and Cameron 1983). Dewsbury (1984, 1988) has shown that subordinate male rats copulate less than dominants, and even those that do copulate typically ejaculate after fewer intromissions, thus reducing the likelihood of successful implantation. This reduction in reproductive opportunity does not necessarily depend upon the threatening presence of the dominant male. When given a choice under laboratory conditions, female rats (Carr et al., 1982), bank voles (hoffmeyer and Lund 1982) and hamsters (White, 1986) associate and copulate with dominant males rather than subordinates males.

Under certain circumstances, these adaptive mechanisms fail. Attempts to cope with social stress are unsuccessful and death may result, sometimes even in the absence of wounding. For example, Von Holst (1985) has decribed how subordinates from a socially intolerant species such as the treeshrew, when forced to live in the presence of the dominant male, show greatly disrupted patterns of feeding, grooming and sleeping. Even when physical contact is prevented, subordinates living under these conditions often die within two weeks. Death as a result to unsuccessful coping with a social stressor has also been reported in rats. Barnett and coworkers showed that 34% of the intruders introduced in a homecage of three other longhaired rats (Rattus villosissimus) died or collapsed under attack, though unwounded, and died when not removed (Barnett et al., 1975).

Physiological responses in submissive animals

Hormonal responses

Historically, dominant and subordinate animals were characterized physiologically by differential patterns of pituitary-adrenal and -gonadal activities. The initial strategy focused on adrenal and gonadal tissue weights in dominant and subordinate animals. For example, wild rats that intruded into the territory of a resident had high adrenal weights and usually died (Barnett 1958). Subordinate mice show decreased androgen-dependent preputial weights in comparison to dominants which may reduce their aggression-provoking characteristics (Bronson 1973, Bronson and Marsden 1973).

A second strategy attempts to correlate endocrine profiles with the social status of the individual. In large, long-established colonies of mice, subordinates respond primarily to social challenge by pituitary-adrenocortical activities, whereas dominants respond mainly with a sympathetic pattern (Ely and Henry 1978). After three weeks of dyadic confrontations, the adrenals of dominant mice contained elevated levels of norepinephrine, while subordinates had elevated epinephrine contents (Gamal-el-Din 1978, Hucklebridge et al. 1981). In response to an acute aggressive encounter, both winners and losers increase their corticosterone levels, but losers recover more slowly than winners to pre-fight levels (e.g., Bronson 1973, Schuurman 1980, Brain 1980). After prolonged exposures to agonistic encounters, low plasma testosterone are measured in the losers, whereas victorious animals show relatively elevated levels (Bronson et al. 1973, Sachser and Proeve 1984). The reduction in androgen production in subordinate animals may provide a means by which reproductive competitors are incapacitated ('psychological castration') and may stabilize group hierarchies (Huntingford and Turner 1987).

A third and classic research strategy involves the removal of endocrine glands followed by replacement therapy in order to influence the probability of submissive or, alternatively, aggressive behavior. While castration markedly reduces aggressive behavior, it has little effect on submissive behavior (Leshner and Moyer 1975, Barfield and Geyer 1972, Maruniak et al. 1977, Leshner and Politch 1979). In adrenalectomized mice, corticosterone or ACTH replacement therapy enhances the probability for displaying submissive behavior (Leshner 1980). Recently, we measured a rapid and large increase in ACTH in rats that have intruded for the first time into a resident's cage and displayed submissive responses in reaction to the resident's attacks and threats; ACTH remains elevate for one hour which the intruder is exposed to threats but not to physical attacks. The rise in plasma corticosterone follows the ACTH response in these indruder rats; both ACTH and coricosterone return to pre-defeat baseline levels within 3-6 hours. These data suggest an immediate activation of the pituitary-adrenal axis in response to social stress; even in the absence of physical stimulation in the form of attacks, the ACTH-corticosterone levels remain elevated while the animal is in the presence of a potential attacker.

In the long-term,display of submissive behavior is linked to prolonged elevated glucocorticoid levels. An increase in glucocorticoid activity may facilitate the display of submissive behavior which in turn reduces the likelihood of being subjected to future aggressive behavior. However, if glucocorticoids remain elevated, hippocampal corticosterone receptors are downeregulated (Sapolsky et al. 1984).

Cardiovascular responses

In addition to the hypothalamic-pituitary-adrenal axis, the cardiovascular system is highly responsive to short-term challenges as well as to prolonged social stresses (Candland et al. 1970, Adams et al. 1971, Cherkovich and Tatoyan 1973, Buettner and Plonait 1980, Stoehr 1988). Increases in systolic blood pressure (BP) are recorded in mice that live as dominants in large population cages (145 mm Hg in dominants vs. 125 mm Hg in solitary mice; Ely 1981). Even larger increases in BP are seen in rival animals to the dominants (165 mm Hg); the subordinate rivals develop cardiac hypertrophy, small infarcts and separation of myocardial fibers. After having lived for 2 months in large rat colonies, subordinates that have the most agonistic interactions with the dominants show the largest elevation in resting BP (Fokkema and Koolhaas 1985).

With the advent of telemetry technology, cardiovascular parameters may be measured during socially relevant interactions in the absence of potential artifacts of the measurement procedures. Telemetered heart rate (HR) of subordinate tree shrews reveal a loss of the typical circadian rhythmicity after two days of exposure to a resident; the HR of intruder animals remains elevated for prolonged periods of time (Stoehr 1988, von Holst 1985). By contrast, resident tree shrews retain the normal variation in HR during day and night even when exposed to intruders for several days. Coincidental to the prolonged elevation in HR, 'actively coping' intruder tree shrews show elevated adrenal tyrosine hydroxylase levels and elevated NE content in the adrenal medulla (Raab and Oswald 1980, von Holst 1985). This prolonged elevation in cardiovascular and adrenal activity in subordinate animals is presumably due to supramedullary influces that override brainstem homeostatic mechanisms (Folkow 1982) and may lead to cardiovascular pathalogies (Rona 1985).

Recently, it has become possible to monitor concurrently HR, core temperature as well as BP in rats that intrude into a resident's home cage and that are maintained for several months (Tornatzky and Miczek 1989, Meehan, Tornatzky and Miczek 1989). Brief confrontations between the intruder and resident lead to large elevations in HR and core temperature (Tc). These physiological parameters remain elevated for several days leading to a long- term disturbance in circadian rhythmicity. Normal periodicity of the cardiovascular and temperature regulatory systems returns after 7-9 days of rest. Figure 2 illustrates the short- and long- term changes in a rat that has been subjected to five consecutive confontations with a resident and displayed submissive or defeat behavior.

In subsequent experiments the duration of the physical contact during the agonistic interaction was limited by protecting the intruder with a wire mesh cage after he was forced by the resident into displaying a prolonged submissive supine posture and was emitting ultrasounds. The mean resting baseline values of these animals were 350 b/min for HR, 36.6 °C for Tc and 125/80 mm HG for diastolic/systolic BP. During the prolonged 1 hour threat period Tc($+1.8$ °C) as well as BP ($+20/+15$) remained elevated above resting baseline values. HR declined to a level which was similar to the elevated values during activity bouts in the animal's homecage ($+50$ b/min), suggesting that the threat of being attacked in rats is sufficient for inducing large and sustained elevations in physiological parameters reflecting stress responses.

Neurochemical correlates and consequences of social stress

Neurochemical alterations in response to social stress have received much less attention than peripheral hormonal and physiological correlates. Much of the data that does exist has generally focused on neurochemical correlates of aggressive behavior, rather than submission/subordination. Brain monoamines as well neuropeptides such as the endogenous opioids have been investigated.

Catecholamines

While dopamine (DA) release and utilization from terminals in the frontal cortex has been implicated in stress reactions (e.g. Thierry et al. 1976, Lavielle et al. 1978), its involvement in reactions to social stress has not been definitively established. In general, animals that are engaging in aggressive and defensive behavior show marked changes in whole brain or

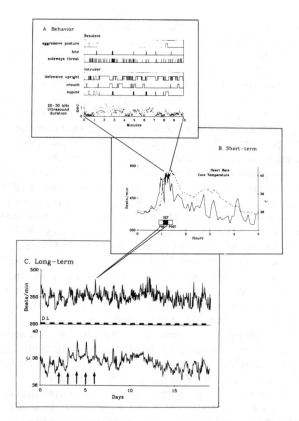

Figure 2. A. Elements of aggressive behavior by a resident rat threatening and attacking an intruder are identified as upward deflections from a timeline (in minutes). Three behavioral elements of the resident are selected, aggressive posture, bite, and sideways threat. The salient elements of the intruder behavior are shown similarly, defensive upright, crouch, and supine posture. Underneath, concurrently recorded ultrasounds in the 20-30 kHz range are portrayed as a time/duration scatter plot. *B.* Short-term effects of being attacked and threatened on the intruder's heart rate (solid line) and core temperature (dashed line) before, during and after defeat (in hours) as obtained via Mini-Mitter Dataquest III system. PRE = Intruder in the resident's home cage, separated by an opaque divider for 10 min; DEF = physical interactions leading to defeat, maximally for 10 min. or until intruder had been attacked 20 times; POST = 10 min of intruder activity in the resident's cage after the defeat with the resident removed; thereafter, the intruder was placed into his own homecage (dark period: 8:00-20:00); heart rate and temperature were sampled every 20 sec during resident-intruder encounters and in the home cage every 5 min. *C.* Long-term effects of 5 consecutive defeats on the magnitude and rhythmicity of heart rate and core temperature as mean values per hour (sampling rate every 5 min). Black horizontal bars denote dark periods; arrows point to the intruder's defeats. Resident-intruder encounters occurred within 1 hour after the onset of the dark period. (From Miczek et al., 1989, In: M.R. Brown et al. (Eds.) Neurobiology and Neuroendocrinology of Stress, New York: Marcel Dekker).

regional determinations of level, turnover, and reuptake of catecholamines. The duration and nature of the social stress, as well as the species under study, appear to be important determinants of the types of changes that are observed.

An early study found a time-dependent pattern of increases and decreases in cortical, hypothalamic and amygdaloid 5-HT and NE in C57BL/6J mice (Eleftheriou and Church

1968). Whereas 5-HT initially was decreased in amygdala and hypothalamus after 2 daily defeat experiences, subsequent determinations following 4,8 and 16 days of defeat found 5-HT to be increased. NE showed a reverse pattern, first an increase then a decrease. Subordinate group housed mice were found to show an increase in the number of reuptake sites for cortical NE as well as a decrease for affinity of reuptake for NE (Hendley et al. 1973). Similar results were found in mice receiving electroconvulsive shock or after attack experience (Welch et al. 1974). Display of attack behavior in Swiss-Webster mice has been shown to result in decreases in NE and NE turnover in olfactory bulb and substantia nigra (Tizabi et al. 1980). Increases in dopamine turnover have also been found in mice following an aggressive encounter using the resident- intruder paradigm (Noda et al. 1985). DOPAC/DA ratios were increased in n. accumbens but not striatum or amygdala of attacking mice after the initial aggressive encounter. In contrast, after 10 days of fighting experience DOPAC/DA ratios were no longer elevated, suggesting parallel neurochemical and behavioral adaptation to the repeated social confrontation. DA metabolites DOPAC and HVA were reported elevated in the olfactory tubercle, but not striatum, of rats exposed to 10 min. of attack. Even just being in the presence of a potential opponent was found to stimulate increases in mesolimbic DA metabolism M_{05} and Van Valkenburg 1979).

In summary, it seems that both aggressive as well as defensive or sumissive behavior can result in significant alterations in catecholamine functioning within the CNS. Species as well as the nature and temporal characteristics of the type of social stress employed are important to the type of neuchemical response that can be observed. Alterations specific to the display of either aggressive or defensive behavior that generalize across studies have yet to be revealed.

Serotonin

Much effort has been expended investigating the proposed link between decreased brain serotonin and heightened aggressive behavior, yet variable results persist (see reviews by Miczek and Donat 1989, Miczek et al. 1989). Even less established is whether or not subordinate or sbumissive behavior is influenced by or correlated with alterations in brain 5-HT.

Defensive reactions in rats exposed to electric footshock have received the most attention. Using this model, defensive behavior has been shown to be increased by manipulations that impair 5-HT neurotransmission, most consistently by chronic admistration of monamine reuptake blockers and MAO inhibitors. Neurotoxic or electrolytic lesions of 5-HT containing neurons or inhinibition of tyrosine hydroxylase also are effective (for reviews see Eichelman 1979, Pucilowski and Kostowski 1983, Prasad and Sheard 1983, Miczek and Donat 1989).

The role of serotonergic function in defensive or submissive behavior in response to social stress has received relatively little attention. Marked changes in raphe cell activity have been shown to be correlated with defensive reactions in treeshrews (Tupaia), with rate of single unit activity of dorsal raphe neurons increasing thee-four times following the display of immobile defense postures in the presence of a dominant conspecific (Walletscheck and Raab 1982). In mice, a doubling in 5-HT utilization, as indicated by increased 5-HIAA/5-HT ratios, was found in attackers, with smaller increases in the intruder, defeated mice, but only after the very first encounter (Noda et al. 1985). Following 10 days of confrontations, no changes in 5- HIAA or 5-HT could be detected.

Benzodiazepine-GABA receptor chloride ionophore complex

Benzodiazepine receptors in the CNS have also been implicated in the mediation of certain aspects of submissive behavior. For example, postnatal treatment with lorazepam resulted in increased submissiveness in male hooded rats intruding into another animal's territory (File and Tucker 1983). More recently, mice subjected to acute defeat showed increased in vivo labelling of benzodiazepine receptor sites by the BDZ antagonist 3H-Ro15-1788, with greatest increases occurring in the hypothalamus, cortex and hippocampus (Miller et al. 1987). These increases were short- term, with no increase observed 4hrs. after defeat, and were only observed in the intruder mice, not the attacking mice. The biological relevance of the stress, i.e. the nature of the stressful event and how the organism is equipped to cope with this event, and not a generalized pituitary-adrenal activation determine the physiological effects observed. Further analyses suggested the defeat-induced alterations in benzodiazepine binding to be due to increased receptor number and not due to changes in affinity. Adrenalectomized intruders were similar to controls, suggesting that adrenal steroids are necessary for the stress-induced increase in receptors. BDZ's may therefore modulate the stress response by indirectly altering circulating concentrations of adrenal steroids or by indirectly altering the number of available BDZ receptors.

Opioid peptides and receptors

The acute display of submissive behavior in mice may activate endogenous pain modulatory systems as evidenced by the display of analgesia (Miczek et al. 1982). Opioid mediation of this analgesic response within the CNS is suggested by several lines of evidence. First, this analgesia is blocked by pretreatment with opioid receptor antagonists such as naloxone or naltrexone (Miczek et al. 1985, 1986, Rodgers and Randall 1985). Second, analgesia is associated with changes in levels of brain opioid peptides a revealed directly by radioimmunoassay determination (Thompson et al. 1986, Kulling et al 1988) or indirectly by in vivo 3H-diprenorphine labelling of opioid receptors (Fig. 3; Thompson et al. 1985, 1986). Third, the analgesic response to morphine is potentiated in rats or mice during the very first exposure to attack and threat, even when the encounter duration or severity is insufficient to alter basal pain response, most likely via the release of endogenous opioids (Donat and Miczek 1988; Miczek and Fier, unpublished observations). This involvement of opioid peptides in submissive behavior has been confirmed in other strains of mice (Teskey et al. 1984; Rodgers and Randall 1985; Siegfried and Frischknecht 1984; Miczek and Thompson 1984) as well as rats (Rodgers et al. 1983).

In contrast to the activation of endogenous opioid systems and analgesia produced by acute exposure to social stress, repeated defeat is associated with the development of tolerance (Miczek et al. 1982). In subsequent encounters with social stress, the intruder mice display less and less analgesia, until eventually they no longer show significant elevations in tailflick latencies in response to defeat. This decline in responsiveness may be attributed to protracted alterations in the endogenous opioid system. For example, at the neurochemical level there is no longer any correlation between analgesia and decreased in in vivo 3H-diprenorphine binding. Instead, following seven days of defeat, binding is significantly enhanced in several brain regions in defeat tolerant mice relative to control mice, suggesting that in defeat-experienced or tolerant mice, acute defeat non longer triggers a large release of endogenous opiods (Fig. 3). Indeed, beta-endorphin levels in tolerant mice assessed

Brain dissection scheme:

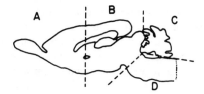

³H–Diprenorphine binding (in vivo)

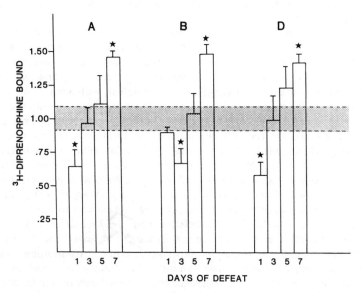

Figure 3. In vivo ³H-diprenorphine binding in B6AF₁/J mice after 1, 3, 5, or 7 days of social stress. Upper panel: Brains were rapidly removed after defeat and divided into four regions as shown (Glowinski and Iversen 1966). Lower panel: Binding in regions A, B, and D expressed as a ratio of binding determined in control mice that were not exposed to social stress. ³H-diprenorphine (4 mCi, 2 mg/kg, s.c.) was injected immediately after defeat, and mice were sacrificed 20 min after injection. Stereospecific in vivo labelling of receptors was determined using a rapid filtration method. Values are means ± SEM; N = at least 7 for each group). Star indicates a significant (P <0.05) difference from the naive controls.

immediately following defeat on Day 7 were similar to those measured in control mice, in contrast to the large increase observed following acute defeat in naive mice (Thompson et al. 1986).

Additionally, in vitro binding studies conducted on brains from mice subjected to daily defeat for 7 days suggest the reduction in analgesic response is associated with mu opiate receptor upregulation. Scatchard analyses utilizing the relatively selective mu receptor ligand 3H-dihydromorphine binding revealed a significant increase (30%) in the number of forebrain binding sites with no change in affinity of receptors for the ligand (Thompson et al. 1985, 1986).

A second manifestation that repeated exposure to defeat stress produces enduring changes in the endogenous opioid system was the finding that defeat tolerant mice show apparent development of a physical dependence-like syndrome similar to that seen after repeated exposure to morphine. Separate groups of mice were injected daily with morphine (5 mg/kg) or subjected to defeat for 8 days. Analgesic response to each treatment showed a similar progressive decline across days of treatment suggesting the development of tolerance. When injected with naloxone immediately after defeat on the 8th day, almost all defeat tolerant mice showed signs of opiate withdrawal such as jumping behavior that resembled that observed in the morphine tolerant mice, while such behavior is rarely observed in defeat naive mice. This finding, combined with the previously discussed data demostrating changes in in vivo and in vitro opioid receptor binding suggest that tolerance-like changes in analgesia with repeated defeat experience may be based on changes in sensitivity or regulation of opioid receptors.

Associated with the decline in analgesic response to defeat stress in the development of cross-tolerance to morphine, wich has been characterized in extensive detail in some of our more recent sudies. In our initial study we found significant tolerance to defeat and morphine cross-tolerance after 14 days of stress. In subsequent experiments, however, we have observed significant cross tolerance after only 3 days of defeat, with a maximum at 7 days. Two studies have attempted to assess the extent of shift in morphine dose response curve. In the first, mice were injected weekly with increasing doses of morphine following 5 daily defeats. Morphine response was found to be shifted 4-8 fold compared to defeat-naive mice, for at least 1 month following the 5 defeats (Miczek and Winslow 1987). Cross- tolerance was very long lasting, with decreased sensitivity to morphine still present 6-8 weeks after the last defeat experience. A second experiment utilized a cumulative dosing protocol to provide dose-effect curves within individuals at specific time points (e.g., Wenger 1980), and assessed responsiveness to both mu and kappa opiate receptor agonists in order to learn which opioid receptor subtypes may be modulated by the defeat experience. Five defeat experiences was sufficient to produce cross-tolerance in mice that lasted for at least 3 months (Miczek in press), and extended to both classes of receptors as determined by morphine (mu) and ethylketocyclazocine (kappa) responsiveness. An interesting finding from both experiments was that development of morphine cross-tolerance encompassed only the analgesic response to morphine but not the suppressant effects of morphine on operant behavior or morphine-induced hypothermia. Functionally selective tolerance limited to the analgesic response has also been demonstrated in a recent study involving rats trained in a drug discrimination paradigm (Miczek, unpublished observations). Rats were trained to discriminate a low morphine dose from saline with better than 95% accuracy, and then exposed to an attacking opponent during five brief encounters during which they displayed submissive behaviors. While all rats subsequently required 4-6 fold higher doses of morphine to induce analgesia, half of the rats actually became more sensitive to the 'cue' or discriminative stimulus properties of morphine, while no rat showed evidence of tolerance to the rate suppressive effects of morphine on operant responding.

Development of cross-tolerance to morphine may be a more sensitive indicator of endogenous opioid response to social stress than the acute display of analgesia. In our attempts to define the minimal conditions of social stress that lead enduring alterations in opioid systems, we found that even a single exposure to defeat could result in time-related reductions in morphine response as early as 3 days after the single defeat on Day 1, with maximum diminished responsiveness by Day 7. Development of dependence like syndrome as measured by naloxone precipitated withdrawal jumping followed a similar time course. A more recent sudy involving rats reveals that such prolonged alterations in morphine responsiveness can occur even under conditions of social stress that are insufficient to induce the display of an acute analgesic response. Rats were exposed as intruders to a brief 3-5 min. confrontation with a resident rat, with the endpoint being the display of submissive-supine or crouch postures and the emission of 20-30 KHz ultrasounds in reaction to attacks by the resident (Miczek and Fier, unpublished alterations). After this brief period of contact, the intruders remained in the presence of the resident for one hour, but behind the protective confines of a wire mesh cage so as to prevent further injurious contact. At no point during this procedure were changes in pain responsiveness in the intruder rat detectable. As mentioned above, during this period and for up to 4 hours after, significant potentiation of morphine analgesic responsiveness was observed. However, subsequent assessments of morphine responsiveness 1,7 or 30 days after the single display of submissive behavior indicated significant and long lasting tolerance development to the analgesic effects of the drug (Fig. 4).

Of the three neurochemical systems discussed here, the endogenous opioid system appears to be most intimately responsive to exposure to social stress. Analgesic response, whether resulting from opioid peptide activation or exogenous morphine challenge, provides a sensitive and easily measured indicator of both acute and long-term alterations in opioid systems in the CNS produced by social stress. At this point it is premature to speculate as to the critical importance of specific opioid peptides or receptor subtypes, or even the critical neural 'pool' of opioid peptide-containing neurons or receptor populations that is most involved in the long-term alterations in analgesic responsiveness produced by defeat. The variety of pain, aversion and coping responses that fall into the category of submissive behavior most undoubtedly involve differential activation of opioid peptides in limbic, diencephalic, and mesencephalic brain regions as well as spinal cord. More detailed examination of the types of pain or aversion responses affected in response to defeat will be required before the complex opioid changes in response to defeat can be further differentiated.

Conclusions:

Exposure to an aggressive or threatening opponent activates species-specific responses at the behavioral, physiological, and neurochemical level. In addition to the previous research on monaminergic mechanisms, our research has idenfied opioid peptides and their receptors,as well as the benzodiazepine-GABA- chloride ionophore complex, as two systems within the brain that are significantly involved in the acute response to a specific social stress. Furthermore, we have demonstrated the persistent alterations in sensivity of opioid receptors that occur in response to social stress that may parallel the long-term behavioral and physiological changes, both adaptive and maladaptive, that are usually associated with subordinate status. These imposed behavioral and physiological constraints, rather than the acute physical severity of the stressor, may be more important to the long-term survival of the individual. An understanding of the feedback relationship between specific social experiences and brain monoamines and neuropeptides, as well as insight into the reasons for

MORPHINE ANALGESIA IN RATS

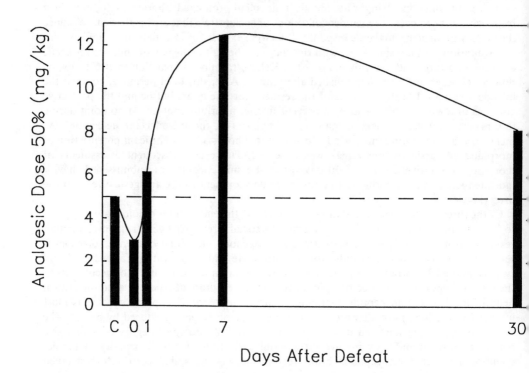

Figure 4. The dose of morphine that elevated tail flick latencies to half of the maximum possible level (AD50) in intruder rats that have displayed submissive behavior once during a 2-5 min encounter with an aggressive resident. Morphine doses were administered cumulatively over 1 hour, either during the confrontation with the resident (0) or 1, 7, or 30 days after the single confrontation. The dotted line indicated the AD50 for undefeated control animals (C). (Miczek and Fier, unpublished observations.)

the varying ability of individuals to cope with specific social stressors, could provide important new pathways for clinical intervention in treating stress pathologies.

Acknowledgements

Preparation of this chapter and our own experimental work was supported by USPHS research grants AA 05122 and DA 02632.

27

References

Adams DB., Baccelli G., Mancia G., Zanchetti A. (1971), 'Relation of cardiovascular changes in fighting to emotion and exercise'. *Journal of Physiology* 212:321-335.
Banks E.M. (1962), 'A time and motion study of prefighting behavior in mice'. *Journal of Genetic Psychology* 101:165-183.
Barfield R.J., Geyer L.A. (1972) 'Sexual behavior: Ultrasonic postejaculatory song of the male rat'. *Science* 176:1349-1350.
Barnett S.A. (1958) 'Physiological effects of 'social stress' in wild rats'. *I.Journal of Phychosomatic Research* 3:1-11.
Barnett S.A., Hocking W.E., Munro K.M.H., Walker K.Z. (1975) 'Socially induced renal pathology of captive wild rats'. *Aggressive Behavior* 1:123-133.
Bishop M.J., Chevins P.F. (1987) 'Urine odours and marking patterns in territorial laboratory mice (Mus musculus)'. *Behavioural Processes* 15:223-248.
Blanchard D.C., Blanchard R.J. (1988) 'Ethoexperimental approaches to the biology of aggression'. *Annual Review of Phychology* 39:43-68.
Brain P.F. (1980) 'Adaptive aspects of hormonal correlates of attack and defence in laboratory mice: A study in ethobiology'. In: McConnell P.S., Boer G.T., Romijn H.J., Van der Poll N.E. (eds), Progress in Brain Research: Adaptive Capabilities of the Nervous System. Elsevier North-Holland Biomedical, Amsterdam, pp. 391-414.
Brain P.F. (1981) 'Differentiating types of attack and defense in rodents'. In: Brain P.F., Benton D. (eds), Multidisciplinary Aproaches to Aggression Research. Elsevier, Amsterdam, pp. 53-78.
Bronson F.H.(1973) 'Establishment of social rank among grouped male mice: Relative effects on circulating FSH, LH, and corticosterone'. *Physiology and Behavior* 10:947-951.
Bronson F.H., Marsden H.M. (1973) 'The preputial gland as an indicator of social dominance in male mice'. *Behavioral Biology* 9:625-628.
Bronson F.H., Stetson M.H., Stiff M.E. (1973) 'Serum FSH and LH in male mice following aggressive and nonaggressive interaction'. *Physiology and Behavior* 10:369-372.
Buettner D., Plonait H. (1980) 'Langfristige Messungen der maximalen, mittleren und Ueberherzfrequen an Laborratten mittels implantierbarer telemetriesender'. *Zentralblatt der Veterinaer Medizin* 27:269-278.
Candland D.K., Bryan D.C., Nazar B.L., Kopf K.J., Sendor M. (1970) 'Squirrel monkey heart rate during formation of status orders'. *Journal of Comparative and Physiological Psychology* 70:417-423.
Carr W.J., Kimmel K.R., Anthony S.L., Schlocker D.E. (1982) 'Female rats prefere to mate with dominant rather than subordinate males'. *Bulletin of the Psychonomic Society* 20(2):89- 91.
Cherkovich G.M., Tatoyan S.K. (1973) 'Heart rate (radiotelemetrical registration) in macaques and baboons according to dominant-submissive rank in a group'. *Folia Primatologica* 20:265-273.
Desjardins S.C., Maruniak J.A., Bronson F.H. (1973) 'Social rank in house mice: Differentiation revealed by ultraviolet visualization of urinary marking patterns'. *Science* 182:939-941.
Dewsbury D.A. (1984) 'Aggression, copulation, and differential reproduction of deer mice (Peromyscus maniculatus) in a semi- natural enclosure'. *Behaviour* 91:1-23.
Dewsbury D.A. (1988) 'Kinship, familiarity, aggression, and dominance in deer mice (Peromyscus maniculatus) in seminatural eclosures'. *Journal of Comparative Psychology* 102:124-128.
Donat P., Miczek K.A. (1988) 'Time dependent hyper-and hyposensitivity to morphine analgesia after defeat in mice'. *Psychopharmacology* 96:S17 (Abstract).
Eibl-Eibesfeld I.B. (1975) 'Ausdrucksformen der Saeugetiere'. In: Kuekenthal (ed), Handbuch der Zoologie, Vol:8, pp.1-26.
Eichelman B. (1979) 'Role of biogenic amines in aggressive behavior'. In: Sandler M. (ed), Psychopharmacology of Aggression. Raven Press, New York, pp.61-93.
Eleftheriou B.E., Church R.L. (1968) 'Brain levels of serotonin and norepinephrine in mice after exposure to aggression and defeat'. *Physiology and Behavior* 3:977-980.
Ely D.L. (1981) 'Hypertension, social rank, and aortic arteriosclerosis in CBA/J mice'. *Physiology and Behavior* 26:655- 661.
Ely D.L., Henry J.P. (1978) 'Neuroendocrine response patterns in dominant and subordinate mice'. *Hormones and Behavior* 10:156-169.
File S.E., Tucker J.C. (1983) 'Lorazepam treatment in the neonatal rat alters submissive behavior in adulthood'.

28

Neurobehavioral Toxicology and Teratology 5:289-294.

Fokkema D.S., Koolhaas J.M. (1985) 'Acute and conditioned blood pressure changes in relation to social and psychosocial stimui in rats'. *Physiology and Behavior* 34:33-38.

Folkow B. (1982) 'Physiological aspects of primary hypertension'. *Physiological Reviews* 62:347-409.

Gamal-el-Din L.A. (1978) 'Some aspects of adrenomedullary function in relation to agonistic behaviour in the mouse (Mus musculus)'. Ph.D.dissertation, Polytechnic of Central London, London.

Ginsburg B., Allee W.C. (1942) 'Some effects of conditioning on social dominance and subordination in inbred strains of mice'. *Physiological Zoology* 15:485-506.

Glowinski J., Iversen L.L. (1966) 'Regional studies of catecholamines in the rat brain. I.The disposition of [3H] norepinepherine, [3H] dopamine, and [3H] DOPA in various regions of the brain'. *Journal of Neurochemistry* 13:655-669.

Grant E.C., MacKintosh J.H. (1963) 'A comparison of the social postures of some common laboratory rodents'. *Behavior* 21:246-259.

Haney M., DeBold J.F., Miczek K.A. (in press) 'Maternal aggression in mice and rats towards male ad female comspecifics'. Aggressive Behavior.

Hendley E.D., Moisset B., Welch B.L. (1973) 'Catecholamine uptake in cerebral cortex: Adaptive change induced by fighting'. *Science* 180:1050-1052.

Hoffmeyer I., Lund H. (1982) 'Responses of female bank voles (Clethrionomys glareolus) to dominant vs subordinate conspecific males and to urine odors from dominant vs subordinate males'. *Behavioral and Neural Biology* 36:178-188.

Hucklebridge F.H., Gamal-el-Din L., Brain P.F. (1981) 'Social status and the adrenal medulla in the house mouse (Mus musculus, L.)' *Behavioral and Neural Biology* 33:345-363.

Huntingford F.A., Turner A.K. (1987) 'Animal Conflict'. Chapman and Hall, London, New York.

Kulling P., Frischknecht H.R., Pasi A., Waser P.G., Siegfried B. (1987) 'Effects of repeated as compared to single aggressive confrontation on nociception and defense behavior in C57BL/6 and DBA/2 mice'. *Physiology and Behavior* 39:01-07.

Kulling P., Frischknecht H-R, Pasi A., Waser P.G., Siegfried B. (1988) 'Social conflict-induced changes in nociception and b- endorphin-like immunoreactivity in pituitary and discrete brain areas of C57BL/6 and DBA/2 mice.' *Brain Research* 450:237-246.

Lavielle S., Tassin J.P., Thierry A.M., Blanc G., Herve D., Barthelemy C., Glowinski J. (1978) 'Blockade by benzodiazepines of the selective high increase in dopamine turnover induced by stress in mesocortical dopaminergic neurons of the rat'. *Brain Research* 168:585-594.

Leshner A.I. (1980) 'The interaction of experience and neuroendocrine factors in determining behavioral adaptations to aggression'. In: McConnell P.S., Boer G.J., Romijn H.J., van der Poll N.E. (eds), Progress in Brain Research. Adaptive Capabilities of the Nervous System. Elsevier/North Holland Biomedical Press, Amsterdam, pp.427-438.

Leshner A.I., Moyer J.A. (1975) 'Androgens and agonistic behavior in mice: Relevance to aggression and irrelevance to avoidance-of- attack'. *Physiology and Behavior* 15:695-699.

Leshner A.I., Politch J.A. (1979) 'Hormonal control of submissiveness in mice: Irrelevance of the androgens and relevance of the pituitary-adrenal hormones'. *Physiology and Behavior* 22:531-534.

Maruniak J.A., Desjardins C., Bronson F.H. (1977) 'Dominant- subordinate relationships in castrated male mice bearing testosterone implants'. *American Journal of Physiology* 233:495- 499.

Meehan W.P., Tornatzky W., Miczek K.A. (1989) 'Blood pressure via telemetry during social stress in freely moving rats'. *Society of Neuroscience Abstracts* 15, in press.

Miczek K.A., Haney M., Tidey J., Vatne T., Weerts E., DeBold J.F. (1989) 'Temporal and sequential patterns of agonistic behavior: Effects of alcohol, anxiolytics and psychomotor stimulants'. *Psychopharmacology* 97:149-151.

Miczek K.A., Thompson M.L., Tornatzky W. (in press): 'Subordinate animals:behavioral and physiological adaptations and opioid tolerance'. In:Brown M.R., Rivier C., Koob G. et al. (eds.). Neurobiology and neuroendocrinology of stress, 1989, New York: Marcel Dekker).

Miczek K.A. (in press) 'Long-lasting functionally specific tolerance to m and k opioid agonists after brief social defeat: Analgesia, hypothermia and suppression of operant performance'. Psychopharmacology.

Miczek K.A., Donat P. (1989) 'Brain 5-HT systems and inhibition of aggressive behavior'. In: Archer T., Bevan P., Cools A. (eds), The Behavioural Pharmacology of 5-HT. Lawrence Erlbaum Associates, Hillsdale, NJ, pp. 117.

Miczek K.A., Mos J., Olivier B. (1989) 'Brain 5-HT and inhibition of aggressive behavior in animals: 5-HIAA and receptor subtypes'. *Psychopharmacology Bulletin* 25:(in press).

Miczek K.A., Thompson M.L. (1984) 'Analgesia resulting from defeat in a social confrontation: The role of endogenous opioids in brain'. In: Bandler R. (ed), Modulation of sensorimotor activity during altered behavioural states. Alan R. Liss, New York, pp, 431-456.

Miczek K.A., Thompson M.L., Shuster L. (1982) 'Opioid-like analgesia in defeated mice'. *Science* 215:1520-1522.

Miczek K.A., Thompson M.L., Shuster L. (1985) 'Naloxone injections into periaqueductal grey area ad arcuate nucleus block analgesia in defeated mice'. *Psychopharmacology* 87:39-42.

Miczek K.A., Thompson M.L., Shuster L. (1986) 'Analgesia following defeat in an aggressive encounter: Development of tolerance and changes in opioid receptors'. In: Kelly D.D. (ed), Annals of the New York Academy of Sciences, v. 467:Stress-induced Analgesia. New York Academy of Sciences, New York, pp. 14-29.

Miczek K.A., Winslow J.T. (1987) 'Analgesia and decrement in operant performance in socially defeated mice: Selective cross- tolerance to morphine and antagonism by naltrexone'. *Psychopharmacology* 92:444-451.

Miller L.G., Thompson M.L., Greenblatt D.J., Deutsch S.I., Shader R.I., Paul S.M. (1987) 'Rapid increase in brain benzodiazepine receptor binding following defeat stress in mice'. *Brain Research* 414:395-400.

Mos J., Van Valkenburg C.F.M. (1979) 'Specific effect on social stress and aggression on regional dopamine metabolism in rat brain'. *Neuroscience Letters* 15:325-327.

Nock B.L., Leshner A.I. (1976) 'Hormonal mediation of the effects of defeat on agonistic responding in mice'. *Physiology and Behavior* 17:111-119.

Noda K., Miczek K.A., Kream R. (1985) 'Regional monoamine activity, sensitivity to amphetamine and aggressive behavior in mice'. *Society for Neuroscience Abstracts* 11:549.

Noirot E., Goyens J., Buhot M-C. (1975) 'Aggressive behavior of pregnant mice towards males'. *Hormones and Behavior* 6:9-17.

Prasad V., Sheard M.H. (1983) 'Time course of chronic desipramine on shock-elicited fighting in rats'. *Agressologie* 24:15-17.

Pucilowski O., Kostowski W. (1983) 'Aggressive behaviour and the central serotonergic systems'. *Behavioural Brain Research* 9:33- 48.

Raab A., Oswald R. (1980) 'Coping with social conflict: Impact on the activity of tyrosine hydroxylase in the limbic system and in the adrenals'. *Physiology and Behavior* 24:387-394.

Rodgers R.J., Hendrie C., Waters A.J. (1983) 'Naloxone partially antagonizes post-encounter analgesia and enhances defensive responding in male rats exposed to attack from lactating conspecifics'. *Physiology and Behavior* 30:781-786.

Rodgers R.J., Randall J.I. (1985) 'Social conflict analgesia: Studies on naloxone antagonism and morphine cross-tolerance in male DBA/2 mice.' *Pharmacology Biochemistry and Behavior* 23:883- 887.

Rona G. (1985) 'Catecholamine cardiotoxicity'. Journal of Molecular and Cellular Cardiology 17:291-306.

Sachser N., Prove E. (1984) 'Short-term effects of residence on the testosterone responses to fighting in alpha male guinea pigs.' *Aggressive Behavior* 10:285-292.

Sales G.D., Sewell N.(1972) 'Ultrasound and aggressive behaviour in rats and other small mammals.' *Animal Behaviour* 20:88-100.

Sapolsky R.M., Krey L.C., McEwen B.S. (1984) 'Stress down- regulates corticosterone receptors in a site-specific manner in the brain.' *Endocrinology* 114:287-292.

Schuurman T. (1980) 'Hormonal correlates of agonistic behavior in adult male rats.' In: McConnel P.S., Boer G.J., Romijn H.J., Van de Poll N.E. (eds), Progress in Brain Research, vol.53:Adaptive Capabilities of the Nervous System. Elsevier Biomedical Press, Amsterdam, pp.415-420.

Scott J.P., Fredericson E. (1951) 'The causes of fighting in mice and rats.' *Physiological Zoology* 24:273-309.

Scott J.P., Marston M.V. (1953) 'Nonadaptive behavior resulting from a series of defeat in fighting mice.' *Journal of Abnormal and Social Phychology* 48:417-428.

Siegfried B., Frischknecht H-R, Waser P.G. (1984) 'Defeat, learned submissiveness, and analgesia in mice: Effect of genotype.' *Behavioral and Neural Biology* 42:91-97.

Siegfried B., Frischknecht H-R, Waser P.G. (1982) 'A new learning model for submissive behavior in mice: Effects of naloxone.' Aggressive Behavior 8:112-115.

Spencer S.R., Cameron G.N. (1983) 'Behavioral dominance and its relationship to habitat patch utilization by the hispid cotton rat (Sigmodon hispidus). 'Behavioral Ecology and Sociobiology 13:27-36.

Stoehr W. (1988) 'Longterm heartrate telemetry in small mammals: A comprehensive approach as a prerequisite for valid results.' *Physiology and Behavior* 43:567-576.

Summerlin C.T., Wolfe J.L. (1971) 'Social influences on exploratory behavior in the cotton rat.' *Communications in Behavioral Biology* 6:105-109.

Teskey G.C., Kavaliers M., Hirst M. (1984) 'Social conflict activates opioid analgesic and ingestive behaviors in male mice.' *Life Sciences* 35:303-315.

Thierry A.M., Tassin J.P., Blank G., Glowinsky G. (1976) 'Topographic and pharmacological study of the mesocortical dopaminergic system.' In:Wauquier A., Rolls E.T. (eds), Brain- stimulation Rewaard: North Holland, Amsterdam.

Thompson M.L:, Shuster L., Hartley J., Hoefler H., Miczek K.A. and Kream R. (1985) 'Changes in opioid receptor binding following acute and chronic defeat in mice.' Paper presented at ASPET Meeting, Fall. 1985

Thompson M.L., Brunner E., Hoefler H., Hartley J., Kumar M.S.A., Shuster L., Kream R. (1986) 'Changes in opioid receptor binding and levels of opioid peptides in the brain following acute and chronic defeat in mice.' *Society for Neuroscience Abstracts* 12:411.

Tizabi Y., Massari V.J., Jacobowitz D.M. (1980) 'Isolation induced aggression and catecholamine variations in discrete brain areas of the mouse.' *Brain Research Bulletin* 5:81-86.

Tornatzky W., Miczek K.A. (1989) 'Alpha- and beta-adrenergic reseptors and etlemetered autonomic response to socal stress.' *Society of Neuroscience Abstracts* 15, in press.

Von Holst D. (1985) 'Coping behaviour and stress physiology in male tree shrews.' In: Hoelldobler B., Lindauer M. (eds), Experimental Behavioral Ecology and Sociobiology. Gustav Fischer, Stuttgart, pp. 461-470.

Walletschek H., Raab A. (1982) 'Spontaneous activity of dorsal raphe neurons during defensive and offensive encounters in the tree-shrew.' *Physiology and Behavior* 28:697-705.

Welch B., Hendley E.D., Turek I. (1974) 'Norepinephrine uptake into cerebral cortical synaptosomes after one fight or electroconvulsive shock.' *Science* 183:220-221.

Wenger G.R. (1980) 'Cumulative dose-response curves in behavioral pharmacology.' *Pharmacology Biochemistry and Behavior* 13:647-651.

White P.J. (1986) 'Female discrimination of male dominance by urine odor cues in hamsters.' *Physiology and Behavior* 37:273-277.

SOCIAL BEHAVIOR OF THE HOUSE MOUSE: A POTENTIAL MODEL FOR PRECLINICAL STUDIES ON STRESS

Simona Cabib and Stefano Puglisi-Allegra
Istituto di Psicobiologia e Psicofarmacologia, C.N.R., via Reno 1, 00198 Roma, Italy

In preclinical psychopharmacology, attention has shifted from models based on superficial similarities between animal behavior and human behavioral pathologies to the mechanisms by which genetic and environmental factors interact to modulate the behavioral output of the organism. An organism can be viewed, in fact, as an open system constantly interacting with its environment so that changes in the environment necessarily bring about changes in the organism (Bertalanffy, 1968).

When the environment is modified, the organism detects discrepancies between observed and expected events and attempts to restore the expected situation. This phase is characterized by a particular picture of physiological changes commonly termed arousal. If attempts to restore the expected situation fail, given the characteristics of the environment, the organism undergoes a crisis we call stress. This crisis produces a number of significant alterations in the organism's functioning which lead either to a new equilibrium between the organism and the environment or to pathology. Thus the relationship between stress and psychic functions is now considered of paramount importance and a number of animal models have been developed to help preclinical research with regard to the pathogenic effects of stress.

Studies carried out since the early seventies, have shown that exposure to unavoidable electric shocks is a potent stressor producing physiological changes such as activation of the hypothalamic-pituitary-adrenal axis, changes in brain neurotransmitters and alterations in autonomic responses (Conner et al., 1971; Weiss et al, 1970). Ability to control the situation prevents physiological stress responses in humans and animals (Jackson et al., 1979). The effects of inescapable shock are drastically reduced if the animal is somehow allowed to act upon the environment, even if the amount of shock received is unchanged (Conner et al., 1971; Weiss et al, 1970). Rats shocked in pairs exhibit lower plasma levels of ACTH than animals individually subjected to shock (Conner et al., 1971). This effect is said to result from the coping response, i.e. the fighting induced by shock attenuates the intensity of the stress response in terms of ACTH release. Moreover, rats fighting in response to electric shock show fewer gastric lesions compared with animals receiving the same shocks alone. Animals which received shock together but do not engage in fighting behavior show no reduction of gastric lesions while gastric lesions were reduced in animals that were allowed to display aggressive behavior even though physical contact was prevented by a barrier (Weiss et al., 1976).

Thus stressful stimulation can markedly alter social behavior by increasing aggressive

S. Puglisi-Allegra and A. Oliverio (eds.), Psychobiology of Stress, 31–40.
© 1990 *Kluwer Academic Publishers. Printed in the Netherlands.*

responses which, in turn, prevent some of the pathogenic effects of these stimulations. Since, in such contexts, aggressive behavior can be considered a successful response to stress, a number of researches have pointed to the study of the factors involved in the expression of this behavior during inescapable shock experiences.

Pharmacological as well as neurochemical studies have shown that a number of neurotransmitter systems modulate aggressive behavior induced by inescapable shock (Valzelli, 1981). In our laboratory, it has been observed that an interaction between age-related and genetic factors is involved in determining the behavioral response of mice to this stressful experience (Puglisi-Allegra et al., 1981; Puglisi-Allegra, 1984; Puglisi-Allegra and Cabib, 1985). Moreover, in collaboration with the group of Paul Mandel, in Strasbourg we have also indicated an involvement of the central GABA system in the modulation of mice behavioral responses to this kind of stress (Puglisi-Allegra et al., 1981; Clement et al., 1987). These results indicate that social responses induced by stress depend on developmental and genetic factors, possibly related to central neurotransmitter functioning and suggest that the interaction between stress and social life may represent a source of models for preclinical research.

The Social Environment as stressor

Social behavior can be viewed as a characteristic of the organism-environment interaction leading to solutions but also precipitating stress.

Isolation may be used to induce stress-related alterations in laboratory animals. Although *Mus Musculus* is considered a territorial species where a few males occupy discrete areas (Brain, 1975), social stimuli such as those produced by sexual or aggressive encounters would not be absent in a natural setting. In laboratory situations, where animals commonly live in small groups, peer contact including aggressive interactions represent important sources of social stimulation (D'Amato, 1986). The experimental procedure known as 'social isolation' commonly used in the laboratory involves individually housing an animal in a cage for some weeks . Such long-term social isolation thus may represent a chronic stressful experience and produces a number of physiological and behavioral alterations including changes in hormonal and neurotransmitter functioning, hyperreactivity, decreased learning skills and marked aggressive or fear behavior in the presence of conspecifics (Brain, 1975; Frischknecht et al., 1985; Siegfried et al., 1981; Simler et al., 1982; Valzelli, 1981). There is evidence that alterations observed following long-term social isolation are not unrelated phenomena (Frischknecht et al., 1985), suggesting that there is a 'syndrome' induced by social isolation in the mouse (Valzelli, 1981).

Adaptation to social isolation involves a number of physiological alterations also related to central neurotransmission. Some of these alterations are similar to those observed following other chronic stressful experiences. We have shown that isolated mice exhibit higher pain thresholds than group-housed mice and that the former animals are tolerant to immobilization-induced analgesia, indicating altered functioning of the endogenous opioid system in these mice (Puglisi-Allegra and Oliverio, 1983). Moreover, a sensitization to the behavioral effects of the dopamine agonist apomorphine has been observed in isolated mice with respect to group-housed animals (Wilmot et al., 1984). This effect has been also observed in group-housed mice subjected to chronic immobilization and it is possibly related to altered sensitivity of dopamine receptors (Cabib et al., 1984; 1985; 1988b).

The permeating physiological alterations produced by this chronic stressful condition allow the organism to establish a new equilibrium with the surrounding environment. Once this new equilibrium has been established, for the isolated mouse,a different social environment will represent a discrepancy between expected (being alone) and the observed (social interaction) events, i.e. a potential stressor. Clear evidence of the stressful properties of social interaction for isolated mice was provided by J.P. Henry and coworkers in the late sixties. He observed that in colonies formed by preisolated males and females, females had fewer pregnancies and increased litter mortality in comparison with socially-experienced females in newly-formed control colonies. Both males and females exhibited higher plasma corticosterone and systolic blood pressure during postisolation social interaction as well as hypertension, cardiovascular and renal pathology compared with mice in control colonies (Ely et al., 1974; Henry, 1980). In the colonies formed by pre-isolated mice a stable social hierarchy was absent and persistent aggression was shown by all males in the group leading a large number of them to die within 40 days of the initial social interaction.

Some alterations of central neurotransmitter functioning , such as those concerning brain GABA system (Puglisi-Allegra et al., 1984; Simler et al., 1982; Puglisi-Allegra and Mandel, 1980) or the endogenous opioid system (Puglisi-Allegra and Oliverio, 1981; Puglisi-Allegra et al., 1982; Lynch et al., 1983, Puglisi-Allegra et al., 1984), seem implicated in the changes of social behavior produced by isolation in the mouse. Genetic factors seem also to modulate social behavior expressed by isolated mice interacting with a conspecific, determining the choice between aggression and defence (Puglisi-Allegra, 1984).

These results show that experimental models based on social isolation describe a true interaction between stress responses and the social life of an organism. Adaptation to isolation may produce, depending on genetic make-up, major alterations of social behavior that may render the organism unfit for different social settings and lead to pathology or even death.

Evidence exists to indicate that experimental defeat is a stressful experience involving the hypothalamic-pituitary-adrenal system as well as the endogenous opioid system (Roche and Leshner, 1979; Miczek et al., 1982; Rodgers and Hendrie, 1983; Siegfried et al., 1984). The experience of being attacked and defeated is not uncommon for laboratory mice living in groups of peers. Some authors suggest that this is, indeed, the main characteristic of social interaction for grouped mice (Brain,1975). However, socially experienced mice present a marked increase in defensive behavior toward non-aggressive opponents after experimental-ly-induced defeat. Moreover, experimental defeat induces analgesia in group housed mice while tolerance to this phenomenon has been described in mice after repeated exposure to defeat (Miczek and Winslow, 1987). These results indicate a lack of cross tolerance between defeat experienced in the social group and experimentally-induced defeat suggesting that these two experiences do not produce equivalent effects. One possible explanation of this difference is the dyadic form of the aggressive interaction used for experimental defeat which does not allow the attacked mice to escape temporarily by merging with the group. This response might give the defeated mice some control over the situation, thus reducing the stressful properties of such an experience.

A number of studies have indicated that the dopamine (DA) system is involved in the expression of defensive and submissive behavior of laboratory animals interacting with aggressive and non-aggressive conspecifics. Rats interacting with an aggressive male conspecific exhibit the greatest increase of DA activity in the nucleus accumbens when compared with animals interacting with non-aggressive males or a female conspecific

(Louilot et al., 1986). Both attack and social stress have been shown to increase 3, 4-dihydroxyphenylacetic acid and homovanillic acid in rat olfactory tubercle (Mos and Van Valkenburg, 1979). Finally, psychostimulants, d-amphetamine and cocaine, were shown to induce defensive behavior in mice confronted with isolated resident conspecifics (Miczek and O'Donnell, 1978).

On the basis of these results it may be suggested that changes in social behavior induced by experimental defeat may depend on the activation of the central DA system due to the stressful properties of this experience.

Dopamine-mediated hyperdefensive behavior of the mouse

C57BL/6 (C57) mice were subjected to three daily defeat experiences (separated by 24h) and then confronted with a non aggressive opponent (testing) 24 h after the last aggressive experience (Puglisi-Allegra and Cabib, 1988). Aggressive confrontation were carried out by introducing a C57 mouse to a Swiss mouse housed with a female. On day 4 the C57 mouse was introduced into the test box (transparent breeding cage with fresh hardwood sawdust on the floor) and confronted with a non aggressive Swiss mouse during a 2-min testing session.

Each test was videotaped and, later, the behavior of the C57 mouse was recorded using a

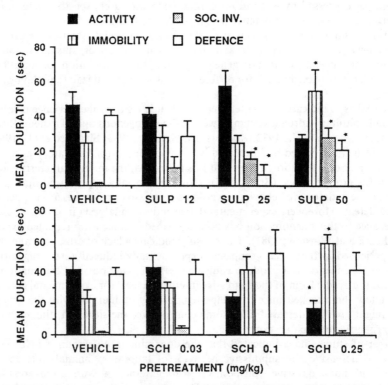

Figure 1. Effects of different doses of (-)-sulpiride and SCH23390 on behavior exhibited by defeted mice during interaction with a nonaggressive opponent.* = Different from vehicle treated group

keyboard system connected to an Apple computer (Puglisi-Allegra and Cabib, 1988). Previously defeated mice exhibited significantly more defensive behavior and less social investigation than mice previously exposed to nonaggressive opponents. The selective D2 DA receptor antagonist (-)-sulpiride administered before confrontation with non-aggressive opponents (4th day) dramatically decreased defensive behavior and increased social investigation in defeated mice. The selective D1 DA receptor antagonist SCH 23390 did not affect either defence or social behavior (Fig. 1).

These results indicated that defensive behavior exhibited by defeat-experienced mice when confronted with non-aggressive opponents involves D2 receptor activation. They also suggest the hypothesis that during the stressful experience of being defeated, D2 receptors are activated producing sustained defensive response in the mouse. Subsequent confrontations with nonaggressive opponents induce a conditioned activation of the same receptor mechanism in defeated mice and the expression of defensive behavioral patterns. An alternative explanation could be that (-)-sulpiride but not SCH23390 has an amnestic effect in mice interfering with recollection of previous defeat experiences during the test. In order to rule out the latter hypothesis, naive C57 mice were treated with the selective D2 agonist LY171555 to induce defensive behavior toward a nonaggressive opponent.

In these experiments mice were moved to a new cage 30 minutes after injection of the selective D2 agonist LY171555 (0.3 to 5 mg/kg), vehicle, or the selective D1 agonist SKF38393. After two minutes' stay in the cage, a nonaggressive but unknown male of the same strain was placed in the same cage. Behavior exhibited by the test mouse was then recorded for 5 minutes and analyzed as previously described.

LY171555 produced a clear dose-dependent increase in defensive behavior and immobility as well as a decrease in social investigation and activity. Low doses of the selective D2 agonist LY171555 produced a behavioral profile which paralleled that of defeat-experienced animals in naive mice interacting with non-aggressive conspecifics (Figure 2).

By contrast, higher doses of this D2 agonist (up to 5 mg/kg) produced hyper-defensiveness (where defence accounted for 75 % of total behavior). The D1 DA agonist SKF 38393 (up to 30 mg/kg) did not increase defensive behaviors.

These results support a major role of the D2 receptor in modulating defensive behavior in the mouse. Furthermore, it may be argued that overstimulation of D2 receptors produces hyper-defensiveness. This could be viewed as a distorted perception of environmental stimuli causing the individual to perceive danger and/or threat when it is actually absent. The survival of both man and animals depends on effective recognition of danger. The failure to discriminate between an aggressive and a non aggressive conspecific can be viewed as a

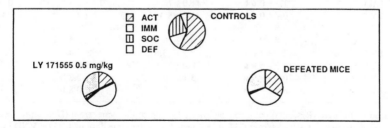

Figure 2. Time spent in behavioral categories by C57 mice interacting with a non aggressive-opponent. Results are expressed in percent mean values of 12 mice per group. ACT = activity; IMM = immobility; SOC.INV = social investigation; DEF = defence

distorted perception of the social context in which the animal lives. Such a perceptive malfunctioning is reminiscent of paranoia. In fact, the paranoid person perceives serious dangers where absent. The alteration of the ability to recognize absence of danger may thus provide a useful way to develop animal models of such pathology (Haber et al., 1977).

A number of investigations on genetic differences related to dopaminergic functions have been carried out on inbred strains of mice (Ingram and Corfman, 1980). Major strain-dependent differences have been evidenced for the behavioral response to the mixed D1, D2 dopaminergic agonist apomorphine, suggesting a different distribution of receptors (Cabib and Puglisi-Allegra, 1985; Cabib and Puglisi-Allegra, 1988; Kendler and Davis, 1984). Finally, it has been shown that genotype plays a major role in the modulation of behavioral responses to repeated defeat experiences (Siegfried et al., 1984). To evaluate the possibility of a genotype-dependent modulation of the effects of the D2 selective agonist LY171555 on defensive behavior, mice of the C57BL/6, DBA/2 (DBA) and BALB/c (BALB) strains were treated with different doses of LY171555 (0.1 to 5 mg/kg s.c.) 30 minutes before interaction with a nonaggressive opponent of the same strain. The behaviors exhibited by the treated mice were recorded and analyzed as described.

While in C57 mice LY171555 induced the typical dose-dependent increase in defense involving all defensive behaviors in a similar way, in mice of the DBA/2 strain none of the doses used affected any of the defensive behaviors. The response of the BALB/c strain was more complex. Defense exhibited by treated mice increased significantly in comparison with

Table I. Effects of different doses of LY 171555 on defensive behavior exhibited by mice of different inbred strains

Strain	Dose	Defensive item			
		Sideways	Crouch (c)	Escape	Upright
C57	Vehicle	1.3 ± 0.5	1.1 ± 0.6	1.1 ± 0.6	4.0 ± 1.3
	LY 0.1	21.8 ± 4.8 a	8.2 ± 3.5	4.7 ± 3.0 a	3.8 ± 1.5
	LY 0.5	28.5 ± 5.3 a	3.5 ± 1.6	26.3 ± 4.9 a	41.5 ± 13.1 a
	LY 1	21.4 ± 4.4 a	6.3 ± 3.2	30.0 ± 10.2 a	34.8 ± 16.0 a
	LY 5	65.5 ± 9.5 a	5.6 ± 1.9	33.1 ± 5.0 a	97.1 ± 14.0 a
DBA	Vehicle	1.2 ± 0.7	2.2 ± 1.5	5.0 ± 2.6	3.1 ± 1.2
	LY 0.1	5.0 ± 0.7 b	8.5 ± 2.9	2.7 ± 1.3	7.2 ± 3.6
	LY 0.5	5.1 ± 2.7 b	7.6 ± 4.2	2.5 ± 1.2 b	3.0 ± 0.9 b
	LY 1	1.3 ± 0.6 b	11.0 ± 7.2	0.4 ± 0.3 b	0.0 ± 0.0 b
	LY 5	0.4 ± 0.2 b	31.2 ± 8.4	1.0 ± 0.6 b	0.0 ± 0.0 b
BALB	Vehicle	1.8 ± 0.7	0.5 ± 0.3	1.0 ± 0.5	1.6 ± 1.6
	LY 0.1	2.7 ± 1.6 b	8.2 ± 3.8	0.6 ± 0.4	9.7 ± 7.7
	LY 0.5	11.7 ± 5.4 a	28.0 ± 9.6	0.2 ± 0.2 b	10.7 ± 9.2
	LY 1	10.4 ± 3.7 a	15.1 ± 8.8	0.3 ± 0.3 b	39.3 ± 25.8 a
	LY 5	8.7 ± 5.9 b	43.2 ± 27.0	2.0 ± 0.8 b	21.2 ± 19.6 b

a = Significantly different (P <0.05) from vehicle-injected group of the same strain (Duncan test).
b = Significantly different (P <0.05) from C57 mice injected with the same dose of LY171555 (Duncan test).
c = Two-way ANOVA showed only a strain main effect (F(2,90)= 5.7; P <0.005) and a treatment main effect (F(4,90) = 4.3; P <0.005). Lack of significant interaction between strain end treatment did not allow individual between groups comparisons.

vehicle-injected controls at the dose of 0.5 ,1 and 5 mg/kg but each doses affected a different group of defensive behaviors (Table I) (Cabib and Puglisi-Allegra, 1989).

These results, indicating strain-dependent differences in the effects of LY171555 on defensive behavior exhibited by mice during interaction with nonaggressive opponents, suggest that genotype may play a major role in the modulation of social behavior by the D2 dopaminergic system.

Chronic treatment with DA-active substances is known to produce changes in DA functioning possibly related to altered DA receptor sensitivity in laboratory animals and psychotic-like syndromes in humans (Kokkinidis and Anisman, 1981). In further experiments we thus decided to subject naive mice to chronic treatment with the psychostimulant cocaine, in order to investigate possible alterations of social behaviors related to altered dopaminergic functioning. Mice (C57BL/6) were injected with a challenge dose (20 mg/kg) of cocaine for ten days 72 h after the end of chronic treatment with cocaine (two daily injections of 20 mg/kg for ten days) and then confronted in 5 min tests in a neutral arena with non aggressive opponents from a different cage (which was docile to the test mouse). Treated mice exhibited a clear cut increase of defensive behavior in comparison with non drugged mice. Administration of the selective D2 antagonist (-)-sulpiride (25 mg/kg) before the challenge dose of cocaine completely antagonized the increase in defensive behavior, while the selective D1 receptor antagonist SCH23390 (0.5 mg/kg) did not significantly affect defensive behavioral patterns (Table II) (Filibeck et al., 1988).

Table II. Effects of different treatments on defensive behavior observed in C57 mice interacting with non-aggressive opponents

	Sideways	Crouch	Escape	Upright
SAL/SAL	5.6 ± 0.8	4.1 ± 2.6	1.8 ± 0.7	0.4 ± 0.3
SAL/COCA	5.0 ± 2.8 b	0.1 ± 0.1	5.3 ± 1.9 b	1.6 ± 1.0 b
COCA/COCA	20.6 ± 3.6 a	0.1 ± 0.1	28.0 ± 6.5 a	71.5 ± 26.4 a
COCA/SAL	5.8 ± 1.2 b	8.3 ± 2.1	5.1 ± 2.2 b	0.8 ± 0.3 b
VEHICLE	25.5 ± 4.1	0.3 ± 0.1	35.5 ± 5.5	72.0 ± 17.1
(-)-SULPIRIDE	3.5 ± 1.4 c	0.8 ± 0.5	2.1 ± 0.9 c	1.6 ± 0.9 c
VEHICLE	22.4 ± 4.1	0.6 ± 0.2	33.5 ± 5.5	67.2 ± 26.2
SCH23390	20.5 ± 4.9	1.6 ± 0.7	16.8 ± 5.8	49.0 ± 14.9

Results are expressed as means + S.E. duration (s) of behaviors.
SAL/SAL = mice chronically treated with saline and injected with saline. SAL/COCA = mice chronically treated with saline and injected with cocaine. COCA/COCA = mice chronically treated with cocaine and injected with cocaine. COCA/SAL = mice chronically treated with cocaine and injected with saline. VEHICLE = mice chronically treated with cocaine and injected with antagonist vehicle before the challenge dose of cocaine. (-)-SULPIRIDE = mice chronically treated with cocaine and injected with (-)-sulpiride (25 mg/kg) before the challenge dose of cocaine. SCH23390 = mice chronically treated with cocaine and injected with SCH23390 (0.2 mg/kg) before the challenge dose of cocaine.
a = different from SAL/SAL ($p < 0.05$ by the Duncan test); b = different from COCA/COCA group ($p < 0.05$ by the Duncan test); c = different from vehicle-treated group ($p < 0.01$ by the Student's t-test, two tailed)

Single injections of cocaine, at doses of up to 75 mg/kg, did not produce changes in defensive behavior in mice chronically treated with saline. The increase in defensive behavior was absent also in mice tested after the last injection of the chronic treatment and in mice receiving the challenge dose of cocaine 12 h after the end of the treatment.

These results suggest that chronic cocaine treatment and the 3-day wash-out period following it modifies the sensitivity of D2 receptors and/or the balance between D1 and D2 receptors. When the challenge dose is administered, the resulting dopamine activation predominantly stimulates D2 receptor, producing the observed hyperdefensiveness.

Recently we have shown that chronic exposure to a nonsocial stressor (immobilization) may change DA receptor sensitivity in the mouse (Cabib et al., 1984, 1985,1988a, b). Moreover, a cross-sensitization has been demonstrated between stress and amphetamine treatments (Antelman et al., 1980). On the basis of these results, the hypothesis emerges that hyper-defensiveness in mice might be also induced by non-social stress via alterations of central dopaminergic functioning.

Taken together, these experiments indicate that the stressful experience of being defeated stimulate defensive behavior in mice by activating the D2 dopamine receptor system. Moreover, D2 receptor overstimulation in the mouse results in a behavioral syndrome characterized by extreme defensive behavior. Genetic factors appear to play a major role in the expression of this syndrome, suggesting a genotype-dependent modulation of the activity of this receptor system. Finally, pharmacological manipulations which supposedly alter D2 dopaminergic function produce hyperdefensiveness in undefeated mice. On the basis of these results one may suggest that the behavioral repertoire presented by mice during social confrontation may depend on their genetic background, as well as on their previous experience of social or non social stress.

Conclusions

Experimental situations allow relative control to be exerted over the environmental variables involved in the modulation of social behavior, thus revealing complex interactions among these variables. We have attempted to show that aggression and defence may be viewed as part of a dynamic interaction between the organism and the environment. Aggressive behavior against conspecifics, for example, may represent an attempt to gain some control over potentially stressful stimuli of a non social nature. Moreover, adaptation to a stressful social environment, as in the case of individual housing or repeated defeats, may alter the perception of common social experiences by changing social interactions in stressful stimuli leading to aggressive or defensive responses.

The stress response involves different physiologial systems depending on the characteristics of the stressor, on genetic make up and on the history of the individual (i.e. previous stressful experiences). By exploring the role of the dopaminergic system in the expression of defensive behavior we have suggested that alterations of social behavior may be produced also as a consequence of adaptation to non social stessors. This may produce a change in the interaction between the organism and the social environment since a stressed mouse,due to altered dopaminergic functioning, might be more prone to exhibit defensive behavior thus assuming a submissive position inside the group (Thiessen and Upchurch, 1981) that would increase the occasions for new stressful experiences (for example, limited access to food).

Finally, the role of stress in the modulation of social behavior of the house mouse appears to be a very promising field for studies in preclinical psychopharmacology since it offers a view of the mechanisms by which physiological and environmental factors may interact to alter the behavioral output of the organism.

References

Antelman, S.M., Eichler, A.J., Black, C.A. and Kican, D. (1980) Interchangeability of stress and amphetamine sensistization, *Science*, 207, 329-331

Bertalanffy, L. von. (1968) General Systems Theory. Essays on Its Foundation and Development. Braziller, New York;

Brain, P.F. (1975) What does individual housing mean to a mouse? *Life Sciences*, 16, 187-200

Cabib, S., Pugisi-Allegra, S. and Oliverio, A. (1984) Chronic stress enhances apomorphine-induced streotyped behavior in mice: involvement of endogenous opioids. *Brain Research*, 298, 138-140

Cabib, S. and Puglisi-Allegra, S. (1985) Different effects of apomorphine on climbing behavior and locomotor activity in three strains of mice, *Pharmacology Biochemistry and Behavior*, 23, 555-557

Cabib, S., Puglisi-Allegra, S. and Oliverio, A. (1985) A genetic analysis of stereotypy in the mouse: dopaminergic plasticity following chronic stress. *Behavavioral Neural Biology*, 44, 239-248

Cabib, S. and Puglisi-Allegra, S. (1988) A classical genetic analysis of two apomorphine-induced behaviors in the mouse. *Pharmacology Biochemestry and Behavior* 30, 143-147

Cabib, S., Kempf, E., Schleef, C., Oliverio, A. and Puglisi-Allegra, S. (1988a) Effects of immobilization stress on dopamine and its metabolites in different brain areas of the mouse: role of genotype and stress duration. *Brain Research*, 441, 153-160

Cabib, S., Kempf, E., Schleef, C., Mele A., and Puglisi-Allegra, S. (1988b) Different effects of acute and chronic stress on two dopamine-mediated behaviors in the mouse. *Physiology Behavior* 43, 223-227

Cabib, S. and Puglisi-Allegra, S. (1989) Genotype-dependent modulation of LY 171555-induced defensive behavior in the mouse. *Psychopharmacology*, 97, 166-168

Clement J., Simler, S., Ciesielski, L., Mandel, P., Cabib, S. and Puglisi-Allegra, S. (1987) Age-dependent changes of brain GABA levels, turnover rates and shock-induced aggressive behavior in inbred strains of mice. *Pharmacology Biochemistry and Behavior*, 26, 83-88

Conner, R.L., Vernikos-Danellis J., and Levine S. (1971) 'Stress, fighting and neuroendocrine function'. *Nature* 234, 564-566.

D'Amato, F.R. (1986) Time budgets and behavioural synchronization in aggregated and isolated male and female mice. *Behavioural Processes*, 13, 385-397

Ely, D.L. and Henry, J.P. (1974) Effects of prolonged social deprivation on murine beaviour patterns, blood pressure, and adrenal weight. *Journal of Comparative and Physiological Psychology*, 4, 733-740

Filibeck, U., Cabib, S., Castellano, C. and Puglisi-Allegra, S. (1988) Chronic cocaine enhances defensive behaviour in the laboratory mouse: Involvement of D2 dopamine receptors. *Psychopharmacology* 96, 437-441

Frischknecht, H.R., Siegfried, B. and Waser, P.G. (1985) Postaggression footshock inhibits aggressive behavior in dominant but not in isolated mice. *Behavioral Neural Biology*, 44, 132-38

Henry, J.P. (1980) Present concept of stress theory. In: E. Usdin, Kvetnansky and Kopin (eds.) Catecholamine and Stress: Recent Advances, Elsevier/North Holland Biomedical Press, Amsterdam, pp. 557-587.

Haber, S., Barchas, P.R. and Barchas, J.D. (1977) Effects of amphetamine on social behavior of rhesus macaques: an animal model of paranoia. In: Hanin, I. and Usdin, E. (eds.) Animal Models in Psychiatry and Neurology, pp 107-115, Pergamon, Oxford

Ingram, D.K. and Corfman, T.P. (1980) An overview of neurobiological comparisons in mouse strains. *Neuroscience Biobehavioral Reviews*, 4, 421-435

Kendler, K.S. and Davis, K. (1984) Genetic control of apomorphine-induced climbing behavior in two strains of mice. *Brain Research* 293, 343-351

Kokkinidis, L. and Anisman; H. (1981) Amphetamine psychosis and schizophrenia: a dual model. *Neuroscience Biobehaviral Reviews*, 5, 449-461

Jackson, R.L., Maier, S.F. and Coon, D.J. (1979) Long-term analgesic effects of inescapable shock and learned helplessness. *Science*, 206, 91-93

Lynch, W.C., Libby, L. and Johnson, H.F. (1983) Naloxone inhibits intermale aggression in isolated mice. *Psychopharmacology* 79, 370-371

Louilot, A., Le Moal, M. and Simon, H. (1986) Differantial reactivity of dopaminergic neurons in the nucleus accumbens in response to different behavioral situations: An in vivo voltametric study in free moving rats. *Brain Research* 397, 395-400

Miczek K.A. and O'Donnell, J.M. (1978) Intruder-evoked aggression in isolated and nonisolated mice: Effects of Psychomotor stimulants and L-Dopa. *Psychopharmacology* 57, 47-55

Miczek, K.A., Thompson,. M.L. and Shuster, L. (1982) Opioid-like analgesia in defeated mice. *Science* 215, 1520-1522

Miczek, K.A. and Winslow, J.T. (1987) Analgesia and decrement in operant performance in socially defeated mice: Selective cross-tolerance to morphine and antagonism by naltrexone. *Psychopharmacology* 92, 444-451

Mos J. and Van Valkenburg C.F. (1979) Specific effects of social stress and aggression on regional dopamine metabolism in rat brain. *Neuroscience Letters* 15, 325-327

Puglisi-Allegra, S. and Mandel, P. (1980) Effects of sodium n--dipropylacetate, muscimolhydrobromide and (R,S) nipecotic acid amide on isolation-induced aggressive behavior in mice. *Psychopharmacology*, 70, 287-290

Puglisi-Allegra, S., Simler, S., Kempf, E. and Mandel, P. (1981) Involvement of the GABAergic system on shock-induced aggressive behavior in two strains of mice. *Pharmacology Biochemistry and Behavior* 14(1), 13-18

Puglisi-Allegra, S., Oliverio, A. and Mandel, P. (1982) Effects of opiate antagonists on social and aggressive behavior of isolated mice. *Pharmacology Biochemistry and Behavior* 17, 691-694

Puglisi-Allegra, S. and Oliverio, A. (1983) Social isolation: Effects on pain treshold and stress-induced analgesia. *Pharmacology Biochemistry and Behavior* 19, 679-681

Puglisi-Allegra, S. (1984) Deux Examples de Comportement d'Agression chez La Souris: Etude Comportementale, Neurochimique et Pharmacologique. These de Doctorat d'Etat, Universite Louis Pasteur, Strasbourg

Puglisi-Allegra, S. Mele, A. and Cabib, S. (1984) Involvement of endogenous opioid systems in social behavior of individually housed mice. Ethopharmacological Aggression Reserach, Alan R. Liss, Inc., New York, pp. 209-225

Puglisi-Allegra, S. and Cabib, S. (1985) The effect of age on two kinds of aggressive behavior in inbred strains of mice. *Developmental Psychobiology*, 18, 477-482

Puglisi-Allegra, S. and Cabib, S. (1988) Pharmacological evidence for a role of D2 dopamine receptors in the defensive behavior of the mouse. *Behavioral Neural Biology* 50, 98-111

Roche, K.E and Leshner, A.I. (1979) ACTH and vasopressine treatments immediately after a defeat increase future submissiveness in male mice. *Science* 204, 1343-1344

Rodgers, R.J.& Hendrie, C.A. (1983) Social conflict activates status-dependent endogenous analgesic or hyperalgesic mechanisms in male mice: Effects of naloxone on nociception and behavior. *Physiology and Behavior* 30, 775-780

Siegfried, B., Alleva, E., Oliverio, A. and Puglisi-Allegra, S. (1981) Effects of isolation on activity, reactivity, excitability and aggressive behavior in two inbred strains of mice *Behavioral Brain Research* 2, 211-218

Siegfried, B., Frischknecht, H.R. and Waser, P.G. (1984) Defeat, learned submissiveness and analgesia in mice: Effect of genotype. *Behavioral Neural Biology* 42, 91-97

Simler,S., Puglisi-Allegra, S. and Mandel, P. (1982) y-Aminobutirric acid in brain areas of isolated aggressive and non-aggressive inbred strains of mice. *Pharmacology Biochemistry and Behavior* 16, 57-61

Thiessen, D.D and Upchurch, M. (1981) Haloperidol and chlonidine increase, and apomorphine decreases ultrasonic vocalizations by gerbils. *Psychopharmacology* 75, 287-290

Valzelli, L. (1981) Psychobiology of Aggression and Violence. Raven Press, New York

Weiss, J.M., Stone, E.A., and Harrel, N. (1970) 'Coping behavior and brain norepinephrine levels in rats'. *Journal of Comparative and Physiological Psychology* 72, 153-160

Weiss, J.M., Pohorecky, L.A., Salman, S. and Gruenthal, M. (1976) Attenuation of gastric lesions by psychological aspects of aggression in rats. *Journal of Comparative and Physiological Psychology* 90, 252-259

Wilmot, C.A., WanderWende C. and Spoerlein, M.T. (1984) Behavioral responses to apomorphine and amphetamine in differentially housed mice. *Psychopharmacology* 84, 105-108

A ROLE FOR AFFECTIVE NEUROSCIENCE IN UNDERSTANDING STRESS: THE CASE OF SEPARATION DISTRESS CIRCUITRY

Jaak Panksepp
Department of Psychology
Bowling Green State University
Bowling Green, Ohio 43403
U.S.A.

In psychological terms, the concept of 'stress' is difficult to define in scientifically useful ways. At a neurobiological level, however, the concept is coming to have a deceptively simple underlying reality. Every environmental event or personal experience which would normally be deemed emotionally 'harsh' or 'overwhelming' by humans (and the imposition of comparable events on animals) activates the pituitary-adrenal axis and often the sympatho-adrenal one as well. Thus, one could posit the existence of a unitary brain 'stress system' which is intimately linked to corticotrophin releasing factor (CRF) circuits and sympathetic (e.g., norepinephrine (NE)) circuits within the brain. Since the discovery of brain CRF and NE systems and the demonstration of their influence on many stress-related behaviors (Swanson et al., 1986), such ideas have become common currency in the field. Indeed, blockade of NE activity is a common way to alleviate the autonomic impact of stressors (see Kelly, 1980). Likewise, it is widely believed that development of a pharmacological antagonist to brain CRF and glucocorticoid receptors, which could gain entry into the brain following peripheral administration, could be of considerable benefit in the treatment of stress disorders.

However, the apparent unity of the so-called 'stress response' may be deceptive. The brain may not really have a specific neural system which was constructed during brain evolution to deal with a unitary psychological process called 'stress'. Rather, the final physical 'stress response', which can be indexed by various peripheral changes (but especially by the classic ones mentioned above), may emerge from many different homeostatic and emotional systems of the brain which require recruitment of rapid patterns of bodily change so that stored energy reserves are made readily available for action. In such situations the cardiovascular system needs to invigorated, while many other bodily processes (such as digestion and anabolic disposition of nutrients) need to be quelled. Such neuro-hormonal reflexes (including secretion of catecholamines and catabolic steroids), which facilitates behavioral readiness in a very generalized way, have traditionally been viewed as the physical manifestations of the stress response. However, these terminal responses appears to be capable of being activated by several distinct brain systems, probably in differential patterns which remain poorly characterized. Accordingly, it is presently essential to identify and properly conceptualize the many brain systems which utilize such final common response pathways. If we make such a shift in stress research, from focussing just on the terminal

S. Puglisi-Allegra and A. Oliverio (eds.), Psychobiology of Stress, 41–57.
© 1990 Kluwer Academic Publishers. Printed in the Netherlands.

'stress responses' to also consider the diversity of psychoneuronal input systems, then I suspect an understanding of brain emotional systems (and the central-states they produce) becomes an essential ingredient for understanding the the manner in which the brain copes with so-called stressful events. Only when the underlying biology of these brain systems is beset by pathologies of overstimulation and exhaustion (e.g., Sapolsky et al., 1985; Sapolsky and Meaney, 1986; Selye, 1980), and the brain substrates for negative feedback control of the aroused systems become compromised, might it be appropriate to begin viewing the changes under some global rubric of 'stress'. Before such disorders of regulation begin to emerge, all the brain may really have is variety of normal emotional and homeostatic processes poised for action.

Unfortunately, the study of emotional processes in the brain has not been an especially popular research strategy in modern behavioral neuroscience. Obviously, emotions, as well as all other central-states elaborated by the brain, cannot be observed directly. They cannot be weighed and measured in totally reliable and valid ways, yet. They are characterized by subjective/felt attributes, and in animals they must be inferred indirectly from behavioral and other organismic changes. However, from an evolutionary perspective, it is likely that the brain controls many ingrained psychological tendencies, which can only be empirically approached with the use of indirect theoretical approaches (outwardly resembling those which have guided the revolutionary insights of chemistry and physics). Because of the difficulties in forging a consensus on how central affective states could ever be studied in animals in reliable ways, most of stress research, indeed most data collection in behavioral neuroscience, remains remarkably free of and disconnected from psychological theory. Although the *Zeitgeist* still commends remaining at a radical empirical level indefinitely-- merely correlating bodily/behavioral changes with imposed influences of various sorts--I think that strategy is rapidly becoming narcissitically self-centered in its presumed rigor (which has more than a passing resemblance to the 'radical behaviorism' which 'cognitive psychology' asserts it killed, but whose philosophy survives relatively intact within 'behavioral neuroscience'). Although not appreciated by its adherents, this radical neurobehaviorism sustains and continues to prolong a dualistic conception of the nature of psychology and the brain.

Modern neuroscience remains rather myopic in its vision of what types of psychic functions actually emerge from the dynamic activities of the brain. To put it another way, within modern neuroscience the behaviorist 'reflex model' of animal nature still prevails over the 'active agent' model, and a host of important questions concerning brain organization remain largely unaddressed. Indeed, proponents of 'the reflex model' have often actively discouraged open discussion of the possibility that the many types of mammals and birds we study may, in fact, possess various sophisticated central-state mechanisms within their brains. If they do have such hidden neuropsychological filters, which symbolize our most important ancestral successes during evolution, then our failure to openly air theoretical possibilities is bound to generate research strategies which ultimately yield an impoverished picture of the brain processes we wish to understand reductionistically. I think much of present-day stress research is actually struggling at the borderline of these two paradigms, not quite ready to accept the 'active agent' model but somewhat disenchanted with an exclusive focus on the conceptually sterile 'reflex model' which may only suffice for studying the simplest types of central mechanisms. I think that one reason that many are presently quite willing to entertain the existence of monolithic integrative 'stress systems' in the brain, but not 'emotional systems', is a lingering consequence of psychology's long infatuation with the behaviorist reflex-model of animal actions. In fact, the brain is probably a much more complex a 'symbolic organ' than is generally assumed.

The present paper is premised on the idea that we will make faster progress if we forthrightly make the 'active agent' model axiomatic in much of psychobiological research. I think it is self-evident and hence a reasonable starting premise that evolution has constructed a variety of adaptation promoting primal process 'central-state circuits' in the nervous system-some of which are designed to mediate emotionality. If so, then we must properly conceptualize these neuroemotional filters before we can make coherent sense of the effects of stressful manipulations on the mind, brain and body. In the parlance of modern artifical intielligence, these 'central states' may resemble the 'hidden neurons' of the neural-net connectionistic models for mental processing which have become the current rage in cognitive psychology and artifical intelligence. It has been found that such 'hidden units' in parallel-distributed-processing (PDP) models of mind can exert considerable power over an information processing system, helping neural nets to simulate cognitive processes in a way resembling how human minds operates in the real world (Pagels, 1988; Rumelhart and McClelland, 1986).

It seems doubtful that anyone could have the evolutionary insight to properly conceptualize all of the hidden 'state control' systems of 'the mammalian brain' without much more psychoneurological research. Thus, for the time being, we must be satisified with various successive, inductively-derived approximations which may yield a sufficient number of theory-driven empirical observations which can lead to more refined conceptions. The overall aim of the present paper is to consider where such a path might empirically take us in stress research.

Paranthetically, I would emphasize that I do not deny the utility of research on stress responsivity that has been and is being collected using the existing 'reflex model' of brain organization. That will continue to be a useful research strategy for a long time to come, especially when we have no prior theoretical insights into a system. I would only suggest that much of that data will eventually have to be re-interpreted within an emotional systems-analysis context. To take one example--in the study of how the brain controls the pituitary-adrenal axis, it is clear that the amygdala and hippocampus provide higher negative feedback control over hypothalamic pituitary-adrenal control mechanisms (Allan and Allen, 1974). For instance, lesions of the central amygdala in rats can markedly reduce stress-induced ACTH secretion (Beaulieu et al., 1986) as well as corticosterone release (Seggie, 1987). Furthermore, the excessive secretion of ACTH in stressed adrenalectomized rats requires intact afferent input into the medial basal hypothalamus (Allen, Allen and Greer, 1974), and is attenuated following bilateral lesions of the ventral amygdalofugal pathway but not by stria terminalis lesions (Allen and Allen, 1975). Additionaly, the amygdala possesses abundant glucocorticoid receptors (Gerlach, et al. 1974; Warembourg, 1975) which may provide end-product inhibition of ACTH secretion following removal of a stressor (Allen and Allen, 1975). By removing the tonic effects of this feedback mechanism, bilateral amygdalectomy can result in tonic plasma coricosterone elevations (Eleftherious et al., 1966; Martin et al., 1958). Similar types of higher negative feedback controls emanate from hippocampal tissues (Angelucci, et al., 1980), which presumably help link emotional issues to associative/cognitive ones. These are well known empirical effects in the field (typically untinged by psychological issues), but in order to make adaptive sense of such effects we will ultimately have to interface the above findings with an understanding of the the emotional/psychic processes of animals which emerge from neural activities in these brain areas.

For instance, what is the normal evolutionary adaptive reason for there being so many glucocorticoid receptors in hippocampal/amygdaloid tissues which have now been demonstrated to be so remarkably susceptible to stress-induced injury (e.g., Sapolsky, et al, 1985)? We do not know for sure, but presently there is an abundant literature that the amygdala and hippocampus participate in the elaboration of several emotional responses including 'anger' and 'fear' (for summary see Gray (1987) and Panksepp et al. (1989)), and it will be of considerable interest to try to determine how those glucocorticoid control systems interface with the emotional processes those brain areas also subserve. Do they increase, dampen, prolong, shorten or deepen emotional and related memorial responses? At present, there is relatively little empirical work which is seeking to clarify such issues, mainly, I would suggest, because we have not yet learned to blend the best aspects of the 'reflex model' of radical neuroscience and the 'active agent' model of evolutionary psychobiology. I suspect that one function of these receptors may be to promote the increased effort, and sustained emotionality, that is so useful in various emergency situations. When higher limbic aspects of this system begin to dysfunction, because of glucocorticoid receptor mediated metabolic 'burnout' in the affected systems, the gateway to chronic despair, depression, and withdrawal from active confrontation with the world may be opened.

In sum, my position is that many of the changes which result from 'stress' are in fact the physiological imbalances which can transpire in primal-process emotional systems of the brain. Thus, to understand how disorders of stress emerge, we may first have to unravel the nature of emotionality. I suspect that presently these 'central-state' emotional circuits are most effectively labeled with common affective labels which have long been part of our linguistic heritage (i.e., terms such as anger, fear, etc) rather than the concept of 'stress'. By viewing certain intrinsic brain systems from such a vantage, I suspect that we may be encouraged to do very different kinds of experimental studies than if we merely view the underlying responsive circuits as 'stress systems.' Accordingly, my specific goals for this essay are to discuss i) how a better understanding of emotionality may highlight key issues of stress research, ii) to suggest some new modes of inquiry which may help identify the psychophysiological patterns which characterize different emotion-specific forms of stress and also iii) to summarize some of our past research concerning the separation-distress response, which may help us better conceptualize how certain certain emotional control systems of the brain such as endorphins, CRF and oxytocin respond to stressors.

A Potential Taxonomy of Emotions and Stress Processes in the Brain.

How many distinct circuits of the brain participate in the genesis of what has been commonly called 'stress responses'? Many, no doubt--but a common denominator may be that these changes typically arise from brain circuits which generate strong affective central states. At a general conceptual level, stress-sensitive systems may be viewed as neurophysiological systems which register major bodily deviations from homeostatic equilibrium into affective consciousness--whether they be deviations from explicit body constancies such as energy storage (i.e., hunger) or implicit brain constancies (such as the tendency of the brain to try to avoid arousal of negative emotions). Accordingly, deviations from resting levels (i.e., homeostatically defended levels) of activity within several major emotional circuits of the brain may generate feelings of distress and displeasure , while a return of these system toward homeostatic equilibrium may be experienced as relief or pleasure (see Cabanac, 1979, as well as in this volume). Of course, each emotional system also has more specific subjective concommitants, none of which can, at the present time, be measured directly in animals. Eventually, however, I suspect we will be able to measure them as changing neurochemistries and electrophysiologies of the brain which are not rigidly linked to momentary stimulus inputs.

From my present research vantage, of trying to understand the basic neural organization of primal emotional systems of the brain, I would still be willing to hope that distinguishable 'stress patterns' may emerge from the arousal of each of the distinct emotion generating circuits which exist in the brain (for a more extensive anatomical, functional and conceptual discussion of emotion circuits, see Panksepp, 1982, 1988, as well as the specific reviews for each emotional system which are cited after each item enumerated below). To the best of our knowledge, the brain has primal-process executive systems (i.e., behavioral state control circuits) for generating at least three distinct negative emotional processes 1) anger-rage (i.e., which on the response side yields offensive behavior patterns), 2) anxiety-fear (yielding defensive behaviors) (Panksepp et al., 1989), and 3) panicy-separation distress (yielding contact and care seeking behaviors) (Panksepp et al., 1988). Although it is more common to think about stress in terms of negative emotional responses, the brain also contains positive emotional systems, for instance those which mediate 1) the central state of anticipatory eagerness and expectation of positive events (expressed as foraging behavior patterns) (Panksepp, 1986), as well as 2) circuits for joyful/playful arousal (expressed as rough-and-tumble ludic activities) (Panksepp et al., 1984), and perhaps 3) circuits which mediate nurturance/acceptance , which may be the source of parental care and the apparent 'love/friendship/affection' that one often sees between animals. Activity in these positive emotional systems may normally help counteract the perceived distress which emerges from the activity of the negative systems. However, it is also possible that excessive activity of some of these positive emotional systems may also yield bodily imbalances which may promote the peripheral 'stress response' as traditionally understood. In addition to specific patterns of bodily change that are aroused by the individual emotions, all systems may also share certain common changes which arise from the fact that generalized background processes have been aroused. All emergency situations need increased sensory arousal, attention, vigilance and general motor readiness for effective behavioral coping. Thus, each of the specific emotional patterns of physiological change may be accompanied by various non-specific arousal patterns, especially within biogenic amine circuits such as those which emanate from the locus coeruleus, which may reflect the generalized fact that a brain in any of a variety of states of homeostatic disequilibrium requires increased vigilance, attention and adaptive behavioral change for re-equilibration to occur.

The conceptual view I would advocate is that there are both specific arousal responses which characterize each of the distinct emotional patterns as well as non-specific adaptive responses shared by most strong emotional tendencies. Of course, in most real-life situations, specific and non-specific responses may blend together in relatively seamless ways as animals shift successively between very different emotional tendencies (especially as homeostatic opponent-processes are recruited during sustained emotional arousal). Abundant psychophysiology/sociophysiology literature collected in humans (e.g., Waid, 1984; Wagner, 1988) bears out the relative futility of attempting to characterize different emotions via different patterns of autonomic change in humans, although some success has been achieved using well-controlled artificial laboratory procedures in humans (Ax, 1953; Ekman et al., 1983). Still, I suspect the the development of a correct natural taxonomy of emotional systems in the brain may be especially useful in guiding future analyses of state-specific autonomic patterns. Such work may eventually clarify the spectra of patho-physiological change observed in individuals confronted by distinct forms of stress, as well as psychosomatic disorders which are correlated with the various affective personalities/ temperaments. For instance, individuals prone to anger have a higher incidence of certain medical problems (Johnson and Broman, 1987) especially hypertension (Julius et al., 1985). If differentiable stress patterns do, in fact, characterize different autonomic disorders (Henry, 1986), then a new generation of theory-guided approaches to animal brain research may yet provide clarity where past approaches have yielded little more than confusion.

One of the best strategies for identifying the autonomic patterns that characterize the primal emotions is to fully deploy the very best animal-models of 'pure emotion' that can be studied under strict stimulus control. My recommendation as the 'best choice' is the induction of different emotions by electrical stimulation of specific areas of the brain (Panksepp, 1982). With judicious placement of electrodes within specific diencephalic areas, one may be able to study the patterned autonomic manifestations of each of the primal emotions in especially productive ways. Promising work along these lines has already been conducted for certain systems such as defensive and aggressive ones (e.g., Mancia and Zanchetti, 1981). What is needed now is a thorough comparative evaluation of the physio-logical and autonomic changes that arise from the different types of brain stimulation in a single global study conducted concurrently for all the emotive circuits. To minimize confusion in such a study, the 'psychological purity' of electrode sites should first be characterized on the basis of behavioral criteria prior to the analysis of physiological changes. Presumably the analysis of many brain sites would be rejected because of mixed affective tendencies, which reflect the considerable overlap of emotive system as they course through the diencephalon. Indeed, these systems may eventually be studied in even more specific ways by direct (cannula guided) manipulation of neurotransmitter activity levels in the brain. I suspect that highly distinct emotion-specific patterns of bodily change will emerge when such work if finally done. However, in the absence of an extensive set of definitive, well-replicated findings in the field at the present time, I will not attempt to summarize the suggestive patterns from existing data in the limited space available here. I will restrict my coverage to the organization of the separation-distress/panic system in the brain, which, along with the study of playfulness, has been the focus of our own empirical work on the brain mechanisms of emotion for the past decade.

The separation distress system may also prove to be an important case-study for analyzing the specific and non-specific role of brain CRF systems in stress. As will be summarized later, brain CRF systems appear to be integrally linked to the separation distress response, at least in domestic chicks. Is the amplification of distress vocalizations that one can see after intracerebral CRF reflect a specific function of this neurochemical system or is it merely a secondary reflection of a generalized stress system of the brain? Electrophysiological analysis of CRF neurons will be needed to discriminate the extent to which components of brain CRF systems evolved primarily for social emotional control functions vs. the extent to which CRF evolved as a generalized system which is tapped by every type of emotional arousal. Surely the function of the paraventricular CRF system which controls pituitary ACTH release (and which, unlike most brain CRF circuits, does receive negative feedback influences from circulating glucocorticoids) is quite non-specific since it can be aroused by so many diverse emotional and motivational situations. However, the main CRF systems which courses through the brain do not appear to be under negative feedback regulation of circulating glucocorticoids (Swanson et al., 1983), and those CRF circuits may be better conceptualized as mediators of a specific type of emotional response rather than simply as a generalized integrative system for mediation of stress. We suspect that this CRF system may be the primal substrate for the experience of social loss, and that it, along with many other emotive systems, can access the generalized bodily stress response which is controlled by the paraventricular, hypophysiotropic CRF system of the hypothalamus. In other words, while the emotional response of separation distress is mediated by the long-axoned CRF systems, the generalized stress response of the body, which is concurrently evoked, may be controlled by the anatomically restricted paraventricular cell group. The reason the higher CRF system is not under negative feedback control may reflect the fact that feedback from the social environment, rather than circulating glucocorticoid levels, is the regulatory endpoint of the system. Namely the system remains active until social comfort is restored. Although the psychological distress which results from social separation may emerge from the higher CRF related circuits, it is reasonable to suppose that the general health consequences of bereavement/social loss may emerge more from the activity of the lower paraventricular systems. Excessive steroid secretion has long been known to have many health consequences (including changes in cardiovascular, reproductive, digestive, growth and immune functions) and it may be largely through those effects that social-loss, a very common stressor indeed, comes to compromise the continuation of good health.

The Health Consequences of Separation Distress

Before discussing the details of the separation distress system, I would briefly highlight a role for the separation-distress system in the broader health context. The degree of social bonding and the consequent perception of social support which one experiences appear to be key vectors in how bereavement affects bodily health. Social loss appears to initiate a cascade of physiological changes in the body which can result in premature morbidity (Rahe, 1975). This appears to be the case in both animals (Rabin et al., 1988) and humans, especially in areas such as immunocompetence and susceptibility to coronary heart disease (for excellent recent reviews see Melnechuk, 1988; Pelletier and Herzing, 1988). Indeed, for the human condition, it has been stated that 'marital disruption is the single most powerful sociodemographic predictor of stress-related physical illness' (Kiecolt-Glaser, et al. 1987). The disasterous consequences of early social deprivation for the development of humans and other primates has long been recognized (Bowlby, 1973; Rutter, 1972. Early social support has life-long consequences for subsequent psychobehavioral competence, and positive/supportive types of early experiences as well as abundant social play may strengthen the neurobiological mechanisms which underlie human health-promoting personality characteristics such as 'heartiness' (Kobasa et al., 1982). Such lines of thought suggest that an understanding of brain mechanisms which mediate the experience of social separation and social loss, as well of the opposite side of this emotional coin, namely the feelings of social connectedness and playfulness, may provide a neurological foundation for understanding how certain 'stress-related' medical disorders emerge.

   It is noteworthy that the autonomic changes which emerge during the acute phase of the separation response (i.e., protest) as well as the more protracted effects (i.e., despair) have now been relatively well characterized in rodents (Hoffer, 1984) and primates (Reite and Capitanio, 1975), and consist of clear cardiovascular, thermoregulatory/metabolic, and vigilance/sleep changes. Many of these effects may be secondarily due to the massive glucocorticoid response that separation triggers. The massive arousal of the pituitary-adrenal system during the 'protest' phase of the separation response (Coe et al., 1985), and the brain changes (e.g., limbic glucocorticoid receptor down-regulation and eventual hippocampal damage) that can result from sustained activation of the pituitary-adrenal axis (Sapolsky, et al, 1985), may establish the chronic brain condition of apathetic-depression which characterizes the protracted 'despair' phase which commonly results from both chronic social separation and social defeat. There is even an outside chance that such depression of the limbic system may in fact be adaptive, especially for the young: The decreased behavioral output may reduce the probability that socially isolated animals will pursue a potentially foolish actions . By remaining inactive, the probability of social reunion with a departed parent may be increased. Some forms of psychiatric depression in adults may be a residual result of this psychobehavioral disposition which might only have been adaptive for younger organisms.

Brain Mechanisms of Separation Distress

Our strategy for understanding the brain mechanisms which elaborate the affective experiences evoked by social loss has been to focus on the neuroanatomical and neuro-chemical controls of the separation call. The over-riding theoretical scheme is that the actual emotional experience of social separation depends on activity of those executive coordinating circuits which provide excitatory control over the separation call (which is a highly conserved primitive social response of all warm-blooded vertebrates), while many of the protracted effects may be due, as already described, to the long-term physiological consequences of an overactive pituitary adrenal axis . The self-evident evolutionary function of the separation response is to sustain social cohesion between infants and care-takers, and hence the neurochemical dynamics within this system may provide the primal 'motivational' mechanism via which social cohesion and social bonding are sustained. This same call, with minor variations, also appears to be used by animal for seeking care when non-social needs emerge which could be alleviated by others--for instance, when animals are hungry, cold or in pain. Still, the least physically intrusive and most effective way to activate this care-seeking system in the brain appears to be via social isolation. Hence we typically call the underlying circuitry the separation distress/panic system, although it may certainly have a more generalized distress function that is activated (perhaps conditionally) by enforced separation of an animal from non-social attachments such as food, possessions and territory. We have primarily used neuropharmacological maneuvers to understand the neurochemical organization of this brain system, and our underlying assumption is that neurochemical influences which inhibit the separation call are ones which may be beneficial not only in alleviating many forms of acute psychic distress and may also be useful for sustaining health following social loss. In addition to summarizing our knowledge of the separation distress system, I will also briefly consider the potential anti-stress functions of intrinsic brain mechanisms which mediate certain natural pro-social behaviors such as play and social feelings such as acceptance (both of which seem to emanate from brain processes diametrically opposite to those which mediate separation distress).

Neuroanatomical Substrates of Separation Distress: The main technique we have used to identify the location of neural circuits which mediate separation distress has been identification of brain sites where localized electrical stimulation of the brain (ESB) produces isolation calls/distress vocalizations (DVs). Extensive analysis along these lines has been done in primates and domestic chicks (see Newman, 1988 for review). In our own work (e.g., Herman & Panksepp, 1981; Panksepp, et al., 1988), much of which remains unpublished, we have used the ESB technique to mapped out the general trajectory of neural systems which mediate separation-type distress vocalizations in guinea pigs and domestic chicks. In guinea pigs, some rostral vocalization sites can be found in the amygdala and infrequently in hippocampus around the subiculum, but these effects are not reliably observed in all animals. Also, with repeated testing, the responses evoked from these higher areas do not remain stable (perhaps because of a concurrent kindling-like processes being recruited). In our experience, marked elevations in thresholds commonly occur after several test sessions, typically to the point where reasonable levels of ESB will no longer evoke the response again.

In guinea pigs, the highest brain area from which highly reliable DVs can be evoked with ESB is contained in a broad zone in the basal forebrain/preoptic area that extends into the septum especially vental parts. In frontal section through the mid-septal area, the zones surrounding and just medial to the horns of the anterior comissure yield highly consistent DVs, most of which are emitted at ESB offset. It is noteworthy that these zones overlap substantially with brain areas which have long been implicated in the organization of sexual and maternal behaviors. The active zone from which such DVs can be evoked continues caudally, with many active sites in the bed-nucleus of the stria terminalis and abundant sites within the dorsomedial zones of the thalamus with some active electrode sites being found in the dorsomedial hypothalmus. At the mesencephalic level, DVs can be evoked from scattered sites running from the medial parts of the central gray, laterally through the nucleus of the posterior comissure and the reticular formation dorsal to the red nucleus, as well as some sites in the ventrolateral pons. At more caudal levels of the midbrain, one tends to get fewer and fewer DVs and obtains more and more pain-like screams. This mesencephalic anatomy suggests that the distress vocalization machinery is rooted within the primitive nonspecific pain circuitry of the pontine periaqueductal gray and reticular formation. This distress/panic circuitry is quite similar in species as divergent as chicks and primates, highlighting how ancient and well-conserved is this system for the maintenance of social cohesion. Analysis of neurochemical controls (see next two sections) further affirms this conclusion, for the effects that have been observed are again quite comparable in diverse vertebrate species.

Separation Distress and Peptide Systems. In our anatomical analysis of the separation distress system with ESB techniques, we have been especially struck by the resemblance of the active fields to the trajectories of two neurochemical circuits--those for B-endorphin and CRF (for summary see Panksepp, et al. 1988). Both of these systems have proved to be very important for inhibiting and promoting natural DVs respectively. It is as if these quintessential pro-distress (CRF) and anti-distress (endorphins) systems are interwoven in their trajectory through the neuroaxis-- exerting opposing modulatory control over DV circuitry as well as over each other. Indeed, our work with central CRF injections into the chick brain (Panksepp et al., 1987) has raised the possiblity that CRF is not just a modulatory factor in DV circuitry but may in fact be a 'command transmitter' itself and hence the very core of the executive circuitry of separation distress. As summarized in Figure 1, CRF can sustain a dramatically elevated level of DVs for up to six hours following 4th ventricular administration.

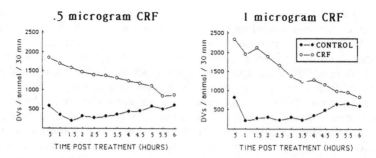

Fig. 1: Frequency of DVs (across 6 continuous hours in two week old birds following i.c.v. CRF into fourth ventricle region.

Certain investigators have raised concerns over whether the reduction of separation induced crying induced by opiates is really a specific emotional effect. For instance, Levine's group has reported that opiates reduced DVs in monkeys without a concurrent reduction of the separation-induced pituitary adrenal activation (Wiener et al., 1988). We, of course, suspect that the behavioral response is a more specific indicator of activity in the emotional systems which subserves separation distress than the glucocorticoid response, especially since the pituitary adrenal system is well known to respond to practically all stessors. It seems likely to us that other negative emotions such as fear and anger could easily be evoked by procedures typically used to produce social separation, and we suspect that the failure of the pituitary adrenal response to be attenuated in certain studies is simply due to the fact that this final common pathay can, in fact, respond to many distinct emotions. Thus opiates may be quite specifically reducing separation distress while leaving other negative emotional responses intact which can still arouse the pituitary adrenal system. In any case, it is noteworthy that other investigators have observed anti-distress effects of opiates with both vocalization and glucocorticoid measures in monkeys (Kalin, et al. 1987).

We have now analyzed the effects of many other centrally administered neuropeptides on separation distress using the avian model, and have observed many modest effects (e.g., slight elevations with arginine vassopressin, and ACTH, and bimodal effects (facilitation followed by inhibition) with a-MSH, and a substantial inhibition of all behaviors with somatostain). In our experience, however, only one other class of peptides beside opioids has yielded a most striking calming of separation distress that as not due to any apparent sedation. These agents were the posterior-pituitary peptides, oxytocin and vasotocin (Panksepp, 1988), which are known to control other social processes such as partruition, lactation, nursing and maternal behaviors. In birds, vasotocin, can inhibit DVs at remarkably low doses for several hours. We have observed reliable reductions with central doses as low as .01 nanomoles which lasts for an hour, with slightly higher doses suppressing DVs for up to 3 hrs (Figure 2). These animals exhibit no inhibition of various other behaviors like feeding and approaching a flock of birds. They locomote more, walking around seemingly 'courageouly' (i.e., they are willing to move farther from a flock than control birds), and they also exhibit heightened levels of several other behaviors, including yawning, lateral head shaking, annd wing-flapping . It is noteworthy that the wing-flapping needs social-facilitation (i.e., the presence of other birds) in order to be vigorously expressed, and we suspect that it reflects an internal state akin to a sense of dominance/confidence.

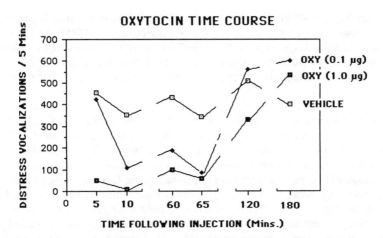

Figure 2: Time course of i.c.v. oxytocin effects on separation-induced DVs in week old chicks.

The ability of oxytocin and vasotocin to powerfully inhibit separation distress is congruent with various other physiological effects of this peptidergic system. I suspect oxytocin will prove to be as powerful a central anti-distress system as opioids, especially stress which arises from social interactions. Indeed, oxytocin secreted during suckling, may not only deliver milk to the child, but may reciprocally and directly alleviate the mother's tensions concerning any distress the child may be experiencing. Perhaps in psychic resonance to the activation of the maternal oxytocin systems, the childs own system becomes aroused, alleviating its distress, promoting relaxation and thereby facilitating feeding. As mentioned before, oxytocin can also promote maternal behavior (Pedersen, et al., 1982) even though this seems to occur in rats only if the effect of olfactory inhibitory processes are first diminished (Wamboldt and Insel, 1987). In any case, it is quite attractive that a system that would promote maternal intent would also modulate the separation responses.

Oxytocin, of course, is also found in high levels in males. Not only does this peptide promote ejaculation, but it can cause spontaneous erections in male rats (Melis et al., 1986). It is also massively released during sexual activity and appears to mediate the post-ejaculatory period of peaceful behavioral relaxation (Hughes et al., 1987), and may be a key ingredient in the erotic feelings which result from sexual activities. Whether the positive emotional response which emerges from this peptide in emotional situations as diverse as maternal care and sex is fundamentally similar, deserves further consideration. In addition to providing the affective background of pleasure and relaxed comfort which emerges from sociosexual activity, high levels of this peptide may directly promote peaceful attitudes and social interactions. For instance, vasotocin, the ancestral precursor of oxytocin, has a very powerful anti-aggressive effects in rats (Brown and King, 1984). Perhaps the high levels of sexual presentation which female primates commonly direct toward dominant males is an instrumental attempt to control male aggression via the inducing release of peace-promoting hormones. Although these are very specualtive lines of thought, further work with oxytocin and induction of pro-social attitudes is warranted. I suspect, in agreement with several other investigators that additional research will reveal that oxytocin and vasotocin directly promotes the bonding that normally occurs as a result of positive social relationships

(Pedersen and Prange, 1986; Kendrick et al., 1987). We have previously suggested that opioids may also subserve related functions (Panksepp et al., 1985), and the fact that oxytocin (and its constituent peptide Pro-Leu-Gly) can modify the speed at which opiate addiction and tolerance occur (Pucilowski et al., 1985; Van Ree and de Wied, 1976) may provide an additional interactive mechanisms for the regulation and chanelling of social attachments.

Pharmacological Approaches: In addition to peptide studies, we have also evaluated the role of most other known neurotransmitter systems using pharmacological strategies. Since many of the key results have been summarized elsewhere (Panksepp, et al., 1988), I will only highlight the major findings here. It would not be surprising that most drugs at sufficiently high doses modify DVs, but we have been impressed by the fact that most pharmacological effects are relatively weak and typically only appear when concurrent symptoms of neurotoxicity are apparent. Several drugs however yield what appear to be specific effects, and our thinking concerning the nature of the underlying substrates is guided more by manipulations which strongly increase rather than those which reduce DVs. Several drugs increase DVs in relatively weak modulatory ways (e.g., mu opiate receptor, serotonin receptor and muscarinic acetylcholine receptor antagonists), but we have identified three agents which turn on vocalizations in an obligatory manner as if fixed-action pattern circuits had been triggered. Beside CRF described above, we have found this effect with central adminsitration of curare, and the glutaminergic receptor agonists, N-methyl-D-aspartate (NMDA) and kainic acid.

With regard to drugs that can reduce DVs markedly and specifically, the opiate receptor agonists have proven to be consistently effective in whatever form they are give. In addition to the alkaloids morphine and oxymorphone, we have tested a dozen opioid peptides which have relatively strong effects on the mu receptor and all reduce DVs in accord with their reported receptor affinities, with DAGO and B-endorphin being most potent. Delta receptor agonists such as DSLET and kappa receptor agonists such as Dynprophin A are essentially ineffective by comparison. In addition to opiates, we have also found that nicotine and the alpha-norepinephrine receptor agonist clonidine are quite powerful. In our experience, clonidine was not exerting its effect pre-synaptically since total brain NE denervation did not attenutate the response (Rossi et al., 1983). Although strong and presumably behaviorally specific reductions of DVs have been reported with benzodiazepines in infant rats (Insel, Hill and Mayor, 1986), in our experience chlordiazepoxide has a comparatively modest effect on DVs of chicks. It is noteworthy that Insel and colleagues obtained their largest effects when animals were under thermoregualtory stress, and hence the effects may not have been specific to separation distress. The difference may also be due to species differences, but in general the cross-species generality of most of the pharmacological effects which have been reported so far has been rather impressive.

A Possible Role for Play in the Modulation of Stress.

Considering how much of every-day stress typically arises from social interactions--given the battles for dominance and the routine politics of every-day life in social animals--one might expect that the brain has evolved mechanisms which may help innoculate organisms against future social stress. For over a decade now we have been working on the puzzling phenomenon of social play--trying to identify the brain mechanisms that control this pro-social psycho-behavioral system of the brain (Panksepp et al., 1987). We have become ever more intrigued by the possibility that one non-obvious function of this brain system may be to provide the types of experiences and behavioral strategies which help animals ward of the effect of future stressors. In addition to the obvious social-learning function that this system may subserve (although that possibility remains to be empirically affirmed), a basic physiological function of play behavior may be to promote central traits such as 'heartiness' which help animals cope not only psychologically but also physiologically with future stressors they encounter. For instance, play may help regulate the speed and vigor with which the pituitary-adrenal response can be recruited and diminished following a stressor. Although speculative, there are some pertinent lines of evidence to consider.

The ability of an excessive and chronic pituitary-adrenal stress response to destroy hippocampal cells in a manner that resembles natural aging (Sapolsky et al., 1985), suggests that long-term emotional homeostasis may be a key vector which ameliorates and counteracts degenerative processes in the brain. Indeed, *if one makes the assumption that a healthy hippocampus has an abundance of receptors which help provide higher cognitive (e.g., hippocampal) feedback over stress responses,* it is noteworthy that early infantile 'handling' selectively increases corticosterone receptors in the hippocampus, while in other areas of the brain such as frontal cortex, this beneficial effect is sustained only if animals have abundant social experiences (play?) throughout their juvenile period of development (Meaney et al., 1985). Thus, it is reasonable to supposed that certain forms of social interaction, such as rough and tumble play, may indeed help invigorate and strengthen the stress-protective systems of the brain.

In summary, the separation distress system may be one of the most common sources of 'distressful' emotional arousal which emanates from primal process affective systems of the brain. Considering the fact that CRF, opioids, oxytocin and alpha-2-noradrenergic systems seem to be powerfully involved in controlling this response, it is noteworthy that all of these systems seem to be anatomically highly interconnected. Two prominent nodal points of this system are the locus coeruleus and the paraventricular nucleus of the hypothalamus. Such anatomies suggest that the system may operate in a relatively homogeneous mood/behavior controlling capacity. Indeed, the existence of pro-social neurochemical components (i.e., opioids and oxytocin) in this brain system, which help reduce distress and promote bonding, makes us wonder whether these brain substrates might not be the primal neural mechanisms for psychic states known as acceptance, trust, and perhaps even love, from which altruistic intent can ultimately arise. The ability of helping behaviors to induce emotional states which may promote health has received considerable attention recently, and it is worth considering whether the sense of well-being and the touted health benefits of helping behaviors are elaborated through the brain circuits described above. It may be through the consideration of such circuit properties that the most effective future strategies for pharmacological alleviation of stress may be constructed.

In Summary

From the perspective that the ultimate psychological meaning of 'stress' is the failure of homeostasis in emotional systems of the brain, it is noteworthy that there appear to be specific systems in the brain which have been designed to promote distress (CRF). Some components of the system (the paraventricular/hypophyseal system and its steroid feedbacks onto hippocampus, etc.) may operate in a very general 'stress' promoting ways, while others (the internal brain CRF systems) may operate in more specific emotional ways. The brain also contains powerful anti-distress systems, some of which appear to have trajectories which parallel brain CRF systems. Perhaps the most comprehensive theoretical encapsulation of opioid functions in the brain is that those systems provide a general mechanisms for re-establishing homeostatic equilibrium whenever major deviations from stable regulated levels have occurred (Panksepp, 1986). The longer the disequilibrium, the more likely it becomes that long-term medical/psychiatric consequnces will emerge not only because of changing sensitivities in the underlying systems, but also because of actual damage to the underlying physiological substrates. Of course, we can be certain that many other peptides will participate in the many emotional changes of the brain which, in the past, have been too facily subsumed under a global 'stress concept'. On the pro-stress side, there are candidates such as brain ACTH, MSH, AVP and perhaps neuropeptide Y. On the anti-stress side, it looks like molecules such as oxytocin and vasotocin will surely have very powerful anti-stress effects and ones like neurotensin (Nemeroff et al., 1982) and delta-sleep-inducing protein (Graf and Kastin, 1986), presently appear to enhance the ability of the brain to cope with stress. The exact adaptive functions of these peptides remain largely unknown, but I suspect that we simply will not be able to make functional sense of many of them without considering the types of affective symbolic processes exist in the brain because of the type of evolutionary history the vertebrate line has passed through. These neurosymbolic affective circuits, designed to promote survival, cannot be easily captured in the non-psychological terminology of neurophysiology. I think we really have no reasonable alternative but to begin using some of the ancient affective concepts that evolved early in the development of human languages as labels for the many emotional systems which we need to understand in much greater detail. Once we have labeled some primal processes relatively adequately, we will be in a much better position to study them empirically and eventually to understand them reductionistically.

References

Allen, J.P. and Allen, C.F. (1974) Role of amygdaloid complexes in the stress-induced release of ACTH in the rat. *Neuroendocrinology 15, 220-230* .

Allen, J.P. and Allen, C.F. (1975) Amygdalar participation in tonic ACTH secretion. *Neuroendocrinology 19*, 115-125.

Allen, C.F., Allen, J.P. and Greer, M.A. (1974) Anterolateral hypothalamic dafferentation prevents compensatory hypersecretion of ACTH folloing adrenalectomy in the rat. *Proc. Soc. exp. Biol. Med.* 146, 840-843.

Angelucci, L., Valeri, P., Grossi, E., Veldhuis, H., Bohus, B. and De Kloet, E. (1980) Involvement of hippocampal corticosterone receptors in behavioral phenomena. In *Progress in Psychoneuroendocrinology,* F. Brambilla, G. Racagani and D. de ied (Eds), Elsevier Biomedical Press, Amsterdam, 177-185.

Ax, A.F. (1953) The physiological differentiation of fear and anger in humans. *Psychosomatic Medicine* 15, 433-442.

Baulieu, S., Paolo, T.D. and Barden, N. (1986) Control of ACTH by the central nucleus of the amygdala: Implication of the serotonergic system and its rlevance to the glucocorticoid delayed negative feedback mechanism. *Neuroendocrinology 44*, 247-254.

Bowlby, J. (1973) *Separation,* New York: Basic Books.

Brown, R. and King, M.G. (1984) Arginine vasotocin and aggression in rats. *Peptides* 5: 1135-1138.

Cabanac, M. (1979) Sensory pleasure. *Quarterly Review of Biology* 54: 1-29.

Coe, C.L., wiener, S.G., Rosenberg, L.T., and Levine, S. (1985) Endocrine and immune responses to separation and maternal loss in nonhuman primates. In. M. Reite and T. Fields (Eds.) *The Psychobiology of Attachment and Separation,* Academic Press, New York, 163-199.

Ekman, P. Levenson, R.W., Friesen, W.V. (1983) Autonomic nervous system activity distinguishes between emotions. *Science* 22, 1208-1210.

Eleftheriou, B.E., Zolovick , A.J. and Pearse R. (1966) Effect of amygdaloid lesions on pituitary-adrenal axis in the deermouse. *Proc. Soc. exp. Biol. Med.* 122, 1259-1262.

Gerlach, et al (1974) Corticosterone, cortisol, and estadiol bind differentially to specific cell groups in rhesus monkey brain and pituitary. *Proc. 4th Ann. Meeting of the Society for Neuroscience* 232: 224.

Graf, M.V. and Kastin, A.J. (1986) Delta sleep-inducing peptide (DSIP): An update. *Peptides* 7: 1165-1187.

Gray, J.A. (1987) *The Psychology of Fear and Stress.* Cambridge: Cambridge University Press.

Henry, J.P. (1986) Neuroendocrine patterns of emotional response In *Emotion: Theory, Research, and Experience, vol. 3, Biological Foundations of Emotions* (R. Plutchik and H. Kellerman, Eds.) Academic Press, Inc, New York, pp 37-60

Herman, B. H. and Panksepp, J. (1981) Ascending endorphinergic inhibition of distress vocalization. *Science* 211, 1060-1062.

Hofer , M.A. (1984) Relationships as regulators: A psychobiologic perspective on bereavement. *Psychosomatic Medicine* 46, 183-197.

Hughes, A.M., Everitt, B.J., Lightman, S.L. and Todd, K. (1987) Oxytocin in the central nervous system and sexual behaviour in male rats. *Brain Research* 414: 133-137.

Insel, T.R., Hill, J.L. and Mayor , R.B. (1986) Rat pup ultrasonic isolation calls: possible mediation by the benzodiazepine receptor complex. *Pharmacology Biochemistry and Behavior* 24: 1263.

Johnson, E.H. and Broman, C.L. (1987) The relationship of anger expression to health problems among black americans in a national survey. *Journal of Behavioral Medicine* 10: 103-116.

Julius, S., Schneider, R., and Egan, B. (1985) Suppressed anger in hypertension: facts and problems. In Chesney, M.A. and Rosenman, R.H. (Eds.) *Anger and Hostility in Cardiovascular and Behavioral Disorders.* Hemisphere, McGraw-Hill, Ne wYork.

Kalin, N.H., Shelton, S.E. and Barksdale, C.M. (1988) Opiate modulation of separation-induced distress in nonhuman primates. *Brain Research* 440: 285-292

Kelly, D. (1980) *Anxiety and Emotions,* Charles C. Thomas, Springfield, Illinois

Kendrick, K.M., Keverne, E.B., and Baldwin, B.A. (1987) Intracerebroventricular oxytocin stimulates maternal behaviour in the sheep. *Neuroendocrinology* 46, 56-61.

Kiecolt-Glaser, et al (1987) Marital quality, marital disruption and immune function. *Psychosomatic Medicine* 49: 13-21.

Kobasa, S.C., Maddi, S.R., and Kahn S. (1982) Hardiness and health: a prospective study. *Journal of Personality and Social Psychology* 42, 168-177.

Mancia, G. and Zanchetti, A. (1981) Hypothalamic control of autonomic functions. In. Morgane, P.J. and Panksepp, J. (Eds.) *Handbook of the Hypothalamus, Volume 3, Part B. Behavioral Studies of the Hypothalamus.* Dekker, New York, 147-202.

Martin, J., Endroczi, E. and Bata, G. (1958) Effects of the removal of amygdaloid nuclei on the secretion of adrenal cortical hormones. *Acta Physiol. Hung.* 14, 131-134.

Meaney, M.J., Aiken, D.H., Bodnoff, S.R., Iny, L.J., Tatarewicz, J.E., and Sapolsky, R.M. (1985) Early postnatal handling alters glucocorticoid receptor concentrations in selected brain regions. *Behavioral Neuroscience* 99, 765-770.

Melenchuk , T. (1988) Emotions, brain, immunity, and health: a review. In. Clynes, M. and Panksepp, J. (Eds.) *Emotions and Psychopathology,* Plenum Press, New York, 181-247.

Melis, M.R., Argiolas, A., and Gessad, G.L. (1986) Oxytocin-induced penile erection and yaning: Sites of action in the brain. *Brain Research* 398: 259-265.

Nemeroff, C.B., Hernandez, D.E., Orlando, R.C. and Prange, A.J. (1982) Cytoprotective effect of centrally administered neurotensin on stress-induced gastric ulcers. *American Journal of Physiology* 242, 342-246.

Newman , J.D. (Ed.) (1988) Neural and neurochemical control of the separation distress call. In. *The Physiological Control of Mammalian Vocalizations.* Plenum Press, New York.

Pagels, H.R. (1988) *The Dreams of Reason: The Computer and the Rise of the Sciences of Compexity.* Simon and Schuster, New York.

Panksepp, J. (1982) Toward a general psychobiological theory of emotions. *The Behavioral and Brain Sciences* 5, 407-467.

Panksepp, J. , Siviy, S. , & Normansell, L. (1984) The psychobiology of play: Theoretical and methodological perspectives. *Neuroscience and Biobehavioral Reviews* 8, 465-492.

Panksepp, J. , Siviy, S. M. , & Normansell, L.A. Brain opioids and social emotions. In *The Psychobiology of Attachment and Separation,* Eds. M. Reite and T. Fields. New York, Academic press, 1985, 3-49.

Panksepp, J. , (1986) The neurochemistry of behavior. *Annual Review of Psychology* 1986, *37,* 77-107.

Panksepp, J., Normansell, L., Cox, J., Crepeau, L., and Sacks, D. (1987) Psychopharmacology of social play in J. Mos (Ed.). *Ethnopharmacology of Social Behavior,* Duphar, Holland, 132-144.

Panksepp, J. , Crepeau, L. , and Clynes, M. Effects of CRF on separation distress and juvenile play. *Neuroscience Abstracts* 1987, *13,* 1320.

Panksepp, J. Posterior pituitary hormones and separation distress in chicks. *Neuroscience Abstracts* 1988, *14,* 287.

Panksepp, J. , Normansell, L. , Herman, B. , Bishop, P. & Crepeau, L. (1988) Neural and neurochemical control of the separation distress call. In. J. D. Newman (Ed.). *The Physiological Control of Mammalian Vocalizations.* Plenum Press, 263-300.

Panksepp, J., Sacks, D.S., Crepeau, L.J., Abbott, B.B. (1989) The psycho- and neuro-biology of fearsystems in the brain. In *Aversive Events and Behavior,* M.R. Denny (Ed.), Lawrence Erlbaum Assocs. Inc. Publisher, 1989, In Press.

Pedersen, C.A. et al. (1982) Oxytocin induces maternal behavior in virgin female rats. *Science* 216: 648-650.

Pedersen, C.A. and Prange, A.J. (1986) Oxytocin and mothering behavior in the rat. In D. de ied, .H. Gispen and T. B. van w imersma Greidanus (Eds.) *Neuropeptides and Behavior Volume 2: The Neurohypophyseal Hormones,* Pergamon Press, Oxford, 141-156.

Pelletier, K.R. and Herzing, D.L. (1988) Psychoneuroimmunology: Toward a mind-body model: A critical review. *Advances, Journal of the Institute for the Advancement of Health* 5: 27-56.

Pucilowski, O., Kostowski, w. and Trzaskowska, E. (1985) The effect of oxytocin and fragment (MIF-1) on the development of tolerance to hypothermic and hynotic action of ethanol in the rat. *Peptides* 6: 7-10.

Rabin, B.S., Cunnick , J.E., and Lysle, D.T. (1988) Alteration of the immune system by housing. *Advances, Journal of the Institute for the Advancement of Health* 5: 15-25.

Rahe, R.H. (1975) Epidemiological studies of life change and illness. *Intern. J. Psychiatry Med.* 6: 133-146.

Reite, M. and Capitanio, J.P. (1985) On the nature of social separation and social attachment, In *The Psychobiology of Attachment and Separation,* Eds. M. Reite and T. Fields. New York, Academic press, 1985, 223-258.

Rossi III, J. Sahley, T.L. and Panksepp, J. (1983) The role of brain norepinephrine in clonidine suppression of isolation-induced distress in the domestic-chick. *Psychopharmacology* 79: 338-342.

Rumelhard, D.D. and McClelland, J.L. and the PDP Research Group (1986) *Parallel Distributed Processing. Explorations in the Microstructure of Cognition* vols 1 and 2. Cambridge, Mass: MIT Press.

Rutter , M. (1972) *Maternal Deprivation Reassessed,* Penguin Books, Middlesex, England.

Sapolsky, R.M., Krey, L.C. and McEwen, B.S. (1985) Prolonged glucocorticoid exposure reduces hippocampal neuron number: implications for aging. *J. Neurosci.* 5, 1222-1227.

Sapolsky, R.M. and Meaney, M.J. (1986) The maturation of the adrenocortical stress response in the rat. *Brain Research Reviews* 11, 65-76

Seggie, J. (1987) Differential responsivity of corticosterone and prolactin to stress folliong lesions of the septum or amygdala: implications for psychoneuroendocrinology. *Prog. Neuro-Psychopharmacol. and Biuol. Psychiat.* 1, 449-461.

Selye, H. (1980) *The Stress of Life.* New York: McGraw-Hill Paperback (Original work publsihed 1976).

Swanson, L.W., Sawachenko, P.E. and Lind, R.W. (1986) Regulation of multiple peptides in CRF parvocellular neurosecretory neurons: implications for the stress response. *Progress in Brain Research* 68, 169-190

Swanson, L.W., Sawachenko, P.E., Rivier, J. and Vale, W.W. (1983) Organization of ovine corticotropin-releasing factor immunoreactive cells and fivers in the rat brain: An immunohistochemical study. *Neuroeneocrinogloy* 36: 165.

van Ree, J.M. and de wied, D. (1976) Prolyl-leucyl-glycinamide (PLG) facilitates morphine dependence. *Life Sciences 19, 1331-1340.*

Wagner , H.L. (Ed.) (1988) *Social Psychophysiology and Emotion.* John wiley & Sons, Chichester.

Waid, W.M. (1984) *Sociophysiology,* Springer-Verlag, New York.

Wambolt, M.Z. and Insel, T.R. (1987) The ability of oxytocin to induce ;short latency maternal behavior is dependent on peripheral anosmia. *Behavioral Neuroscience 101:* 439-441.

Warembourg, M. (1975) Radioatographic study of the rat brain after injection of (1,2³H)corticosterone. *Brain Research* 89, 61-70.

Weiner, S., Levine , S., and Coe, C. (1988) Neural and neurochemical control of the separation distress call. In. J. D. Newman (Ed.). *The Physiological Control of Mammalian Vocalizations.* Plenum Press, New York.

BEHAVIORAL EFFECTS OF MANIPULATIONS OF THE OLFACTORY ENVIRONMENT IN DEVELOPING MICE: INVOLVEMENT OF THE DOPAMINERGIC SYSTEM.

Francesca R. D'Amato and Simona Cabib
Istituto di Psicobiologia e Psicofarmacologia del C.N.R., Roma, Italy

Interest in stress has greatly increased in the last few years, not only because it has been shown to produce, or at least to be associated with, neurochemical alterations, affective disorders (Anisman and Zacharko, 1982; Willner, 1984) and alterations in the immune system functioning (Maier et al., 1985), but also because it represents an useful paradigm to investigate the capability of the organism to respond to new situations.

It has become increasingly clear that stress, or the deleterious consequences of exposure to a stressor, is not merely the result of exposure to a given physical event or psychological condition. Stress itself is the result of a complex interaction of psychological factors with the actual physical event, and coping processes have come to be viewed as central in this regard (Levine and Ursin, 1979). The term coping generally refers to problem solving efforts to adapt to important environmental demands. Stress is said to result from exposure to an aversive event only if the individual cannot cope, or anticipates that it cannot cope with the event (Lazarus et al., 1974).

Exposure to inescapable and unavoidable electric shock can result in decreased unconditioned activity, interference with subsequent learning, enhanced reactivity to stimulant drugs, and alterations in a variety of neurochemical and hormonal systems. When the animal is allowed to escape or to produce some behavioral responses to the stressful stimulation these effects will be markedly reduced even if the amount of shock the animal is submitted to is unchanged (Weiss et al., 1970; Conner et al., 1971).

All kinds of repeated and consistent alterations of an individual's environment that cannot be coped with, can thus represent stress situations, leading to modifications of the internal (physiological) environment of the subject. When the stress experience involves the developing organism, implications and scopes of the research are different from studies conducted on adult subjects. In fact plasticity is at its maximum during the early stages of life and represents the mechanism through which each organism can adapt to its environment. The process of development is led by the genetic program, but the feedback from the environment is absolutely necessary to allow each organism to be adapted to the environment it is going to face. For this reason, stress experienced during the developmental phase will interact with maturation processes possibly leading to life-long effects in terms of histological modifications, hormonal and neurochemical functioning, as well as in the behavioral outcome.

Both mice and rats are of great relevance for the study of long-lasting effects of neonatal stress because these species are characterized by offspring not mature at birth. This signifies that the process of development has not been completed during the fetal life and that

S. Puglisi-Allegra and A. Oliverio (eds.), Psychobiology of Stress, 59–71.
© 1990 *Kluwer Academic Publishers. Printed in the Netherlands.*

plasticity is still present during the first weeks postpartum.

Among rodents, not only altricial but also precocious species can be found, as for example the spiny mouse and the guinea pig. The different development timing characterizing these different reproductive strategies represents an useful tool for investigating maturation times and plasticity of specific neurotransmitter and hormonal systems in genetically close species or within the same species, by using inbred strains (Oliverio et al., 1979).

Early stimulation in rodents

Stress can be applied to the developing organism during different phases of its development, according to the timing of maturation of the system one is interested in. If the system is already mature at birth, stress must be applied during the prenatal life, and the subject under stress will be, of course, the mother, during gestation. The physiological response to stress expressed by the mother can be experienced by the offspring through the exchange of information that takes place via the placenta. In this case, pups have no experience of the stressor, but they share with the mother the physiological response generally associated with it; we can hypothesize that they 'experience' stress. This can occur if the system in the developing organism is already mature and can perceive the messages coming from the mother. Probably no attempt will be made to cope with the stressful situation if no 'normal' (baseline) situation has been already experienced. These prenatally stressed pups are characterized by an abnormal internal environment that can affect the maturation processes of those systems needing feedback from the pups' fetal environment. In this case, parameters that should represent the basal levels on which the system must be set, have been modified by maternal response to an hostile situation.

It is important to underline the adaptive meaning in terms of individual survival of this regulation of the ontogenetical program allowing the growing organism to respond to, and face a certain amount of variation in its environment. Of course, limits exist to the variability the organism is programmed to face, and these are determined by the genetic material of the individual, and by the natural history of the species.

Postnatal Stress

After birth stressors can be applied directly to the pups, according to the sensorimotor capacities that characterized the different phases of their development (Fox, 1965). Pups can try to cope with the stressful event according to their sensorimotor maturity. In this case the definition of stress is the same for adults as for developing animals: an alteration of the individual's environment the subject is unable to cope with. Generally, because of the relative immaturity of young rodents at birth, any change, whatsoever in their environment can represent a stress, once their sensory system is mature enough to detect that alteration.

Since the 1950's, many studies have investigated the mechanisms by which profound effects on behavior and physiology of adult rodents emerge following stimulation applied during development (Daly, 1973). Hypotheses emerging from these studies have been reviewed by Russell (1971) and included: (1) the Stress hypothesis (Levine, 1956); (2) the Direct Action hypothesis (Levine, 1962); (3) the Hypothermia hypothesis (Schaefer and Weingarten, 1962); and (4) the Maternal Mediation hypothesis (Barnett and Burn, 1967; Meier and Schutzman, 1968).

The Stress hypothesis suggests that whatever the stimulation the pup is subjected to, it is in some way noxious or stressful. This definition suffers from circularity, until an independent measure of stress, possibly based on physiological correlates at the time of the treatment, is provided. The Direct Action hypothesis indicates additional stimulation deriving from treatment (whether mere handling or handling plus gentling or shock) to be responsible for modifying physiological systems in the neonate.

The third hypothesis proposed by Schaefer and Weingarten (1962), views the effects of treatment as the result of incidentally cooling the pups, during treatment. The last hypothesis suggests that the effects of treatment may be mediated not by the direct action on the pup, but undirectly, through the maternal behavior. Young (1965) suggested that infantile treatments may change the stimulus properties of the neonate in such a way as to influence maternal behavior. Alterations of maternal behavior can be obtained in several ways and seem to be responsible for affecting offspring behavior (Cohen-Salmon et al., 1985; D'Amato et al., 1988; D'Amato and Populin, 1987; Denenberg and Whimbey, 1963, Hudgens et al., 1972; Joffe, 1965; Muir et al., 1985; Ottinger et al., 1963; Villescas et al., 1977).

The Developing Organism and its Environment.

It has become increasingly clear that the intensity of the stressor is as important as its quality. Exposure to a stressful event a pup might meet in its natural environment can result in a response that has been positively reinforced by natural selection. On the contrary, exposure to a completely artificial stressor could correspond only in a disorganization of the system. The infant differs from its adult counterpart in anatomical form, physiological function and behavioral repertoire. Live-born mammalian offspring undergo a postnatal period during which they experience intense and specialized interactions with their parents. Stimuli that regulate parent-offspring interactions, and mechanisms that mediate them, are characteristic of the species under investigation. The reproductive strategy of rodents, like that of most mammals, does not extensively involve paternal behavior; the female takes care of her offspring, nursing, transporting, nest-building and licking the young, until weaning (Rheingold, 1963). Recent literature emphasizes the active role of the offspring in regulating the amount and the quality of parental behavior (Trivers, 1972). The mother-offspring relationship is now viewed as the result of mutual control and reciprocal relations; synchrony between mother's activities and the offspring's maturation level represents continuous adaptation to different developmental stages the infant goes through.

The strong interdependence of infant and mother stimuli in modifying each other's physiology and behavior, represent an additional variable to be kept in mind when analyzing the effect of specific stimulations in the mother or in the infant. In fact, we can hypothesize that if the pup is able to communicate with the mother about the nature of the stressor (for example, through ultrasonic communication), the mother could try to solve its problem. The quality of the stressor is particularly important in this regard: if the nature of the stressor is artificial, natural selection could not provide the pup with the ability to inform the mother about it; in this case the stressor cannot be removed by the caretaker.

Experimental stress in developing rodents

Separation from the mother appears to be the most powerful stressor in the developing organism. In fact, the system of attachment is critical in terms of infant survival (Bowlby, 1969), and even short events of maternal separation can result in long-lasting (behavioral, hormonal, neurochemical, immunological) effects in monkeys (Hinde and Spencer-Booth, 1971; Hinde et al., 1978; Laudenslager et al., 1985; Kraemer et al., 1984). The same seems to be true in rodents, even if separation from the mother was usually combined with other stressing factors as isolation, cooling, handling, and undernurishing (Plaut, 1970; McCarthy et al., 1981; Michaut et al, 1981; Raymond et al, 1986).

Albert (1986) stressed that pups inhabit a different niche than the adults; as a consequence, he suggested that perhaps pups should be investigated with different procedures. Rodents' sensorimotor system is relatively immature at birth, and so stressor they are to be submitted to must be in phase with the pups' development. The study of maturing reflexes can provide indications of the perception of the external stimuli the pups have at different ages (Fox, 1965.). Due to the sensory characteristics of rodent species, olfactory cues seem to represent the main channel of environmental investigation by pups during the early postpartum period. In fact pups show behavioral modifications to olfactory stimulation. For example the odor of the male elicits different ultrasonic responses than the female scent (Conely and Bell, 1978).

Another problem concerning developing organisms refers to the change of the stressful content of a stressor, as the pups grow up. In fact, an event may have a stressful meaning for a 8-day, but not for a 15-day, old pup (for example, temperature; Oswalt and Meier, 1975). When chronic stress is applied to developing animals, the phenomenon of habituation may not only be viewed as a desensitization to that stressor (behavioral adaptation) or the result to changes in biochemical systems, but also explained in terms of maturation processes that allow the growing organism to cope with the stressor. For example, Richardson, Siegel and Campbell (1988) reported that adult rats (60-day old) showed a rapid decline in heart rate during 90-min exposure in an unfamiliar environment, while 16-day old animals did not adapt to the situation and maintained a high level of fear. Rats of intermediate ages displayed intermediate responses.

Effects of acute and chronic exposure to a novel olfactory environment on developing mice

In order to investigate the effects of stress experience in developing mice, we have evaluated the behavioral response to acute and chronic treatment at different stages of the pups' maturation.

We decided to use as stressor exposure to clean bedding, which may represent the most powerful stressor, that is to say, separation from the mother. The stressing procedure consisted in exposing the entire litter of outbred Swiss mice to clean bedding for a 15 min period. Once a day (chronic treatment) each whole litter was transferred to a new cage, the floor of which was covered with either clean bedding (C-B group) or with their home-cage bedding (HC-B group). During the entire 15 min of the procedure's duration, the experimental cage was placed on a hot plate at a temperature of 35 °C. The stress sessions were conducted in a soundproof cabin with constant temperature (25 ± 1 °C). According to the description of the stress procedure above provided, isolation, low temperature and handling cannot explain differences between our stressed group (C-B) and its control group (HC-B). Isolating and submitting pups to one single and definite stressor is important because different stressors may have different physiological and behavioral responses in adults as well as in pups (Denenberg, 1971; Kant et al., 1985).

Experimental investigations have shown that manipulations of the pups' environment can stimulate consistent emission of ultrasounds in rodent pups (Oswalt and Meier, 1975). Thus, the monitoring of ultrasonic signals by infant rodents provides a useful tool in the simultaneous assessment of the infant responsiveness to an environmental event. The number of ultrasonic vocalizations (70 $\pm$ kHz) emitted by the whole litter was recorded for 5 min, between the 5th and the 10th min of the 15 total min of stress.

On the basis of ultrasonic emission (distress vocalizations), six day-old pups reveal the capacity to discriminate between clean and home-cage bedding (H $= 11.08$, p $< .05$)(Fig. 1). This difference is statistically significant for both acute (U $= 5.00$, $n_1 = n_2 = 5$, p $< .05$) and chronically treated pups (U $= 25.50$, $n_1 = 10$, $n_2 = 12$, p $< .05$). Thirteen day-old pups continue to differ in the amount of ultrasonic calls emitted following alteration of their olfactory environment (H $= 12.25$, p $< .01$). Pups exposed for the first time to clean bedding at 13 days show the same ultrasonic response as home cage bedding exposed pups. This would suggest that exposure to clean bedding is not a stressful event for a 13 day-old pup. Alternatively, this olfactory manipulation may still be stressful, but the 13 day-old pups do not respond with ultrasounds to this stressor, but show a different reaction. Chronically treated mice instead continue to show the ultrasonic response at 13 days (U $= 23.00$, $n_1 = n_2 = 10$, p $< .03$), as they do when they are younger. This indicates that there is no habituation to this stressor, as the behavioral response is still evident in the older pups, and no statistical difference discriminates between pups at different ages. In addition, chronically stressed pups continue to emit a significantly greater amount of ultrasounds than acute stressed pups (U $= 4.5$, $n_1 = 10$, $n_2 = 6$, p $< .01$).

Effects of stressing mouse pups on the mothers' behavior

The presence of the mother until weaning and the strong relationship she has with the pups during the first weeks postpartum, represent a variable that cannot be ignored when stress is applied on young animals. The relation between mother and offspring is so close during the first weeks after birth, that the pups can modulate the mother's behavior in order to signal their needs, and the mother can regulate pups' development and limit their behavior. In addition, through the milk they feed on, pups can acquire further information about the environment (for example, about the food the mother has eaten, Galef and Henderson, 1972).

The presence of the mother could either emphasize or reduce the effectiveness of a stressor. Data from the literature suggest that stressing the pups results in an increase in the physiological correlates of stress in the mother (Smotherman et al., 1977). Returning shocked pups to their home cage resulted in different mother's behavioral responses than those in response to returning handled pups (Brown et al., 1977). Moreover, handling the pups resulted in alterations of mother's behavior in an open-field test (Hudgens et al., 1972) and towards pups (Villescas et al., 1977). These changes in the mother's physiology and behavior, which may result in further alterations of mother-offspring relationship (Bell, 1979), can represent an additional stressor from the pups' viewpoint. Otherwise, physiological correlates of mother's stress can be transferred to the pups via the milk they feed on (Angelucci et al., 1983).

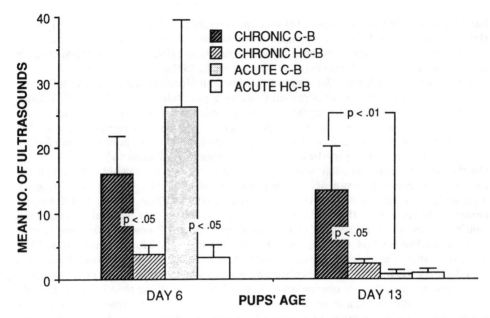

Fig. 1 Ultrasonic calls (mean ± S.E.) emitted by the entire litter during the 5 central minutes of the total 15 minutes of chronic and acute episodes of separation from the mother and exposure to clean bedding (C-B) or to home-cage bedding (HC-B).

However, the presence of the mother may also reduce the stressful experience when she can block the power of the stressor with her presence, her behavior or something else. The presence of an anesthetized lactating dam is sufficient to reduce the increase in the heart rate of 16-day old rats following exposure to an unfamiliar environment (Richardson et al., 1988). In addition, recent studies emphasize mother's capacity to moderate the pups' plasma corticosterone response in new situations (Hennessy et al., 1988; Stanton et al., 1988).

Changes in maternal behavior, according to the severity of the treatment the pups were subjected to, can be explained in terms of the amount of ultrasonic vocalizations emitted by the stressed pups (Noirot, 1972; Smotherman et al., 1974). As the post-treatment mother-pup interaction proceeded, a reduction in both active maternal behavior and ultrasonic calls occurred (Bell et al., 1974). Bell (1979) reviewed ultrasound-induced alterations in maternal behavior and reported that, following several successive days of exposure to augmented ultrasounds, the dams continued to show altered responsiveness to the pups' cues which can best be described as increased maternal responsiveness. However, to our knowledge, no data are available on maternal behavior far removed in time from stressful events involving pups.

Mothers' behavior outside the stressful events was evaluated in pups exposed to chronic stress, as previously described. The behavior of the females in their home cages with the pups was observed for 30 min, on Days 3, and 11 of the pups' life. Maternal behavior tests were conducted prior to the daily stress session in a soundproof cabin, 45 min after the cages had been transferred. Three experimental mothers' groups were compared: 1) mothers of chronically stressed pups, 2) mothers of pups chronically exposed to home cage bedding, and 3) mothers of unhandled pups. Behavioral observations for this last group were conducted on different subjects in the two sessions.

In a previous report we showed that more maternal behavior was found in mothers of litters exposed to clean bedding than in mothers of pups exposed to home-cage bedding (D'Amato and Cabib, 1987). These data suggest a strong modification of mother-offspring interaction in concomitance with chronic stress on pups. Table 1 reports the amount of some maternal behaviors of not only C-B and HC-B mothers but also Control mothers.

Mothers of pups exposed to home-cage bedding more closely resembled control mothers than mothers of pups exposed daily to clean bedding. In fact, the latter scored highest for nursing posture and time spent in nest, and lowest for grooming pups. Generally, mothers of the stressed litters spent most of their time in the nest; this could indicate that they were trying to compensate for stress effects on pups by displaying more maternal behavior.

It is well known that maternal behavior is stimulated by pups' ultrasonic calls (Bell et al., 1974). Our experimental design, with separate sessions for recording maternal or pups' behavior, does not allow any direct relation to be established between pups' behavior and mothers' response. However, one could hypothesize that stressed pups continue to emit more ultrasounds also outside the stress sessions, thus eliciting maternal behaviors (Bell, 1979; Terkel et al., 1979; but also Stern et al., 1984) and possibly reducing rough manipulation by the mother (Bell, 1974).

Effects of chronic exposure to stress during development on dopamine-mediated behavior

As previously stated, interest in neonatal stress consists in inducing stable modifications that can persist throughout the life of the individual. In adult subjects stress has been shown to activate dopaminergic, noradrenergic, GABAergic, and endorphinergic systems in various experimental paradigms (Amir et al., 1980; Dunn and File 1983; Nakagawa et al., 1981; Oliverio et al., 1984; Yoneda et al., 1983). Moreover, alterations of brain neurotransmitter functioning have been found following chronic exposure to stress (Antelman et al, 1980; Stone, 1983). More recently, it has been demostrated that, in adult mice, repeated exposure to immobilization stress modifies apomorphine-induced wall climbing, an effect which suggests an altered functioning of the central dopaminergic system (Cabib et al., 1984).

In particular, we were interested in evaluating the effects of chronic stress on the dopaminergic system during ontogeny. In fact, the persistence of neonatal stress effects on the dopaminergic functioning until adulthood can provide useful information and represents a good model for investigating the related problem in humans, according to the genetic-environmental hypothesis of schizophrenia.

Table 1. Behaviors displayed by mothers whose pups have been subjected to different experimental manipulations.

		C-B	HC-B	Control	H
IN NEST	(day 3)	29.37	29.12	24.57	1.69
	(day 11)	28.37	21.87	16.87	6.34*
NP	(day 3)	26.37	22.00	14.28	7.23*
	(day 11)	25.25	17.12	15.25	4.26°
GRP	(day 3)	1.00	2.12	2.00	3.08
	(day 11)	0.37	0.50	0.75	0.61

* $p < 0.05$; ° $p < 0.10$ (Kruskall-Wallis analysis of variance).
NP: nursing posture; GRP: grooming and licking pups.

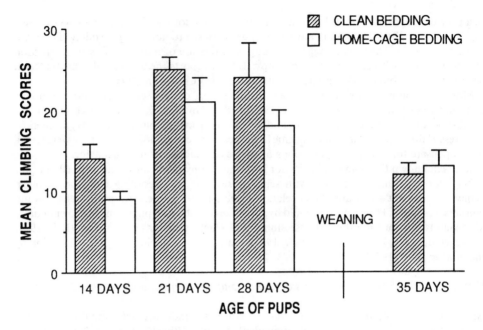

Fig. 2. Climbing scores (mean ± S.E.), following apomorphine injection (0.5 mg/kg sc.), in mouse pups chronically exposed 15 min daily to clean bedding or to home-cage bedding from day 1 to13 of life. Pups were weaned when 29 days old and tested again at 35 days of age.

The effect of chronic stress on apomorphine-induced climbing behavior (0.5 mg/kg) has been investigated in developing mice. As previously described, stress procedure consisted in exposing the entire litter to clean bedding for a 15 min period once a day. Pups submitted to chronic stress from day 1 to 13 of life, were tested at different ages during development. For this purpose, pups of each litter (each litter was culled to six pups) were individually marked. Two pups were tested on day 14, a further two on day 21, and the last two on day 28. On day 29 pups were weaned, but left with their littermates. All pups were tested again on day 35. A two-way ANOVA and a t-test were used to evaluate behavioral differences between C-B and HC-B treated subjects from day 14 to 28 and on day 35, respectively.

The results of this experiment confirmed previous findings with regard to the behavior of mice the day after the last stress session (D'Amato and Cabib, 1987): apomorphine induced wall climbing scores of pups belonging to litters exposed to clean bedding were significantly higher than those of pups from control litters and from a second group of litters, the pups of which were left undisturbed. This result indicates that clean bedding exposure increases apomorphine-induced climbing behavior, while handling and maternal separation per se do not induce this behavioral alteration. This finding suggests an alteration in dopaminergic functioning of mouse pups as has been observed in adult mice following chronic exposure to other kinds of stressors (Cabib et al., 1985).

The results of the long-term effects of 13 days of postnatal stress on climbing behavior are reported in Fig. 2. A significant effect of chronic stress was revealed by the ANOVA that indicated a treatment ($F = 7.63$, $df = 1/83$, $p < 0.01$) and an effect of pup's age ($F = 15.90$,

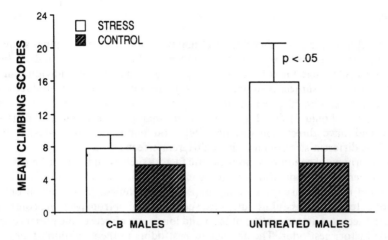

Fig. 3. Climbing scores (mean ± S.E.) of adult mice, either stressed during postnatal development (C-B males) or control (untreated males), following apomorphine injection (0.25 mg/kg sc.) 24 hrs after the last session of chronic stress treatment (10 consecutive days of 2 hrs of immobilization).

df = 2/83 p < .001): C-B group exhibited higher scores of climbing behavior from day 14 to 28. No difference persisted after weaning, as revealed by the t-test (t = -0.46, df = 86, ns). These results are open to different interpretations. The first explanation suggests a disappearance of the effect of 13 days of postnatal chronic stress on DA functioning with increasing age, i.e. at 35 days of age. However, another possible hypothesis is related to the concomitant occurrence of the weaning process. It has been shown that mothers' behavior of stressed pups is modified by chronic treatment. Thus, C-B pups, in addition to the experimental treatment, were also exposed to a different kind of mothering, and to an individual (the mother) in a different physiological condition. We can hypothesize that the mother represents the real or the crucial disturbing factor in our experimental situation. In this case, the weaning process, that is to say, the removal of the mother, might correspond to the elimination of the factor responsible for the maintenance of the altered apomorphine-induced response.

On the basis of the above experiments, the alteration of the dopaminergic functioning induced by postnatal stress disappeared in 35 day-old mice. However, one possibility could be that stressed and control animals did not differ in the dopaminergic functioning in basal condition but that they differed when again submitted to chronic stress in adulthood. Adult mice submitted to chronic stress (10 consecutive days of 120 min of immobilization) show altered sensivity to the behavioral effects of apomorphine if tested 24 hrs after the last stress session (Cabib et al., 1984; 1985). On the contrary, postnatally stressed male mice fail to show behavioral sensitisation to a challenge dose of apomorphine (Fig. 3). Two different interpretations could be formulated: the first one suggests that postnatally stressed animals are less sensitive to stressful stimulation in adulthood; alternatively, these animals, due to altered dopaminergic functioning, could lack the capability of adapting to a chronic stress procedure.

Conclusions

This study explores the possibility that brief manipulations of the individual's environment throughout the first weeks of life can produce life-long functional and/or behavioral modifications. We suggest that the adaptive significance (in terms of individual survival) of the manipulation the subject is subjected to, is important in order to reproduce a situation the animal is programmed to encounter, even if it cannot cope with it. In addition, the stressor selected should 1) avoid long and severe manipulations that might affect infant mortality and have direct non-specific effects on maturation processes 2) be clearly identified, as different stressors may have different effects.

In our experimental conditions, non-specific factors such as undernourishment, reduction of body temperature and painful or invasive manipulations, which could affect maturation processes and increase casualties among pups, were excluded. Moreover, in our model a number of potentially stressful stimulations, such as isolation from peers are controlled, thus allowing the accurate determination of brief and long-term effects of the presence or absence of a single factor: nest odor. The absence of nest odors in the condition of clean bedding exposure may simulate separation from the mother that, especially for species characterized by immature offspring, represents a critical situation in terms of survival which the infant, because of its immaturity, cannot cope with. Otherwise handling (picking up each pup and putting it into the new cage) may be considered as the critical factor; in this case, the absence of stress effects in the group exposed to home-cage bedding might depend on the capability of this familiar stimulus to reduce stress response (Kehoe and Blass, 1986; Richardson et al., 1988). In a recent study Siegel et al. (1988) demonstrated that tactile and thermal, but nor olfactory, stimulus dimension of the anesthetized dam are important in reducing the emotional response of young rats between 16 and 30 days, in an unfamiliar environment. This data question the role of nest odor in our experiments. It is possible that animals refer to olfaction rather than to other stimulus when they are yonger, as in our experiments; in this case home cage odor is sufficient to reduced stress effect due to handling procedure. Otherwise, it might be suggested that exposure to clean bedding, not handling per se, is the stressful event.

As far as the behavioral effects of chronic exposure to stress during development are concerned, our results suggest that stress produces alterations in some developmental processes. Chronically stressed mice, in fact, still present stress-induced ultrasonic callings at an age when this response has disappeared in mice exposed to clean bedding for the first time. Moreover, at the end of the chronic stress treatment, mice present a behavioral sensitization to the DA agonist apomorphine which suggests altered dopaminergic functioning. This last effect is no longer evident in adult mice; however adult mice submitted to chronic stress during development differ from controls in that they do not exhibit classic behavioral sensitization when reexposed to chronic stress treatment.

The absence of this response to chronic stress may depend on a reduced sensitivity to stress or, alternatively, to an alteration of central dopaminergic functioning that is not evident in the basal condition since it involves mechanisms of adaptation. In both cases, chronic exposure to stress during early postnatal life appears to produce changes in the subsequent responses of the organism to environmental pressure.

In conclusion, our model of postnatal stress seems very interesting because it does not produce non-specific effects on maturation, while resulting in 'at risk' subjects. In fact, only exposure to stress again, in adulthood, is able to evidence the functional alteration that characterizes these animals as, on the surface, they resemble normal subjects.

References

Angelucci, L., Patacchioli, F.R., Chierichetti, C., and Laureti, S. (1983) 'Perinatal mother-offspring pituitary-adrenal interrelationship in rats: corticosterone in milk may affect in adult life', *Endocrinologia Experimentalis* 17, 1921-205.

Alberts, J.R. (1986) 'New views of parent-offspring relationships', in W.T. Greenought and J.M. Juraska (eds.), Developmental Neuropsychobiology, Academic Press, pp. 449-478.

Amir, S., Brown, Z.W., and Amit, Z. (1980) 'The role of endorphins in stress: evidence and speculations', *Neurosciences and Biobehavioral Reviews* 4, 77-86.

Anisman, H.A. and Zacharko, R.M. (1982) 'Depression: The predisposing influence of stress', Behavior and *Brain Science* 5, 89-137.

Antelman, S.M., Eichler, A.J., Black, N.J., and Kolan, D. (1980) 'Interchangeability of stress and amphetamine in sensitization', *Science* 207, 329-331.

Barnett, S.A. and Burn, J. (1967) 'Early stimulation and maternal behavior', Nature 213, 150-152.

Bell, R.W. (1974) 'Ultrasounds in small rodents: Arousal-produced ans arousal-producing', *Developmental Psychobiology* 7, 39-42.

Bell, R.W. (1979) 'Ultrasonic control of maternal behavior: Developmental implications', *American Zoologist* 19, 413-419.

Bell, R.W., Nitschke, W., Bell, N.J., and Zachman, T. (1974) 'Early experience, ultrasonic vocalizations, and maternal responsiveness in rats', *Developmental Psychobiology* 7, 235-242.

Bowlby, J. (1969) Attachment and Loss, vol.1, Attachment, Hogarth Press, London.

Brown, C.P., Smotherman, W.P., and Levine, S. (1977) 'Interaction-induced reduction in differential maternal responsiveness: An effect of cue-reduction or behavior?', *Developmental Psychobiology* 10, 273-280.

Cabib, S. Puglisi-Allegra, S., and Oliverio, A. (1984) 'Chronic stress enhances apomorphine-induced climbing behavior: Involvment of endogenous opioids', *Brain Research* 298, 138-140.

Cabib, S., Puglisi-Allegra, S., and Oliverio, A. (1985) 'A genetic analysis of stereotypy in the mouse: Dopaminergic plasticity following chronic stress', *Behavioral and Neural Biology* 44, 239-248.

Cohen-Salmon, C., Carlier, M., Roubertoux, P., Jouhaneau, J., Semal, C., and Paillette, M. (1985) 'Differences in patterns of pup care in mice V-Pup ultrasonic emissions and pup care behavior', *Physiology and Behavior* 35, 167-174.

Conely, L. and Bell, R.W. (1978) 'Neonatal ultrasounds elicited by odor cues', *Developmental Psychobiology* 11, 193-197.

Conner, R.L., Vernikos-Danellis J., and Levine S. (1971) 'Stress, fighting and neuroendocrine function', *Nature* 234, 564-566.

Daly, M. (1973) 'Early stimulation of rodents: A critical review of present interpretations', *British Journal of Psychology* 64, 435-460.

D'Amato, F.R. and Cabib S. (1987) 'Chronic exposure to a novel odor increases pups' vocalizations, maternal care, and alters dopaminergic functioning in developing mice', *Behavioral and Neural Biology* 48, 197-205.

D'Amato F.R. and Populin R. (1987) 'Mother-offspring interaction and pup development in genetically deaf mice', *Behavior Genetics* 17, 465-475.

D'Amato, F.R., Castellano, C., Ammassari-Teule, M., and Oliverio, A. (1988) 'Prenatal antagonism of stress by naltrexone administration: Early and long-lasting effects on emotional behaviors in mice', *Developmental Psychobiology* 21, 283-292.

Denenberg, V.H. (1971) 'Effects of exposure to stressors in early life upon later behavioural and biological processes', in L. Levi (ed.), Society, Stress and Desease, Oxford University Press, London.

Denenberg, V.H. and Whimbey, A.E. (1963) 'Behaviour of adult rats is modified by the experience their mother had as infants', *Science* 142, 1192-1193.

Dunn, A. and File, S. (1983) 'Cold restraint alters dopaminergic metabolism in frontal cortex, nucleus accumbens, and neostriatum', *Physiology and Behavior* 3, 511-513.

Fox, M.W. (1965) 'Reflex ontogeny and behavioral development of the mouse', *Animal Behaviour* 13, 234-241.

Galef, B.G.Jr. and Henderson, P.W. (1972) 'Mother's milk: A determinant of the feeding preferences of weaning rat pups', *Journal of Comparative and Physiological Psychology* 78, 213-219.

Hennessy, M.B., Collier, A.C., Griffin, A.C., and Schwaiger, S. (1988) 'Plasma corticosterone fluctuations in an infant-learning paradigm', *Behavioral Neuroscience* 102, 701-705.

Hennessy, M.B., Li, J., Lowe, E.L., and Levine S. (1980) 'Maternal behavior, pup vocalizations, and pup temperature changes following handling in mice of 2 inbred strains', *Developmental Psychobiology* 13, 573-584.

Hinde, R.A., Leighton-Shapiro, M.E., and McGinnis, L. (1978) 'Effects of various types of separation experience on rhesus monkeys 5 months later', *Journal of Child Psychology and Psychiatry* 19, 199-211.

Hinde, R.A. and Spencer-Booth, Y. (1971) 'Effects of brief separation from mother on rhesus monkeys', *Science* 173, 111-118.

Hudgens, G.A., Chilgren, J.D., and Palardy, D.D. (1972) 'Mother-infant interactions: Effects of early handling of offspring on rat mothers' open-field behavior', *Developmental Psychobiology* 5, 61-70.

Joffe, J.M. (1965) 'Effect of fostet-mother's strain and pre-natal experience on adult behaviour in rats', *Nature* 208, 815-816.

Kant, G.J., Eggleston, T., Landman-Roberts, L., Kenion, C.C., Driver, G.C., and Meyerhoff, J.L. (1985) 'Habituation to repeated stress is stressor specific', *Pharmacology, Biochemistry and Behavior* 22, 631-634.

Kehoe, P. and Blass, E.M. (1986) 'Opioid-mediation of separation distress in 10-day-old rats: Reversal of stress with maternal stimuli', *Developmental Psychobiology* 19, 385-398.

Kraemer, G.W., Ebert, M.H., Lake, C.R., and McKinney, W.T. (1984) 'Hypersensitivity to d-amphetamine several years after early social deprivation in rhesus monkeys', *Psychopharmacology* 82, 266-271.

Laudenslager, M.L., Capitanio , J.P., and Reite, M.L. (1985) 'Some possible consequences of early separation on subsequent immune function', *American Journal of Psychiatry* 142, 862-864.

Lazarus, R.S., Averill, J.R., and Opton, E.M. (1974) 'The psychology of coping: Issues of research and assessment', in G.V. Coelho, D.A. Hamburg, and J.E. Adams (eds.), Coping and Adaptation, Basik Books, New York, pp. 192-219

Levine, S. (1956) 'A further study of infantile handling and adult avoidasnce learning', *Journal of Personality* 25, 70-80.

Levine, S. (1962) 'The effect of infantile experience on adult behavior', in A.J. Bachrach (ed.), Experimental Foundations of Clinical Psychology, Basic Book, New York.

Levine, S. and Ursin, H. (1979) Coping and Health, Plenum Press, New York.

Maier, S.F., Laudenslager, M.L., and Ryan, S.M. (1985) 'Stressor controllability, immune function, and endogenous opiates', in F.R. Brush, and J.B. Overmier (eds.), Affect, Conditioning, and Cognition: Essays on the Determinants of Behavior, Erlbaum Associates, pp.183-201.

McCarty, R., Horbaly, W.G., Brown, M.S., and Baucom, K. (1981) 'Effects of handling during infancy on the sympathetic-adrenal medullary system of rats', *Developmental Psychobiology* 14, 533-539.

Meier, G.W. and Schutzman, L.H. (1968) 'Mother-infant interactions and experimental manipulation: Confounding or misidentification?', *Developmental Psychobiology* 1, 141-145.

Michaut, R.-J., Dechambre, R.-P., Doumerc, S., Lesourd, B., Devillechabrolle, A., and Moulias, R. (1981) 'Influence of early maternal deprivation on adult humoral immune response in mice', *Physiology and Behavior* 26, 189-191.

Muir, J.L., Pfister, H.P., and Ivinskis, A. (1985) 'Effects of prepartum stress and postpartum enrichment on mother-infant interaction and offspring problem-solving ability in *Rattus norvegicus*', *Journal of Comparative Psychology* 99, 468-478.

Nakagawa, R., Tanaka, M., Kohno, Y., Noda, Y., and Nagasaki, N. (1981) 'Regional responses of rat brain noradrenergic neurons to acute intense stress', Pharmacology, *Biochemistry and Behavior* 14, 729-732.

Noirot, E. (1972) 'Ultrasounds and maternal behavior in small rodents', *Developmental Psychobiology* 5, 371-387.

Plaut, S.M. (1970) 'Studies of undernutrition in the young rat: Methodological considerations', *Developmental Psychobiology* 3, 157-167.

Oliverio, A., Castellano, C., and Puglisi-Allegra, S. (1979) 'A genetic approach to behavioral plasticity and rigidity', in J.R. Royce and L. Moos (eds.), Theoretical Advances in Behavior Genetics, NATO-ASI, Sijttoff & Nordhoff, Amsterdam, pp. 139-165.

Oliverio, A., Castellano, C., and Puglisi-Allegra, S. (1984) 'Psychobiology of Opioids', *International Review of Neurobiology* 25, 277-337

Oswalt, G.L. and Meier, G.W. (1975) 'Olfactory, thermal, and tactual influences on infantuile ultrasonic vocalization in rats', *Developmental Psychobiology* 8, 129-135.

Ottinger, D.R., Denenberg, V.H., and Stephens, M.W. (1963) 'Maternal emotionality, multiple mothering, and emotionality in maturity', *Journal of Comparative and Physiological Psychology* 56, 313-317.

Raymond, L.N., Reyes, E., Tokuda, S., andJones, B.C. (1986) 'Differential immune response in two handled inbred strains of mice', *Physiology and Behavior* 37, 295-297.

Rheingold, H.L. (1963) Maternal Behavior in Mammals, Wiley, New York.

Richardson, R., Siegel, M.A., and Campbell, B.A. (1988) 'Effect of maternal presence on the fear response to an unfamiliar environment as measured by heart rate in rats as a function of age', *Developmental Psychobiology* 21, 613-633.

Russell, P.A. (1971) ''Infantile stimulation' in rodents: A consideration of possible mechanisms', *Psychological Bullettin* 75, 192-202.

Schaefer, T. and Weingarten, F.S. (1962) 'Temperature change: The basic variable in early handling phenomenon?' *Science* 135, 41-42.

Siegel, M.A., Richardson, R., and Campbell, B.A. (1988) 'Effects of home nest stimuli on the emotional response of preweanling rats to an unfamiliar environment', *Psychobiology* 16, 236-242.

Smotherman, W.P., Bell, R.W., Starzec, J., and Zachman, T. (1974) 'Maternal responses to infant vocalizations and olfactory cues in rats and mice', *Behavioral Biology* 12, 55-66.

Smotherman, W.P., Wiener, S.G., Mendoza, S.P., and Levine S. (1977) 'Maternal pituitary-adrenal responsiveness as a function of differential treatment of rat pups', *Developmental Psychobiology* 10, 113-122.

Stanton, M.E., Gutierrez, Y.R., and Levine S. (1988) 'Maternal deprivation potentiates pituitary-adrenal stress responses in infant rats', *Behavioral Neuroscience* 102, 692-700.

Stern, J.M., Thomas, D.A., Rabii, J, and Barfield, R.J. (1984) 'Ultrasonic cries provoke prolactin secretion in lactating rats?', *Hormones and Behavior* 18, 86-94.

Stone, E.A. (1983) 'Adaptation to stress and brain noradrenergic receptors', *Neuroscience and Biobehavioral Reviews* 7, 503-509.

Terkel, J. Damassa, D.A., and Sawyer, C.H. (1979) 'Ultrasonic cries from infant rats stimulate prolactin release in lactating mothers', *Hormones and Behavior* 12, 95-102.

Trivers, R.L. (1972) 'Parental investment and sexual selection', in B. Campbell (ed.), Sexual Selection and the Descent of Man, Aldine, Chicago, pp.136-178.

Villescas, R., Bell, R.W., Wright, L., and Kufner, M. (1977) 'Effect of handling on maternal behavior following return of pups to the nest', *Developmental Psychobiology* 10, 323-329.

Yoneda, Y., Kanmori, K., Ida, S., and Kuriyama, K. (1983) 'Stress-induced alteration in metabolism of -aminobutirric acid in ratr brain', *Journal of Neurochemistry* 40, 350-356.

Young, R.D. (1965) 'Influence of neonatal treatment on maternal behaviour', *Psychonomic Science* 3, 295-296.

Weiss, J.M., Stone, E.A., and Harrel, N. (1970) 'Coping behavior and brain norepinephrine levels in rats', *Journal of Comparative and Physiological Psychology* 72, 153-160.

Willner, P. (1984) 'The validity of animal models of depression', *Psychopharmacology* 83, 1-16.

ORDER AND DISORDER IN THE HYPOTHALAMO-PITUITARY-ADRENOCORTICAL STRESS ACTIVATION

L. Angelucci and R. Montez*
Istituto di Farmacologia 2a, Facoltà di Medicina, Università di Roma 'La Sapienza'
* Visiting scientist from Departimiento de Fisiologia, Universidad de Granada

Whatsoever the verbal definition of stress, the concept at present is that of a physiological mechanism enabling the organism to produce adaptation to an environmental change. Apart situations of unavoidable physical offense, where stress stimuli (stressors) impinge on the hipothalamo-pituitary-adrenocortical axis independently from a substantial mediation by higher structures in the central nervous system, it has been clearly shown that stimuli signalling changes in the environment, and perceived as the actual experience, undergo comparison with previous experiences, after which a stress or no-stress response is actuated (Mason, 1971). So a cognitive element is inseparable from stress activation in the situations of ordinary life, and is a prerequisite for any emotional activation such as arousal, anxiety, fear, and for motor activation such as exploration, flight, fight. On these behavioral neurology grounds it is conceivable that some order regulates qualitatively and quantitatively the physiology of the stress response, and this order be assured by informations, nervous and humoral, reaching the central nervous system. The most likely structure involved in this regulation is the hippocampus for which is known a controlling role on stress responsiveness (Knigge and Hous, 1963; Shadé, 1970; Casady and Taylor, 1976; Conforti and Feldman, 1976; Wilson et al., 1980), as well as a role as a cognitive comparator (Vinogradova, 1975; Gray, 1984). It is remarkable that in the performing of both roles an adrenocorticoid receptor system (McEwen, 1982) is involved; for instance, lack of this receptor in the homozigous Brattleboro rat (De Kloet and Veldhuis, 1980), or its inoccupancy in the adrenalectomized rat impedes forced extinction (Bohus et al., 1975; Bohus and De Kloet, 1981), and its age-dependent loss (Angelucci et al., 1980; Sapolsky et al., 1983) is accompanied by disinhibition of the stress response from the hypothalamo-pituitary-adrenocortical axis (Angelucci et al., 1987). It is to be added that the serotoninergic innervation, so important for the activation of hypothalamic release of Corticotropin Releasing Hormone (CRH), exerts an inhibitory control on the adrenocorticoid binding capacity in the hippocampus (Patacchioli and Angelucci, 1984), and a relationship exist between serotonin turnover in this structure and the level of adrenocortical secretion (De Kloet et al., 1982). So, a disorder is also conceivable in the regulation of stress responsiveness, as a consequence of a disfunction of the hippocampus in its cognitive and/or hormonal competence, due to disturbances either in the intrinsic activities of its neurons or in their neurotransmitter input, chiefly the serotoninergic one.

We have previously investigated especially the hormonal competence with regard to its

S. Puglisi-Allegra and A. Oliverio (eds.), Psychobiology of Stress, 73–80.

74

standing in both order and disorder in the regulation of stress responsiveness, and found that the adrenocorticoid binding capacity in the hippocampus comes to be down-regulated in conjunction with increases in circulating corticosterone (Patacchioli et al., 1983). During this stage, when produced by stress, a refractoriness to a second psychic stressor is present. This phenomenon, that we have indicated as stress-feedback, is the opposite of that consisting in the ability of the body to maintain its responsiveness to repeated psychic stressors. It is possible that these phenomena occur in strict dependence on temporal parameters, as the duration of single stressor-exposure, or the length of intervals between repeated stressor-exposures, as well as on on the magnitude of the adrenocortical response in different individuals or in different types of stress. Eventually, the cognitive-neuroendocrine component of stress and the hormonal ones would determine habituation to, or endurance of repeated stressors, in the frame of a reciprocal influence between nervous input to the hypothalamus from higher brain structure, expecially the hippocampus, and the glucocorticoid feedback at these levels. With regard to a negative feedback from a previous stress on a second one, this possibility has been advocated (Jones and Stockam, 1966; Plotsky et al., 1986) or demonstrated (Sakakura et al., 1976; Raff et al., 1984) in few circumstances, or, on the opposite, excluded of any actuality (Keller-Wood and Dallman, 1984), possibly because of its being counteracted by a prolonged period of hypersesitiveness in either the central nervous system or anterior pituitary components of the adrenocortical system, left over by the previous stress (Dallman and Jones, 1973).

The stress-feedback

An instance of this phenomenon in the rat – a reduced responsiveness to a stressor-exposure subsequent to a stress – is shown in table 1: fortyfive min after a cold stress (90 min at 4-6 °C), a pure psychic stress procedure (change of cage and room for 1 min) is unable to elicit an adrenocortical response. However, if ether vapor is instead applied as a stressor, a full adrenocortical response is obtained. To be noted that at the time of exposure to the second stressor, the plasma corticosterone concentration has regained basal values. In table 1 it is also shown that the failure to produce a second response is not due to insensitiveness of the adrenal cortex to adrenocorticotropin (ACTH) stimulation (the quotient plasma corticosterone/plasma ACTH remains unchanged), but to an absence of pituitary activation. The psychic stress-feedback appears as a highly reproducible consistent phenomenon: gathering data from replicated experiments, in 106 control rats (not prestressed) the plasma adrenocortical response to the pure psychic stressor has been 14.16 + 0.8, from a basal value of 2.07 + 0.23 ug/100 ml, whereas, following the above model, in 113 prestressed rats, it has been 5.07 + 0.53, from a post-stress recovered value of 2.34 + 0.34 ug/100 ml, with a statistically significant difference at a 1 percent level (non-parametric ANOVA test). It is not surprising how this regulatory phenomenon of the psychic stress responsiveness could have been passed unnoticed by the authors who, using repeated somatic stressors, have looked into the physiology of the glucocorticoid negative feedback. One has to consider, in fact, that the feedback appears specifically operant concerning psychic stress in which activation of the hypothalamic CRH releasing activity is operated by suprahypothalamic brain structures provided with negative glucocorticoid feedback sites. On the contrary, somatic stressors (hemorrhage, histamine, laparatomy, etc) impinge on lower brain structures, devoid of such sites (Keller-Wood and Dallman, 1984), and activation of the hypothalamus by ether can be

Table 1. The temporal course of the negative feedback exerted by a previous stress on the responsiveness to a subsequent psychic stressor (psychic stress-feedback) in the rat.

	plasma corticosterone ug/100 ml	plasma ACTH pg/ml
Basal	1.10 ± 0.4 (23)	36.50 ± 3.50 (6)
Psychic stress[a]:		
at 5 min		65.00 ± 5.10g (7)
at 15 min	6.99 ± 0.41 (20)	52.00 ± 3.70 (7)
Cold stress[b]:		
at 90 min	21.99 ± 1.83 (15)	
Cold stress:		
45 min after	0.94 ± 0.05 (15)	
Psychic stressor[c] 45 min after cold stress:		
at 5 min		44.00 ± 2.50h (8)
at 15 min	1.75 ± 0.24e (24)	44.00 ± 3.40 (7)
Ether stressor[d] 45 min after cold stress :		
at 15 min	26.30 ± 3.04 (5)	
Psychic stressor 180 min after cold stress :		
at 15 min	7.70 ± 1.87f (6)	
Ether stressor 180 min after cold stress :		
at 15 min	23.17 ± 2.92 (6)	

[a]: change of cage and room for 1 min; [b]: 90 min at 6 °C; [c]: as in [a]; [d]: 1 min exposure to vapor. [e]: 1 percent significant difference v. psychic stress; [f]: 1 percent significant difference v. psychic stressor 45 min after cold stress; ANOVA and Mann Whitney 'U' test. [g]: 1 percent significant difference v. basal; [h]: 1 percent significant difference v. psychic stress at 5 min; ANOVA and Duncan's New Multiple Range test.

obtained even after its total deafferentation (Makara et al., 1970). Taking into account what already mentioned: inhibitory control of the hippocampus on stress responsiveness of the hypothalamo-pituitary-adrenocortical axis, existence of adrenocorticoid receptors in this structure, and neurotransmitter regulation of their binding capacity, we have investigated the role of serotoninergic innervation in the production of the psychic stress-feedback. As shown in table 2, eight days after the neurotoxic lesion (5,7-dihydroxytryptamine) of the serotoninergic system at the dorsal raphe level in the rat, basal and stress plasma corticosterone levels are the same as in blank and sham lesioned animals. However, the psychic stress-feedback is substantially suppressed. At this time, as shown in table 3, serotonin (5-HT) and 5-hydroxy-indolacetic acid (5-HIAA) concentrations are strongly reduced in both hippocampus and hipothalamus. So, in this situation no distiction between the two structures is possible with regard to their potential participation in the production of the psychic stress-feedback. To be noticed that 21 days after the neurotoxic lesion in the dorsal raphe the psychic stress-feedback in almost fully regained (table 2), in spite of a still great reduction in the hypothalamic serotonin, while in the hippocampus the concentration

Table 2. The psychic stress-feedback (psychic stressor 45 min after a 90 min cold stress) in rats after a 5,7-dihydroxytryptamine lesion (5,7-DHT) in two areas of the brain serotoninergic system. Mean ± S.E.M. of plasma corticosterone in ug/100ml.

DAY		BASAL	PSYCHIC STRESS	PSYCHIC STRESS -FEEDBCK
Dorsal raphe lesion[aa]				
8th	Lesioned (5,7-DHT)	1.29 ± 0.13 (10)	7.95 ± 0.81 (20)	6.81 ± 1.02[c] (18)
	Sham + blank[b]	2.40 ± 0.50 (5)	9.73 ± 1.35 (8)	3.65 ± 0.62[e] (8)
8th	Lesioned	1.35 ± 0.15 (2)	12.32 ± 2.36 (4)	7.76 ± 2.30 (5)
	Sham	2.70 ± 1.30 (2)	11.02 ± 2.47 (4)	4.05 ± 1.24[d] (4)
21st	Lesioned	2.11 ± 0.73 (19)	7.00 ± 1.11 (14)	3.87 ± 0.76[d] (19)
	Sham	1.49 ± 0.49 (11)	9.14 ± 1.58 (13)	4.50 ± 1.45[d] (13)
Fimbria-fornix + cingulum bundle lesion[a]				
8th	Lesioned	3.32 ± 1.30 (7)	22.09 ± 2.91 (7)	11.11 ± 2.48[d] (9)
	Sham	2.77 ± 1.04 (8)	29.20 ± 2.74 (9)	10.31 ± 2.85[e] (9)
30th	Lesioned	3.38 ± 1.54 (8)	21.65 ± 1.62 (10)	9.46 ± 1.40[e] (9)
	Sham	1.82 ± 0.78 (10)	19.63 ± 2.55 (8)	8.25 ± 2.64[e] (9)

[aa]: as in a but hyphotalamic concentration. [a]: considered as such when individual hippocampal concentration of 5-HT was lower than the mean minus two S.D. in the sham animals. [b]: cumulated data; no difference was found between these two groups . [c]: no statistically significant difference v. psychic stress. [d] and [e]: 5 and 1 percent difference v. psychic stress; Duncan's New Multiple Range Test.

of serotonin is in effect recovered (table 3). At 35 days after the dorsal raphe lesion serotonin in the hippocampus and hypothalamus is almost fully regained, likely due to the sprouting of residual intact neurons, as well as to collateral regeneration (Nobin et al., 1973; Bjorklund et al., 1973). A more selective neurotoxic lesion of the serotoninergic innervation to the hippocampus, produced at the level of the fimbria-fornix plus cingulum bundle, is after 8 days unable, as shown in table 2, to suppress the psychic stress-feedback, in spite of a reduction in the concentration of serotonin and 5-hydroxy-indolacetic acid in the hippocampus, as large as after dorsal raphe lesion, as shown in table 3. A reduction in the serotonin concentration, accompanied by a reduction in 5-hydroxy-indolacetic acid is unexpectedly found in the hypothalamus, half of that as following the dorsal raphe lesion; for which no explanation is readily available on anatomical grounds, while a strong functional, compensatory reduction of serotoninergic activity in this structure might be thought of. To be noted that 30 days after the neurotoxic lesion in the fimbria-fornix plus cingulum bundle, while serotonin in the hippocampus is still reduced (table 3), in presence of a fully active psychic stress-feedback (table 2), the concentration in the hypothalamus is substantially normal. This would indicate that there is no immediate relationship between serotoninergic innervation of the hippocampus and regulation of responsiveness to psychic stress. However, because the increase in plasma corticosterone during stress produces, as already mentioned, occupation of the adrenocorticoid receptor in the hippocampus (Patacchioli and Angelucci, 1984), specifically of the glucocorticoid preferring type (type II, Reul and De Kloet, 1985), followed by a transitory reduction in cytosolic binding capacity in

Table 3. Residual 5-HT and 5-HIAA in the hippocampus and hypothalamus of rats at various days after 5,7-DHT lesion. Mean percent values S.E.M. v. control + blank animals.

DAY		5-HT	5-HIAA
Dorsal raphe lesion[aa]			
8th	Hippocampus (9)	39.07 ± 4.21[d]	34.34 ± 5.97[d]
	Hypothalamus (9)	34.59 ± 3.66[c]	33.02 ± 4.75[c]
21st	Hippocampus (20)	86.27 ± 11.86	73.38 ± 11.68
	Hypothalamnus (20)	45.98 ± 3.13[d]	42.37 ± 3.92[d]
35th	Hippocampus (15)	72.88 ± 2.87[d]	68.28 ± 3.66[c]
	Hypothalamus (15)	73.32 ± 1.98[d]	79.22 ± 3.22[c]
Fimbria-fornix + cingulum bundle lesion[a]			
3rd	Hippcampus (6)	68.39 ± 5.01[d]	65.12 ± 3.81[d]
	Hypothalamus (6)	82.72 ± 3.96[b]	82.25 ± 4.38[d]
8th	Hippocampus (14)	32.57 ± 2.26[d]	47.05 ± 4.67[d]
	Hypothalamus (14)	65.98 ± 3.25[d]	83.30 ± 4.23[c]
30th	Hippocampus (8)	57.24 ± 3.88[d]	46.54 ± 2.91[d]
	Hypothalamus (8)	99.35 ± 4.33	82.33 ± 3.87[c]

[aa]: as in [a] but hypothalamic concentration.

[a] : considered effective when individual hippocampal concentration of 5-HT at day 8th was lower than the mean minus two S.D. in the sham animals. [b], [c] and [d]: 5, 1 and 0.1 percent difference; Duncan's New Multiple Range test.

this structure (down regulation: Angelucci et al., 1980; Patacchioli et al., 1983), it would appear that the glucocorticoid signal and changes in brain serotoninergic activity concur in determining an adjustment of the sensitiveness to repeated psychic stress in the rat. A contributory factor to this adjustment could be pituitary desensitization to Corticotropin Releasing Hormone (CRH), demonstrated to occur both *in vitro,* with a down regulation of the relevant receptors (Holmes et al., 1984), and *in vivo* (Rivier and Vale, 1983b; Wynn et al., 1988). A typical situation of disorder in the regulation of stress responsiveness is that encountered in the aging rat (Wistar and Sprague-Dawley): a disinhibition of the hypothalamo-pituitary-adrenal axis is evident in the form of increased basal plasma levels of corticosterone, strongly reduced dexamethasone suppression, shift in circadian rythmicity, aberrant adrenocortical response to stressors, strong reduction in psychic stress-feedback (Angelucci et al., 1987). We were able to exclude from the determinism of this disorder primary disturbances of the adrenal cortex and pituitary activities, while finding an hyperactivity in CRH function of the hypothalamus. This disorder goes in parallel with a strong morphological damage of the hippocampus, especially in pyramidal cells and mossy fibers, and its divestiture of adrenocorticoid receptors, which makes itself evident with cognitive inability and reduction in hormonal competence (Angelucci et al., 1988). It is plausible the participation of functional hypercorticism in the immunitary and metabolic debilities of the old rat.

Having in mind the damaging action of hypercorticism on the hippocampus, especially following exposure to stress (Sapolsky et al., 1988), one can ask about the relevance of our findings for human. Certainly among the many sources of hypercorticism, medicines, drugs (especially alcohol), undernutrition, various pathologies and overwhelming psychic burden have their place and can find their ways to impact on the brain aging through the vicious circle of hippocampus damage and disorder in the regulation of stress responsiveness.

Endurance of repeated stressors

As an example of how repeates stress activation of the pituitary-adrenocortical axis is modulated, we report an instance in which the pituitary maintain its responsiveness to short interval repeated stressors, in spite of the fact that, as already mentioned, pituitary desensitization to repeated CRH stimulation occurs *in vivo* (River and Vale, 1983b; Wynn et al., 1988) and *in vitro* (Holmes et al., 1984). Factors other than CRH then must concur to maintain this responsiveness. As shown in table 4, the application of a second restraint as a stressor 15 min after a first one is able to fully produce an adrenocortical response. However, in this condition a refractoriness of the pituitary to the CRH releasing signal should be present. In fact, the administration of CRH is unable in the chlorpromazine-morphine-nembutal hypothalamic blocked rat to activate the pituitary-adrenocortical secretion 15 min after a first dose chosen from a dose-response curve as that producing maximal activation. At this moment, a dose of arginin-vasopressin (AVP) chosen from a dose-response curve as a subeffective one its own account, is able, 15 min after CRH administration, to activate the pituitary-adrenocortical secretion. Considering that in the stress procedure (2 min in plexiglass chamber) used for this experiment the psychic component prevails over the somatic one, it can be envisaged that the responsiveness of the pituitary-adrenocortical axis to short interval-repeated psychic stress procedures may be maintained through the conjunct action of CRH and AVP. In the lack of a reliable anti-AVP antiserum, we could not obtain the effective demonstration of this evenience. At any rate, our findings are in agreement with the demonstration of the complex role played by the interaction between CRH and AVP on ACTH release in vivo (Rivier and Vale, 1983a; Linton et al., 1985).

Table 4. The pituitary-adrenocortical response to repeated 2 min-restraint in the rat, and the sensitiveness of pituitary to CRH (100 ng) intravenous administration in the chlorpromazine-morphine-nembutal hypothalamus blocked rat. Mean values ± S.E.M.

TREATMENT		PLASMA CORTICOSTERONE ug/100 ml
0 min	15 min	30 min
nothing	nothing	1.38 ± 0.19 (6)
restraint	nothing	13.21 ± 3.20 (6)
restraint	restraint	53.87 ± 14.89[a] (5)
vehicle	vehicle	0.64 ± 0.36 (4)
CRH	vehicle	23.01 ± 5.11 (7)
CRH	CRH	27.65 ± 3.73 (6)
vehicle	AVP	1.20 ± 0.45 (4)
CRH	AVP	46.50 ± 4.52[b] (6)

[a] and [b]: 0.01 percent significant difference v. restraint/nothing, or v. CRH/CRH; Duncan's New Multiple Range test.

In conclusion, the ability of the rat's brain to maintain its neuroendocrine responsiveness to shortly subsequent psychic stressors, and, on the opposite, to dampen it to repeated presentation of stressful situations at relatively longer intervals, reveals the existence of an order in the regulation of stress responsiveness which can be disturbed if there is an improper function of the serotoninergic system. This notion can be intriguing when considering that in endogenous depressive patiens, in the majority of whom a disorder in the regulation of the HPAA is present, a serotoninergic disfunction has been postulated, and that in this condition serotoninergic drugs are largely resorted to.

Acknowledgements

Supported by C.N.R. Grant n° 87.01408.04 and M.P.I. Fund for 1987.

References

Angelucci L., Amenta F., Ghirardi O., Lorentz G. and Ramacci M.T. (1988) The hippocampus as a site of integration of morphological, endocrine and behavioral markers of aging in the rat: an experimental model for acetyl-l- carnitine. In: G. Nappi, Q. Hornykiewicz, R.G. Fariello, A. Aguagli and H. Kawans (Eds), Neuro-degenerative Disorders: The Role Played by Endotoxins and Xenobiotics. New York, Raven Press, pp 23-31.

Angelucci L., Patacchioli F.R., Scaccianoce S., Di Sciullo A., Catalani A., Taglialatela G. and Ramacci M.T. (1987) Hipothalamo-pituitary-adrenocortical function and process of brain aging. In: D. Nerozzi, F.K. Goodwin and E. Costa (Eds), Hypothalamic Disfunction in Neuropsychiatric Disorders. New York, Raven Press, pp. 293-304.

Angelucci L., Valeri P., Grossi E., Veldhuis H.D., Bohus B. and De Kloet E.R. (1980) Involvement of hippocampal corticosterone receptors in behavioral phenomena. In: F. Brambilla, G. Racagni and D. De Wied (Eds), Progress in Psychoneuroendocrinology. Amsterdam, Elsevier, pp. 177-185.

Bjorklund A., Nobin A. and Stenevi U. (1973) Regeneration of central serotonin neurons after axonal degeneration induced by 5,6-dihydroxytryptamine. Brain Res., 50: 214-220.

Bohus B. and De Kloet E.R. (1981) Adrenal steroids and extintion behavior: antagonism by progesterone, deoxycorticosterone and dexamethasone of a specific effect of corticosterone. Life Sciences, 28 : 433-440.

Bohus B., Van Wimersma Greidanus T.J.B. and De Wied D. (1975) Behavioral and endocrine responses of rats with hereditary hypothalamic diabetes insipidus (Brattleboro strain). Physiol. Behav., 14 : 609-615.

Casady R.L. and Taylor A.N. (1976) Effect of electrical stimulation of the hippocampus upon corticosterone levels in the freely-behaving, non-stressed rat. Neuroendocrinology, 20: 68-78.

Conforti N. and Feldman S. (1976) Effects of dorsal fornix section and hyppocampectomy on adrenocortical responses to sensory stimulation in the rat. Neuroendocrinology, 22: 1-7.

Dallman M.F. and Jones M.T. (1973) Corticosteroid feedback control of ACTH secretion: effect of stress-induced corticosterone secretion on subsequent stress responses in the rat. Endocrinology, 92 : 1367-1375.

De Kloet E.R., Kovacs G.L., Szabo G., Teledgy G., Bohus B. and Versteeg D.H.G. (1982) Decreased serotonin turnover in the dorsal hippocampus of rat brain shortly after adrenalectomy: selective normalization after corticosterone substitution. Brain Res., 239: 659-663.

De Kloet E.R. and Veldhuis H.D. (1980) The hippocampal corticosterone receptor system of the homozygous diabetes insipidus (Brattleboro) rat. Neurosci. Lett., 9: 249-254.

Gray G.A. (1984) The hippocampus as an interface between cognition and emotion. In: H.L. Roitblat, T.G. Bever and H.A. Terrace (Eds), Animal Cognition. Hillsdale, N.J. Erlbaum, pp. 607-626.

Holmes M.G., Antoni F.A. and Szentendrei I. (1984) Pituitary receptors for corticotropin-releasing factor: no effect of vasopressin on binding or activation of adenylate cyclase. Neuroendocrinology, 39: 162-169.

Jones M.T. and Stokam M.A. (1986) The effect of previous stimulation of the adrenal cortex by adrenocorticotropin on the function of the pituitary- adrenocortical axis in response to stress. J. Physiol., 184: 741-750.

Keller-Wood M.E. and Dallman M.F. (1984) Corticosteroid inhibition of ACTH secretion. *Endocr. Rev.*, 5: 1-23.

Knigge K. M. and Hous M. (1963) Evidence of inhibitive role of hippocampus in neural regulation of ACTH release. *Proc. Soc. Exp. Biol. Med.*, 114: 67-69.

Linton E.A., Tilders F.J.H., Hodgkinson S., Berkenbosh F., Vermes I. and Lowry P.J. (1985) Stress-induced secretion of adrenocorticotropin in rats is inhibited by administration of antisera to ovine corticotropin-releasing factor and vasopressin. *Endocrinology*, 116: 966-970.

Makara G.B., Stark E. and Palkovits M. (1970) Afferent pathways of stressful stimuli: corticotropin release after hypothalamic deafferentation. *Journal Endocrinol.*, 47: 411-416.

Mason J. W. (1971) A re-evaluation of the concept of 'non specificity' in stress theory. *J. Psychiatr. Res.*, 8: 323-333.

McEwen B.S. (1982) Glucocorticoids and hippocampus. In: D. Ganten and D. Pfaff (Eds), Adrenal Actions on Brain. Berlin, Springer, pp. 1-22.

Nobin A., Baumgarten H.G., Bjorklund A., Lachenmayer L. and Stenevi U. (1973) Axonal degeneration and regeneration of the bulbo-spinal indolamine neurons after 5,6-dihydroxytryptamine treatment. *Brain Res.*, 56: 1-24.

Patacchioli F.R. and Angelucci L. (1984) The serotoninergic raphe-hippocampal glucocorticoid receptor link in the regulation of the stress response. In : G. Biggio, P.F. Spano, G. Toffano and G.L. Gessa (Eds), Neuromodulation and Brain Function, Advances in the Biosciences vol. 48. Oxford, Pergamon Press, pp. 327-334.

Plotsky P.M., Otto S. and Sapolsky R.M. (1986) Inhibition of immunoreactive corticotropin-releasing factor secretion into the hypophysial-portal circulation by delayed glucocorticoid feed-back. *Endocrinology*, 119: 1126- 1130.

Raff R., Shinsako J., Keil L.C. and Dallman M.F. (1984) Feedback inhibition of adrenocorticotropin increases in endogenous plasma corticosterone in dogs. *Endocrinology*, 114: 1245-1249.

Reul J.M.H.M. and De Kloet E.R. (1985) Two receptor systems for corticosterone in rat brain: microdistribution and differential occupation. *Endocrinology*, 117: 2505-2511.

Rivier C. and Vale W. (1983a) Interaction of corticotropin-releasing factor and arginine vasopressin on adrenocorticotropin secretion in vivo. *Endocrinology*, 113: 930-942.

Rivier C. and Vale A. (1983b) Influence of the frequency of ovine corticotropin and corticosterone secretion in the rat. *Endocrinology*, 113: 1422-1426.

Sakakura M., Saito Y., Takebe K. and Ishii K. (1976) Studies on fast feedback mechanism by endogenous glucocorticoids. *Endocrinology*, 98: 954-957.

Sapolsky R.M., Krey L.C. and McEwen B.S. (1983) Corticosterone receptors decline in a site-specific manner in the aged rat brain. *Brain Res.*, 289: 235- 240.

Sapolsky R.M., Krey L.C. and McEwen B.S. (1988) Prolonged glucocorticoid exposure reduces hippocampal neuron number: implications for aging. *J. Neurosci.*, 5: 1222-1227.

Shad__ J.P. (1970) The limbic system and the pituitary-adrenal axis. In: D. De Wied and J.A.M. Weijnen (Eds) Pituitary, Adrenal and the Brain. Amsterdam, Elsevier, pp. 1-11.

Vinogradova O.S. (1975) Functional organization of the limbic system in the process of registration of information: facts and hipotheses. In: R.L. Isaacson and K.H. Pribram (Eds), The Hippocampus, vol. 2, New York, Plenum Press, pp. 1-70.

Wilson M.M., Greer S.E., Greer M.A. and Roberts L. (1980) Hippocampal inhibition of pituitary adrenocortical function in female rats. *Brain Res.*, 197: 133-141.

Wynn P.C., Harwood J.P., Catt K.J. and Aguilera G. (1988) Corticotropin releasing factor (CRF) induces desensitization of the rat pituitary CRF receptor- adenylate cyclase complex. *Endocrinology*, 122: 351-358.

CORTICOTROPIN-RELEASING FACTOR AS THE MEDIATOR OF STRESS RESPONSES

Adrian J. Dunn and Craig W. Berridge
Department of Neuroscience, University of Florida College of Medicine, Gainesville, Florida 32610, USA

Corticotropin-releasing factor (CRF) is a 41-amino acid peptide isolated and characterized by Vale et al. (1981). It is now believed to be the major active principle that can elicit the secretion of corticotropin (ACTH) from the pituitary (Rivier and Plotsky, 1986). CRF is synthesized in the hypothalamus, especially in neurons in the paraventricular nucleus (PVN). These neurons have terminals in the median eminence region, which can secrete CRF into the portal blood supply. Thus transported to the anterior pituitary, CRF can stimulate the release of ACTH into the general circulation, so that it subsequently elicits the secretion of glucocorticoids from the adrenal cortex. This cascade (Figure 1) constitutes the hypothalamic-pituitary-adrenal (HPA) axis, the activation of which is considered to be characteristic, if not diagnostic, of stress (Selye, 1950; Mason, 1968; Dunn and Kramarcy, 1984).

However, CRF-like immunoreactivity and bioactive CRF have been identified in many extrahypothalamic regions of the brain (Merchenthaler, 1984; Sawchenko and Swanson, 1985; Nakane et al., 1986). The highest concentration of extrahypothalamic immunoreactive CRF is found in neocortex, areas of the limbic system, and regions involved in the regulation of the autonomic nervous system. High-affinity binding sites for CRF have been observed with a similar regional distribution, using both quantitative autoradiographic and biochemical techniques (Wynn et al., 1984; De Souza et al., 1985). Moreover, CRF-like immunoreactivity has been shown to be released from samples of fresh brain tissue by $K+$ in a Ca^{2+}-dependent manner (Smith et al., 1986). These observations suggest that cerebral CRF has a neurotransmitter function in brain, separate from its ability to activate the HPA axis. Regionally specific changes in the cerebral concentrations of CRF occur following both acute and chronic stressful treatments (Chappell et al., 1986), and one preliminary report indicates an increase in the CSF concentration of CRF in stressed rats (Britton et al., 1984), suggesting that release of extrahypothalamic CRF may be related to stressful situations. For these reasons we have investigated the neurochemical and behavioral effects of intracerebroventricular (ICV) administration of CRF.

A number of behavioral and physiological responses has been observed following ICV injection of CRF. Many of these responses resemble those observed in stress. For example, ICV CRF elicits an activation of sympathetic and adrenomedullary output, resulting in increases in the circulating concentrations of catecholamines, as well as classic indices of sympathetic activation, such as increases in mean arterial pressure and heart rate (Brown and Fisher, 1985). ICV CRF also increases firing of locus coeruleus (LC) neurons in anesthetized

S. Puglisi-Allegra and A. Oliverio (eds.), Psychobiology of Stress, 81–93.
© 1990 *Kluwer Academic Publishers. Printed in the Netherlands.*

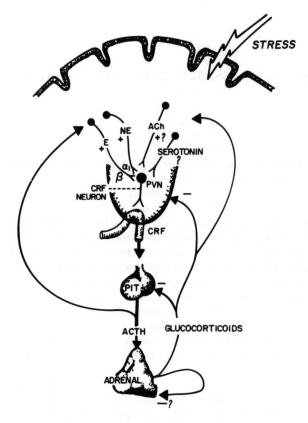

Figure 1. A schematic of the hypothalamic-pituitary-adrenal (HPA) system. Various neurotransmitter systems within the brain (cholinergic – ACh, noradrenergic – NE, adrenergic – E, and serotonergic) regulate the activation of CRF-containing neurons in the paraventricular nucleus (PVN) of the hypothalamus. CRF secreted from terminals in the median eminence region of the hypothalamus is carried in portal blood vessels to the anterior pituitary (PIT) where it stimulates the release of adrenocorticotropin (ACTH) into the general circulation. Circulating ACTH stimulates the release of glucocorticoids (corticosterone in the rat and mouse, cortisol in most other species) from the adrenal cortex. The glucocorticoids exert inhibitory feedback on the brain, pituitary, and possibly the adrenal.

and awake rats (Valentino et al., 1983). Intracerebral administration of CRF elicits changes of locomotor activity (Sutton et al., 1982), increases in grooming (Morley and Levine, 1982; Sutton et al., 1982), decreases in feeding but not drinking (Morley and Levine, 1982), and decreases in sexual behavior (Sirinathsinghji et al., 1983). Some of the observed effects were opposite to those seen following administration of benzodiazepines (Britton et al., 1982, 1985), suggesting that CRF can be regarded as anxiogenic.

These observations have prompted the suggestion that CRF may coordinate a whole body response in stress (Gold and Chrousos, 1985; Koob and Bloom, 1985). We have investigated the strong hypothesis that the release of CRF from brain neurons is both necessary and sufficient to initiate the stress response. Our observations outlined in this chapter provide some support for this hypothesis.

83

Methods

Animals

Male CD-1 mice (25-35 g) were obtained from Charles River (Wilmington, MA) and were maintained on a 12-12 light-dark cycle (lights on 07:00).
Materials Human/rat CRF and α-helical-CRF$_{9-41}$ (ahCRF) were donated by Jean Rivier (Salk Institute). Peptides were dissolved in 0.14 M NaCl (containing 10^{-3} M HCl in the case of CRF, but not ahCRF). Aliquots of the required concentration were stored frozen and used within one hour of thawing. Clonidine and phenylephrine were obtained from Sigma Chemical Co. (St. Louis, MO), DSP-4 was a gift from Astra, and idazoxan was a gift from Reckitt and Coleman (Hull, England).

Surgery

Mice requiring ICV injections were implanted with polyethylene cannulae as described by Guild and Dunn (1982).

Behavioral Testing

The multicompartment testing chamber (MCC) was based on that designed by Arnsten and Segal (1979). For mice this chamber (38 x 38 x 23 cm) consisted of nine interconnecting compartments within each of which a wire mesh stimulus (3.0 cm sphere) was recessed in a 2.5 cm hole below the floor. The animal was placed in this chamber and observed for a 25-30 min period. The MCC was brightly lit, and background sounds were masked using a white noise generator (70 db). Testing was conducted between the hours of 09:00 and 15:30. During the observation period the duration and frequency of a number of behaviors were recorded using an event recorder. Behaviors recorded include: Measures of contact with the stimuli: total number of contacts; total duration of contacts; mean time per contact (total duration of contacts/total number of contacts); Measures of locomotor activity: total number of compartment entries; total number of rears (both forepaws off the floor).

ICV injection was performed in unanesthetized mice using a 10 ml syringe. ICV injections consisted of a total volume of 4 μl divided equally between the two cannulae and given over a 30-45 sec period. Restraint was administered by allowing the mouse to enter a darkened polyethylene tube 10 cm in length and 3 cm in diameter (50-ml centrifuge tubes in which airholes had been punched), preventing their exit by taping the ends of the tube with laboratory labeling tape (Berridge and Dunn, 1986).

Neurochemistry

After decapitation trunk blood was collected into heparinized 1.5 ml Eppendorf tubes for assay of plasma corticosterone, and the brain was rapidly removed from the skull. Brain regions were dissected as previously described (Dunn, 1988a) and tissue samples weighed frozen on dry ice and homogenized by brief ultrasonication in 0.1 M HClO4 containing 0.1 mM EDTA. The samples were thawed and centrifuged briefly immediately prior to HPLC analysis with electrochemical detection as described previously (Dunn and Berridge, 1987; Dunn, 1988a).

Results

Neurochemical Effects of Stress and CRF

There is widespread evidence for the activation of cerebral catecholaminergic systems in stress. The evidence for a general activation of the metabolism of norepinephrine (NE) is overwhelming (Stone, 1975; Anisman, 1978; Glavin, 1985), but there is also extensive evidence for the activation of both dopaminergic (DA) and adrenergic systems in stress (Roth et al., 1982; Bannon and Roth, 1983; Dunn and Kramarcy 1984). We have measured the concentrations of catecholamines, indoleamines, and metabolites associated with a variety of stressful treatments. Our results (Dunn, 1988a,c) indicate that following footshock or restraint treatment there are increases in the concentrations of catabolites of both NE (i.e. 3-methoxy,4-hydroxyphenylethyleneglycol, MHPG) and DA (i.e. 3,4-dihydroxyphenylacetic acid, DOPAC and homovanillic acid, HVA), as well as a general increase in the concentration of free tryptophan (Dunn, 1988a,c). This suggests that there is not only an activation of cerebral noradrenergic systems, but also of dopaminergic systems. The activation of dopaminergic systems is general, occurring in all brain regions studied although there are regional differences in the magnitude of the response. A similar pattern of responses is observed following footshock, restraint, conditioned footshock (Dunn, 1988a,c), or during the training or testing of mice in passive avoidance behavior (Dunn et al., 1986). In addition, when mice get sick with influenza virus or other opportunistic infections, increases in plasma concentrations of corticosterone and of cerebral MHPG and tryptophan similar to those observed following footshock or restraint are observed (Dunn et al., 1989), suggesting that these experiences can be regarded as physiologically stressful. Interestingly, in these cases the dopaminergic component of the response is minor or absent. The mediator of the noradrenergic and tryptophan changes may well be interleukin-1, a polypeptide produced by macrophages following antigenic stimulation, because nanogram doses of this molecule administered peripherally result in similar changes (Dunn, 1988b).

We have investigated the potential role of hormones of the HPA axis in these neurochemical responses. Prior adrenalectomy of the mice used did not substantially alter any of these responses, nor did we detect any activation of cerebral catecholaminergic systems by administration of corticosterone (Dunn, 1988a). However, we have observed substantial changes following administration of CRF. ICV injections of 1 μg CRF caused changes in brain catecholamine metabolites that closely resembled those we observed following footshock or restraint. DOPAC:DA ratios were significantly increased in prefrontal cortex, septum, hypothalamus, and brain stem, and MHPG:NE ratios were increased in prefrontal cortex, hypothalamus, and brain stem (Figure 2: Dunn and Berridge, 1987). However, tryptophan was not altered in any brain region. The MHPG results are consistent with the observations of Valentino et al. (1983) indicating that ICV CRF increases the electrophysiological activity of NE-containing LC neurons. Moreover, Butler et al. (1988) have recently shown an increase in cerebral 3,4-dihydroxyphenylethyleneglycol (DHPG) following ICV or direct local application of CRF in the rat. These results raise the possibility that intracerebral CRF may mediate changes in catecholamine metabolism during stress.

Behavioral Effects of CRF

Because ICV administration of CRF had been reported to elicit a number of behaviors

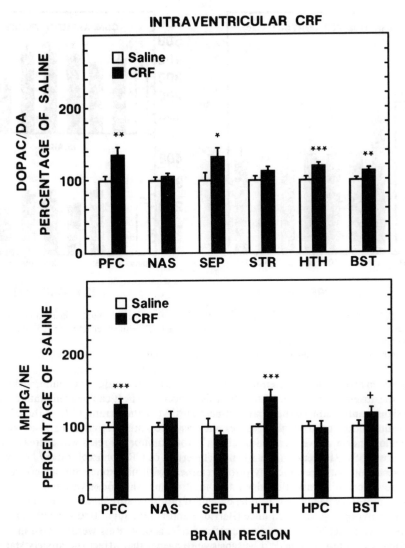

Figure 2. The effect of ICV CRF on MHPG:NE and DOPAC:DA ratios in various brain regions. CRF (1 mg) was injected ICV into mice and brain samples collected 30 minutes later. PFC: prefrontal cortex; NAS: nucleus accumbens septi; SEP: septum; STR: neotsriatum; HTH: hypothalamus; HPC: hippocampus; BST: brain stem. *Significantly different from saline-controls (*P < 0.05, **P < 0.01, ***P < 0.001, +0.05 < P < 0.1).

characteristic of stress, we investigated its effect on exploratory behavior. Arnsten et al. (1985) had previously shown that the time rats spent investigating novel objects in holes in the floor of a multicompartment chamber (MCC) was decreased by prior stressful treatments (restraint, noise, or tail pinch). Using mice, we first showed that 30 minutes of restraint had a similar effect on the mean time they made contact with similar 'stimuli' in a scaled down version of the same apparatus (Berridge and Dunn, 1986). We then showed that ICV CRF (75

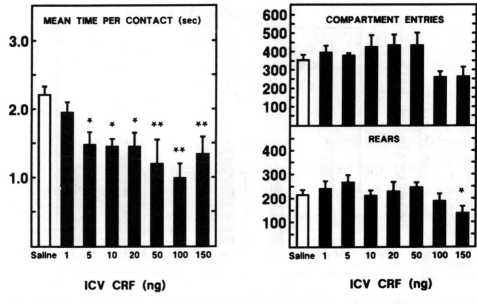

Figure 3. Effect of varying doses of CRF injected ICV on the mean time per contact, the total number of rears and the total number of compartment entries. Mice were injected ICV with either saline or 1, 5, 10, 20, 50, 100, or 150 ng CRF 15 min prior to testing.
*Significantly different from saline-controls (*P < 0.05, **P < 0.01).

ng) resulted in effects closely resembling those of restraint (Berridge and Dunn, 1986). Figure 3 shows the dose-response relationship for this effect of CRF. It can be seen that doses as low as 5 ng significantly reduced the mean contact times. Most important, CRF at these low doses did not alter other behavioral measures such as locomotor activity, rears, or grooming, although locomotor activity as measured by the total number of rears was decreased in some mice at doses of CRF greater than 100 ng. The effect of CRF on exploratory behavior, like that of restraint, was prevented by prior treatment with naloxone. Similar effects of ICV CRF were observed in rats (Berridge and Dunn, unpublished observations). This effect of CRF on exploratory behavior was independent of pituitary-adrenal activation because it was observed in hypophysectomized mice (Berridge and Dunn, submitted for publication).

A stress-like effect of CRF was not confined to this task. Rats were studied in the social interaction test, a test designed to be sensitive to agents that affect the anxiety state of the animals. ICV doses of 100 or 300 ng significantly decreased the social interaction (Dunn and File, 1987), an effect characteristic of anxiogenic agents, and opposite to that observed with anxiolytic drugs, such as the benzodiazepines. We also found that rats tested in the elevated +-maze (Handley and Mithani, 1984) showed an anxiogenic response to ICV CRF, i.e. they spent less time on the open (threatening) arms of the maze, and more time on the closed (safe) arms (Baldwin, File and Dunn, unpublished observations).

CRF as a Mediator of Stress-Related Changes in Behavior

To test whether or not the effect of stress on exploratory behavior in the MCC was mediated by CRF, we employed the CRF-antagonist, a-helical CRF9-41 (ahCRF). This peptide has been shown to inhibit the CRF-induced activation of ACTH release from the isolated anterior pituitaries (Rivier et al., 1984). Unfortunately, its affinity for CRF-receptors is markedly less than that of CRF itself, and a IV dose of 3 mg/kg was necessary to inhibit stress-related ACTH and corticosterone secretion in intact rats. We injected 10, 20, 50 or 100 μg of ahCRF ICV into mice immediately prior to 40 minutes restraint. The peptide had no statistically significant effect on mean stimulus-contact times in unrestrained mice, but in restrained animals there was a dose-dependent effect of the ahCRF, such that doses of 10, 20, and 50 μg significantly reversed the restraint-induced decrease in contact times (Berridge and Dunn, 1987a). This result provides strong evidence that endogenous CRF in the brain mediates this restraint-induced behavioral change.

NE as a Mediator of Stress-Related Behavior

As indicated earlier, a large literature suggests that cerebral noradrenergic systems are activated in stress (Stone, 1975; Dunn and Kramarcy, 1984), and there is some evidence that stress-related behaviors may be noradrenergically mediated. We therefore investigated the involvement of NE in exploratory behavior in the MCC. Drugs active on α2-receptors are known to influence firing of noradrenergic cells in the LC (Svensson et al., 1975). Idazoxan (1 mg/kg SC), an α2-antagonist known to increase LC firing, decreased stimulus-contact times, and clonidine (25 μg/kg SC), an a2-agonist known to decrease LC firing, increased stimulus-contact times (Berridge and Dunn, 1987b). These results are consistent with a noradrenergic mediation of the behavioral response to restraint. This hypothesis was strengthened when we observed that treatment with clonidine (25 μg/kg SC, 5 minutes before restraint for 40 minutes) or the NE-specific neurotoxin, DSP-4 (50 mg/kg IP, 3 days before restraint for 40 minutes) increased the stimulus-contact (exploration) times in restrained mice (Berridge and Dunn, submitted for publication). However, both treatments significantly elevated the stimulus-contact times of unrestrained animals, rendering interpretation of the results ambiguous. Because clonidine does not completely prevent release of NE, or the stress-induced increase, and DSP-4 does not completely deplete NE, we pretreated mice with both drugs. This combined treatment completely reversed the effects of restraint. Likewise, the α1-antagonist, prazosin (200 μg/kg SC 45 minutes before testing), completely prevented the effect of restraint in the MCC (Berridge and Dunn, unpublished observations). Finally, the α1-selective agonist, phenylephrine (50 or 100 ng ICV 10 minutes before testing) had a restraint-like effect on behavior in the MCC.

Interactions between NE and CRF

These latter results strongly support the hypothesis that noradrenergic systems in the brain are involved in the restraint-induced decrease in exploratory behavior. Thus we must conclude that both CRF- and NE-containing cells in the brain can influence exploratory behavior in the MCC. The question that remains is whether the effects are independent or related. As indicated above, CRF administration can activate noradrenergic neurons in the LC. Moreover, there is an extensive literature suggesting that NE-terminals in the PVN can influence the release of median eminence CRF. Current thinking, contradicting some earlier findings, suggests that the primary effect of NE in the PVN is excitatory on CRF release, and is exerted through an α1-receptor (Plotsky, 1987; Szafarczyk et al., 1987).

We tested these potential interactions with the appropriate neuropharmacological experiments. First we showed that the effects of CRF on exploratory behavior in the MCC were unaffected by DSP-4 or prazosin pretreatment. Then we determined that ahCRF at a dose of 20 mg ICV completely reversed the effects of ICV phenylephrine (50 or 100 ng). These results appear to exclude an activation of noradrenergic systems by CRF as responsible for the behavioral effects of restraint, and strongly suggest that the restraint-induced activation of noradrenergic neurons stimulates the release of CRF via an α1-receptor. The CRF released into the ventricles or certain brain sites then affects the behavior. This schema is consistent with the arrangement of these neurotransmitter systems in the activation of the HPA axis as indicated above (Figure 1).

Summary and Discussion

The above has indicated that CRF can elicit endocrine, behavioral, electrophysiological, and neurochemical effects resembling those observed in stress. The effects can be summarized as follows:

Neuroendocrine

CRF activates the HPA axis (Vale et al., 1981). It inhibits the secretion of growth hormone and luteinizing hormone (Ono et al., 1985b; Rivier and Vale, 1986; Rivier et al., 1986).

Physiologic

CRF activates sympathetic and adrenomedullary output, increasing the circulating concentrations of NE, epinephrine, glucagon, and glucose, as well as increasing mean arterial pressure and heart rate (Brown and Fisher, 1985). ICV CRF also increases the electrophysiological activity of the adrenal sympathetic nerve (Kurosawa et al., 1986). Intracisternal CRF decreases gastric acid secretion (Tache et al., 1987), along with decreases in gastric emptying and small intestinal transit, and increases in large intestine transit and fecal excretion (Lenz et al., 1987).

Electrophysiological

ICV CRF produces a dose-dependent activation of the electroencephalogram with signs of both EEG and behavioral arousal predominating at low doses (10-100 ng) and epileptiform activity and seizures observed at higher doses (1-20 μg) (Ehlers et al., 1983). ICV CRF also increases firing of locus coeruleus (LC) neurons in anesthetized and awake rats (Valentino et al., 1983). Similar effects are observed following local injection of CRF on to LC neurons (Valentino and Foote, 1988). In the latter study, CRF was found to increase the basal firing rate, without affecting that due to noxious stimuli. Responses to iontophoretic application of CRF vary with the region; inhibition of cell firing has been recorded in the thalamus and lateral septum, whereas excitation occurred on cortical and hypothalamic neurons (Eberly et al., 1983).

Neurochemical

As discussed above, CRF activates cerebral NE and DA systems as indicated by the increased production of DHPG, MHPG and DOPAC in various brain regions (Dunn and Berridge, 1987; Butler et al., 1988).

Behavioral

CRF increases grooming in mice and rats (Britton et al., 1982; Dunn et al., 1987; Morley and Levine, 1982; Sutton et al., 1982; Veldhuis and de Wied, 1984), and decreases feeding but not drinking in rats (Morley and Levine, 1982; Krahn et al., 1986). Sexual behavior is reduced in both male and female rats (Sirinathsinghji, 1985, 1987). Kalin (1985) reported depression-like effects of ICV CRF in monkeys.

Intracerebral administration of CRF to rats produced a dose-dependent activation of locomotor behavior in a familiar environment (Sutton et al., 1982), whereas in a novel open field it decreased locomotion and rears and increased freezing, results that were interpreted to reflect an increase in the aversive nature of the novel environment (Sutton et al., 1982). CRF decreased approaches to food in a novel open-field (Britton et al., 1982). This effect was opposite to that seen following administration of benzodiazepines, suggesting that CRF enhances the anxiogenic nature of the novel environment. In a conflict test, CRF decreased punished responding, whereas benzodiazepines increased this response (Britton et al., 1985). CRF also exacerbates the acoustic startle response (Swerdlow et al., 1986)and shock-induced freezing (Sherman and Kalin, 1988). CRF has complex effects on avoidance behavior (Sahgal et al., 1983; Veldhuis and de Wied, 1984). As indicated above, we have shown that ICV CRF reduces a measure of exploratory behavior in both rats and mice (Berridge and Dunn, 1986), and decreases social interaction in rats (Dunn and File, 1987).

In every case where it has been tested, these behavioral effects appear to be independent of the the pituitary-adrenal axis. The responses to CRF occur in hypophysectomized animals (Eaves et al., 1985; Lenz et al., 1988; Morley and Levine, 1982; Berridge and Dunn, submitted), and are not prevented by dexamethasone treatment (Britton et al., 1986a; Dunn et al., 1987). Moreover, the actions of CRF appear to be directly on the brain; they are not elicited by peripheral injections at similar doses.

These stress-like effects of CRF suggest, but do not prove that CRF may be an endogenous mediator of stress responses. The CRF administration could merely be stressful, activating the endogenous stress systems. However, a number of stress-related phenomena have now been shown to be attenuated or reversed by the intracerebral administration of the CRF-antagonist, ahCRF. Rivier et al. (1984) initially demonstrated the ability of ahCRF to reverse the ether stress-induced secretion of ACTH. Subsequently, the same group showed that ahCRF (100 μg ICV) could reverse the electric shock-induced decreases in growth hormone (Rivier and Vale, 1985) and LH secretion in male rats (Rivier et al., 1986). Brown et al. (1985) found that ICV ahCRF could reverse the ether-induced increase in plasma epinephrine (but curiously not NE). Williams et al. (1987) showed that ICV ahCRF (50 μg IV or ICV) prevented the restraint-related increases of large intestinal transit time and fecal secretion, and Stephens et al. (1988) showed that intracisternal ahCRF (10 or 50 μg) reversed the surgery-related inhibition of gastric acid secretion. So far as the behavioral effects are concerned, Krahn et al. (1986) demonstrated that 50 or 100 μg ahCRF (ICV) attenuated the shock-induced decrease in feeding. As discussed above, we found that ICV ahCRF (20 or 50 µg) reversed the effect of restraint on exploratory behavior in mice (Berridge and Dunn, 1987a). Tazi et al. (1987) found that 5 or 25 µg ahCRF blocked footshock-induced fighting in rats, and, more recently, Kalin et al. (1988) found that ahCRF (25 µg ICV) prevented footshock-induced freezing behavior in rats.

All these results provide compelling evidence that endogenous CRF mediates at least some of the behavioral and physiological changes that occur in stress. Nevertheless, we do not know the sites of action for any of these effects. In all cases reported, the responses to CRF are elicited with intracerebroventricular injections. While this suggests a periventricular site of action for the peptide, this is not necessarily the case. Ono et al. (1985a) have provided evidence that cerebroventricular CRF can stimulate the activation of endogenous CRF systems, providing a positive ultrashort loop feedback. Such a system would explain the activation by ICV CRF of the pituitary-adrenal axis (Veldhuis and de Wied, 1984; Kalin, 1985; Dunn and Berridge, 1987).

We have preliminary evidence that the effect of ICV CRF on exploratory behavior is exerted via a third ventricle rather than a fourth ventricle site (Berridge, Spadaro and Dunn, unpublished). But our experiments would not reveal the final site of action of the CRF which could be in the cortex. Likewise, we do not know the location of the NE-CRF interaction involved in the behavioral changes in the MCC (if it occurs directly). Whereas it is possible that it is in the PVN, such that collaterals of CRF neurons projecting to the median eminence are responsible for the behavioral effects, it could just as easily be in the cortex, or anywhere else in the brain that both NE terminals and CRF-containing neurons appear.

The currently available evidence reviewed above suggests that CRF administration can reproduce many of the well characterized endocrine, physiological, neurochemical, and behavioral responses in stress. Experiments with CRF antagonists administered to stressed animals indicate that several of the stress-related changes can be attenuated or reversed, suggesting that endogenous CRF may indeed mediate certain stress responses. Thus we can consider seriously the hypothesis that release of endogenous CRF from brain sites may be both necessary and sufficient for a stress response.

References

Anisman, H. (1978) Neurochemical changes elicited by stress. In: Psychopharmacology of Aversively Motivated Behavior, (Ed. Anisman, H. and Bignami, G.), New York: Plenum Press pp.119-172.

Arnsten, A.F.T. and Segal, D.S. (1979) Naloxone alters locomotion and interaction with environmental stimuli. Life Sci. 25: 1035-1042.

Arnsten, A.F.T., Berridge, C.W. and Segal, D.S. (1985) Stress produces opioid-like effects on investigatory behavior. Pharmacol. Biochem. Behav. 22: 803-809.

Bannon, M.J. and Roth, R.H. (1983) Pharmacology of mesocortical dopamine neurons. Pharmacol. Rev. 35: 53-67.

Berridge, C.W. and Dunn, A.J. (1986) Corticotropin-releasing factor elicits naloxone-sensitive stress-like alterations in exploratory behavior in mice. Regulat. Peptides 16: 83-93.

Berridge, C.W. and Dunn, A.J. (1987a) A corticotropin-releasing factor antagonist reverses the stress-induced changes of exploratory behavior in mice. Horm. Behav. 21: 393-401.

Berridge, C.W. and Dunn, A.J. (1987b) a2-noradrenergic agonists and antagonists alter exploratory behavior in mice. Neurosci. Res. Commun. 1: 97-103.

Britton, D.R., Koob, G.F., Rivier, J. and Vale, W. (1982) Intraventricular corticotropin-releasing factor enhances behavioral effects of novelty. Life Sci. 31: 363-367.

Britton, K.T., Lyon, M., Vale, W. and Koob, G.F. (1984) Stress-induced secretion of corticotropin-releasing factor immunoreactivity in rat cerebrospinal fluid. *Soc. Neurosci. Abstr.* 10: 95.

Britton, K.T., Morgan, J., Rivier, J., Vale, W. and Koob, G.F. (1985) Chlordiazepoxide attenuates response suppression induced by corticotropin-releasing factor in the conflict test. *Psychopharmacol.* 86: 170-174.

Britton, D.R., Varela, M., Garcia, A. and Rosenthal, M. (1986a) Dexamethasone suppresses pituitary-adrenal but not behavioral effects of centrally administered CRF. *Life Sci.* 38: 211-216.

Britton, K.T., Lee, G., Vale, W., Rivier, J. and Koob, G.F. (1986b) Corticotropin releasing factor (CRF) receptor antagonist blocks activating and 'anxiogenic' actions of CRF in the rat. *Brain Res.* 369: 303-306.

Brown, M.R. and Fisher, L.A. (1985) Corticotropin-releasing factor: effects on the autonomic nervous system and visceral systems. *Fed. Proc.* 44: 243-248.

Brown, M.R., Fisher, L.A., Webb, V., Vale, W.W. and Rivier, J.E. (1985) Corticotropin-releasing factor: a physiologic regulator of adrenal epinephrine secretion. *Brain Res.* 328: 355-357.

Butler, P.D., Weiss, J.M., Stout, J.C., Kilts, C.D., Coook, L.L. and Nemeroff, C.B (1988) Corticotropin-releasing factor produces anxiogenic and behavioral activating effects following microinfusion into the locus coeruleus. *Soc. Neurosci. Abstr.* 114.14.

Chappell, P.B., Smith, M.A., Kilts, C.D., Bissette, G., Ritchie, J., Anderson, C. and Nemeroff, C.B. (1986) Alterations in corticotropin-releasing factor-like immunoreactivity in discrete rat brain regions after acute and chronic stress. *J. Neurosci.* 6: 2908-2914.

De Souza, E.B., Insel, T., Perrin, M., Rivier, J., Vale, W. and Kuhar, M. (1985) Corticotropin-releasing factor receptors are widely distributed within the rat central nervous system: an autoradiographic study. *J. Neurosci.* 5: 3189-3203.

Dunn, A.J. (1988a) Changes in plasma and brain tryptophan and brain serotonin and 5-hydroxyindoleacetic acid after footshock stress. *Life Sci.* 42: 1847-1853.

Dunn, A.J. (1988b) Systemic interleukin-1 administration stimulates hypothalamic norepinephrine metabolism parallelling the increased plasma corticosterone. *Life Sci.* 43: 429-435.

Dunn, A.J. (1988c) Stress-related activation of cerebral dopaminergic systems. *Ann. N.Y. Acad. Sci.* 537: 188-205.

Dunn, A.J. and Berridge, C.W. (1987) Corticotropin-releasing factor administration elicits a stress-like activation of cerebral catecholaminergic systems. Pharmacol. *Biochem. Behav.* 27: 685-691.

Dunn, A.J. and File, S.E. (1987) Corticotropin-releasing factor has an anxiogenic action in the social interaction test. *Horm. Behav.* 21: 193-202.

Dunn, A. J. and Kramarcy, N. R. (1984) Neurochemical responses in stress: relationships between the hypothalamic-pituitary-adrenal and catecholamine systems. In: *Handbook of Psychopharmacology, Vol. 18* (Eds. L. L. Iversen, S. D. Iversen and S. H. Snyder), New York: Plenum Press. pp.455-515

Dunn, A.J., Elfvin, K.L. and Berridge, C.W. (1986) Changes in plasma corticosterone and cerebral biogenic amines and their catabolites during training and testing of mice in passive avoidance behavior. *Behav. Neural Biol.* 46: 410-423.

Dunn, A.J., Berridge, C.W., Lai, Y.I. and Yachabach, T.L. (1987) CRF-induced excessive grooming behavior in rats and mice. *Peptides* 8: 841-844.

Dunn, A.J., Powell, M.L., Meitin, C. and Small, P.A. (1989) Virus infection as a stressor: influenza virus elevates plasma concentrations of corticosterone, and brain concentrations of MHPG and tryptophan. *Physiol. Behav.* 35: in press.

Eaves, M., Thatcher-Britton, K., Rivier, J., Vale, W. and Koob, G.F. (1985) Effects of corticotropin releasing factor on locomotor activity in hypoph.sectomized rats. *Peptides* 6: 923-926.

Eberly, L.B., Dudley, C.A. and Moss, R.L. (1983) Iontophoretic mapping of corticotropin-releasing factor (CRF) sensitive neurons in the rat forebrain. *Peptides* 4: 837-841.

Ehlers, C.L., Henriksen, S.J., Wang, M., Rivier, J., Vale, W. and Bloom, F.E. (1983) Corticotropin-releasing factor produces increases in brain excitability and convulsive seizures in rats. *Brain Res.* 278: 332-336.

Glavin, G.B. (1985) Stress and brain noradrenaline: A review. *Neurosci. Biobehav. Rev.* 9: 233-243.

Gold, P.W. and Chrousos, G. (1985) Clinical studies with corticotropin-releasing factor: implications for the diagnosis and pathophysiology of depression, Cushing's disease, and adrenal insufficiency. *Psychoneuroendocrinol.* 10: 401-419.

Guild, A.L. and Dunn, A.J. (1982) Dopamine involvement in ACTH-induced grooming behavior. *Pharmacol. Biochem. Behav.* 17: 31-36.

Handley, S. L. and Mithani, S. (1984) Effects of alpha-adrenoceptor agonists and antagonists in a maze-exploration model of 'fear'-motivated behavior. *N-S. Archs Pharmacol.* 327: 1-5.

Kalin, N.H., (1985) Behavioral effects of ovine corticotropin-releasing factor administered to rhesus monkeys. *Fed. Proc.* 44: 249-254.

Kalin, N.H., Sherman, J.E. and Takahashi, L.K. (1988) Antagonism of endogenous CRH systems attenuates stress-induced freezing behavior in rats. *Brain Res.* 457: 130-135.

Koob, G.F. and Bloom, F.E. (1985) Corticotropin-releasing factor and behavior. *Fed. Proc.* 44: 259-263.

Krahn, D.D., Gosnell, B.A., Grace, M. and Levine, A.S. (1986) CRF antagonist partially reverses CRF- and stress-induced effects on feeding. *Brain Res. Bull.* 17: 285-289.

Kurosawa, M., Sato, A., (1986) Swenson, R.S. and Takahashi, Y. Sympatho-adrenal medullary functions in response to intracerebroventricularly injected corticotropin-releasing factor in anesthetized rats. *Brain Res.* 367: 250-257.

Lenz, H.J., Raedler, A. Greten, H. and Brown, M. (1987) CRF inititates biological actions within the brain that are observed in response to stress. *Am. J. Physiol.* 252: R34-39.

Lenz, H.J., Burlage, M., Raedler, A. and Greten, H. (1988) Central nervous system effects of corticotropin-releasing factor on gastrointestinal transit in the rat. *Gastroenterol.* 94: 598-602.

Mason, J.W. (1968) A review of psychoendocrine research on the pituitary-adrenocortical system. *Psychosom. Med.* 30: 576-607.

Merchenthaler, I. (1984) Corticotropin releasing factor (CRF)-like immunoreactivity in the rat central nervous system. Extrahypothalamic distribution. *Peptides* 5 (Suppl. 1): 53-69.

Morley, J.E. and Levine, A.S. (1982) Corticotrophin releasing factor, grooming and ingestive behavior. *Life Sci.* 31: 1459-1464.

Nakane, T., Audhya, T., Hollander, C.S., Schlesinger, D.H., Kardos, P., Brown, C. and Passarelli, J. (1986) Corticotrophin-releasing factor in extra-hypothalamic brain of the mouse: demonstration by immunoassay and immunoneutralization of bioassayable activity. *J. Endocrinol.* 111: 143-149.

Ono, N., Bedran de Castro, J.C. and McCann, S.M. (1985a) Ultrashort-loop positive feedback of corticotropin (ACTH)-releasing factor to enhance ACTH release in stress. *Proc. Natl. Acad. Sci. USA.* 82: 3528-3531.

Ono, N., Samson, W.K., McDonald, J.K., Lumpkin, M.D., Bedran de Castro, J.C. and McCann, S.M. (1985b) Effects of intravenous and intraventricular injections of antisera directed against corticotropin-releasing factor on the secretion of naterior pituitary hormones. *Proc. Natl. Acad. Sci. USA.* 82: 7787-7790.

Plotsky, P. M. (1987) Facilitation of immunoreactive corticotropin-releasing factor secretion into the hypophysial-portal circulation after activation of catecholaminergic pathways or central norepinephrine injection. *Endocrinology* 121: 924-930.

Rivier, C. and Plotsky, P.M. (1986) Mediation by corticotropin releasing factor (CRF) of adenohypophysial hormone secretion. *Ann. Rev Physiol.* 48: 475-494.

Rivier, C. and Vale, W. (1986) Involvement of corticotropin-releasing factor and somatostatin in stress-induced inhibition of growth hormone secretion in the rat. *Endcorinol.* 117: 2478-2482.

Rivier, C., Rivier, J. and Vale, W. (1986) Stress-induced inhibition of reproductive functions: role of endogenous corticotropin-releasing factor. *Science* 231: 607-609.

Rivier, J., Rivier, C. and Vale, W. (1984) Synthetic competitive antagonists of corticotropin-releasing factor: effects on ACTH secretion in the rat. *Science* 224: 889-891.

Roth, K.A., Mefford, I.M. and Barchas, J.D. (1982) Epinephrine, norepinephrine, dopamine and serotonin: differential effects of acute and chronic stress on regional brain amines. *Brain Res.* 239: 417-424.

Sahgal, A., Wright, C., Edwardson, J.A. and Keith, A.B. (1983) Corticotropin releasing factor is more potent than some corticotropin-related peptides in affecting passive avoidance behavior in rats. *Neurosci. Letts* 36: 81-86.

Sawchenko, P.E. and Swanson, L.W. (1985) Localization, colocalization, and plasticity of corticotropin-releasing factor immunoreactivity in rat brain. *Fed. Proc.* 44: 221-227.

Selye, H. (1950) The Physiology and Pathology of Exposure to Stress. *Montreal: Acta Med. Publ.*

Sherman, J.E. and Kalin, N.H. (1988) ICV-CRH alters stress-induced freezing behavior without affecting pain sensitivity. *Pharmacol. Biochem. Behav.* 30: 801-807.

Sirinathsinghji, D.J.S. (1985) Modulation of lordosis behavior in the female rat by corticotropin releasing factor, b-endorphin and gonadotropin releasing hormone in the mesencephalic central gray. *Brain Res.* 336: 45-55.

Sirinathsinghji, D.J.S. (1987) Inhibitory influence of coricotropin releasing factor on components of sexual behavior in the male rat. *Brain Res.* 407: 185-190.

Sirinathsinghji, D.J.S., Rees, L.H., Rivier, J. and Vale, W. (1983) Corticotropin-releasing factor is a potent inhibitor of sexual receptivity in the female rat. *Nature* 305: 232-235.

Smith, M.A., Slotkin, T.A., Knight, D.L. and Nemeroff, C.B. (1986) Release of corticotropin-releasing factor from rat brain regions in vitro. *Endocrinology* 118: 1997-2001.

Stephens, R.L., Yang, H., Rivier, J. and Tache, Y. (1988) Intracisternal injection of CRF antagonist blocks surgical stress-induced inhibition of gastric secretion in the rat. *Peptides* 9: 1067-1070.

Stone, E.A. (1975) Stress and catecholamines. In: Catecholamines and Behavior. Vol. 2 Neuropsychopharmacology (Ed. Friedhoff, A.J.), New York: Plenum Press pp.31-72.

Sutton, R.E., Koob, G.F., Le Moal, M., Rivier, J. and Vale, W. (1982) Corticotropin releasing factor produces behavioral activation in rats. *Nature* 297: 331-333.

Svensson, T.H., Bunney, B.S. and Aghajanian, G.K. (1975) Inhibition of both noradrenergic and serotonergic neurones in brain by the a-adrenoceptor agonist clonidine. *Brain Res.* 92: 291-306.

Swerdlow, N.R., Geyer, M.A., Vale, W.W. and Koob, G.F. (1986) Corticotropin-releasing factor potentiates acoustic startle in rats: blockade by chlordiazepoxide. *Psychopharmacol.* 88: 147-152.

Szafarczyk, A., Malaval, F., Laurent, A., Gibaud, R. and Assenmacher, I. (1987) Further evidence for a central stimulatory action of catecholamines on adrenocorticotropin release in the rat. *Endocrinology* 121: 883-892.

Tache, Y. and Gunion, M. (1985) Corticotropin-releasing factor: central action to influence gastric secretion. *Fed. Proc.* 44: 255-258.

Tazi, A., Dantzer, R., Le Moal, M., Rivier, J., Vale, W. and Koob, G. F. (1987) Corticotropin-releasing factor antagonist blocks stress-induced fighting in rats. *Regulat. Peptides* 18: 37-42.

Vale, W., Spiess, J., Rivier, C. and Rivier, J. (1981) Characterization of a 41-residue ovine hypothalamic peptide that stimulates secretion of corticotropin and B-endorphin. *Science* 213: 1394-1397.

Valentino, R.J. and Foote, S.L. (1988) Corticotropin-releasing factor increases tonic but not sensory-evoked activity of noradrenergic locus coeruleus neurons in unanesthetized rats. *J. Neurosci.* 8: 1016-1025.

Valentino, R.J., Foote, S.L. and Aston-Jones, G. (1983) Corticotropin-releasing factor activates noradrenergic neurons of the locus coeruleus. *Brain Res.* 270: 363-367.

Veldhuis, H.D. and de Wied, D. (1984) Differential behavioral actions of corticotropin-releasing factor (CRF). Pharmacol. *Biochem. Behav.* 21: 707-713.

Williams, C.L., Peterson, J.M., Villar, R.G. and Burks, T.F. (1987) Corticotropin-releasing factor directly mediates colonic responses to stress. *Amer. J. Physiol.* 253: G582-G586.

Wynn, P.C., Hauger, R.L., Holmes, M.C., Millan, M.A., Catt, K.J. and Aguilera, G. (1984) Brain and pituitary receptors for corticotropin releasing factor: localization and differential regulation after adrenalectomy. *Peptides* 5: 1077-1084.

PSYCHOBIOLOGY OF STRESS AND IMMUNE FUNCTIONS

Pierre Mormède
Laboratoire de Psychobiologie des Comportements Adaptatifs
INRA - INSERM U259
Rue Camille St-Saëns, 33077 BORDEAUX, France.

It has been known for a long time that stress has an influence on the susceptibility to disease. As early as the nineteenth century, Louis Pasteur found that chickens could be made susceptible to anthrax by immersing their legs in cold water. In his 1936 paper published in Nature, Hans Selye included involvement of the thymus as one of the three criteria of the stress response, together with adrenal enlargement and gastric erosions. In the past few years a considerable body of data has been accumulated on the effects of environmental factors on the immune system functions, which substantiate these early findings.

Clinical studies have shown that stressful life events, such as space flight, examinations in college students, poor marital relationships, bereavement, unemployment, work overload in accountants, and can alter immune functions. Moreover, the importance of mood states and personality factors on various measures of immune system function as well as on the course of diseases such as cancer has been pointed out by many authors (see Dorian and Garfinkel, 1987, for review).

These data are clearly crucial to an understanding of the individual's receptivity to disease, and have prompted many investigators to study immune functions in experimental models of stress, and to analyze the mechanisms by which environmental and personality factors influence the immune system. Although a great deal of data is now available, a simple overview of this research is not really feasible at the present time due to the great variety of experimental animals, stress protocols, immune function measures, etc. employed in such studies. Instead of giving one more review of this literature, I intend to emphasize a few points which may worth considering when trying to understand the somewhat discordant results available, and which may help develop new approaches to these issues.

These points can be summarized as follows:
– Stress needs to be regarded as a psychobiological response.
– The immune system is diverse.
– The neuroendocrine response to stress is multivariant.

Stress as a psychobiological response

The term stress has been used loosely for any physiological state where blood corticosteroid levels are high or assumed to be so. This use derives from the emphasis put by Selye on activation of the adrenocortical axis. However the original definition of the term by Selye

S. Puglisi-Allegra and A. Oliverio (eds.), Psychobiology of Stress, 95–102.
© 1990 *Kluwer Academic Publishers. Printed in the Netherlands.*

(1950) was 'the non-specific response of the organism to any demand made upon it'. Since the work of Mason (1971) and others, we know that this non-specificity of the response is related to the undifferentiated emotional arousal induced by the stressor. Besides this emotional response, each stimulus triggers a specific array of physiological changes which are related to the nature of the stimulus. The relative importance of the specific and non-specific aspects of the response changes with time, since the initial non-specific arousal tends to wane after exposure to the stimulus, while the response becomes more relevant to the nature of the stimulus and the efficiency of psychological and physiological adaptation mechanisms (Dantzer and Mormède, 1983). This concept, although widely accepted, is sometimes neglected by many physiologists and immunologists. It is not surprising for instance to find different effects of exposure to a cold or a warm atmosphere on immune system function (Kelley, 1985). After an initial phase of non-specific activation of the adrenocortical axis (the stress response *sensu stricto*), exposure to cold chronically increases corticosteroid levels, whereas exposure to warmth chronically decreases them. Many other physiological and behavioral responses are also shaped by the stimulus characteristics. It is therefore misleading to use terms such as cold stress and heat stress, since both stimuli have complex neuroendocrine influences, some of which are diametrically opposite after a period of exposure. It is thus important to distinguish between the non-specific response which is common to both stimuli, and the specific physiological response which is related to the physical nature of the stimulus.

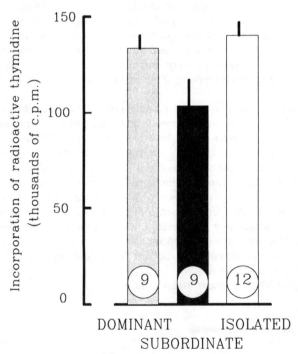

Figure 1. Resident – intruder dyads. Mitogenic response of spleen lymphocytes to pokeweed mitogen after 10 days common housing for dominant and subordinate animals of the dyads, compared to isolated controls. The response was only reduced significantly in the subordinate animals (from Raab et al., 1986).

In order to gain more understanding of the effects of environmental factors on the immune system, it would seem important to sort out the relative influence of each aspect of the response. Another approach is the study of experimental models where the specific physiological response is minimized by using 'pure' psychological stimuli. An elegant example is given by the work of Coe and coworkers (1988) on the influence of separation distress in young monkeys. In the case of psychological stimuli, we can also find complex biological responses which depend on the duration of the stimulus. For example, acute exposure to such a mild stressor as a novel environment or gentle handling can elicit a typical neuroendocrine response, with a surge of circulating levels of ACTH, β-endorphin, corticosteroids, prolactin (Mormède, 1983; Mormède et al., 1986), catecholamines and associated peptides (Castagné et al., 1987). However, when chronic situations are studied, divergent responses emerge which depend on the characteristics of the situation, the efficiency of psychological adaptation mechanisms and probably on poorly defined individual factors shaped by genetic characteristics and previous experience (Dantzer and Mormède, 1983). For instance, we have studied the influence of social factors in the resident-intruder dyad model (Raab et al., 1986). After 10 days of common housing, the proliferative response of spleen lymphocytes was only depressed in the submissive animal, the response of the dominant animals was the same as that of isolated controls (figure 1).

In the case of physical stimuli which are the most widely used, it is particularly important to consider the psychological aspects of the experimental protocol such as stressor predictability and the availability of behavioral control over the stimulus. For example, we have studied the influence of these factors in the effects of footshock in rats (Mormède et al., 1988). When rats were submitted to two sessions of unavoidable and unpredictable footshocks, the in vitro mitogenic response of spleen lymphocytes to lectins was reduced, but the depression of this immune function was completely reversed when a warning signal preceded every impending shock (figure 2). These results show that the physical characteristics of the aversive stimulus (here the electric shock) are not as crucial as 'subtle' procedural factors which are able to modify the way the animal perceives the situation. In a similar design, Laudenslager and coworkers (1983) showed that the effect of footshocks could be abolished when the shock was escapable. In another experiment, rats were trained in a shuttle-box to avoid brief shocks in a schedule of continuous avoidance (Mormède et al., 1988). Yoked animals were wired to animals of the avoidance group, and therefore received the same shocks, but independently from their behavior. In fact, although avoiding animals shuttled actively, yoked rats became less and less active over the sessions. After ten daily sessions, only the yoked rats displayed a reduced mitogenic response of spleen lymphocytes to mitogens (figure 3). This is a further demonstration of the fact that the experimental design is often more important than the actual physical characteristics of the aversive stimulus for induction of functional changes in the immune system. These experiments further support the concept that stress is not a simple reflex response but an elaborate psychobiological reaction. This has been shown to be the case for neuroendocrine activation (Hennessy and Levine, 1979), stress-induced gastric ulceration (Weiss, 1968), tumor growth (Sklar and Anisman, 1979), to only mention a few examples. This must also apply to the influence of stressors on immune functions.

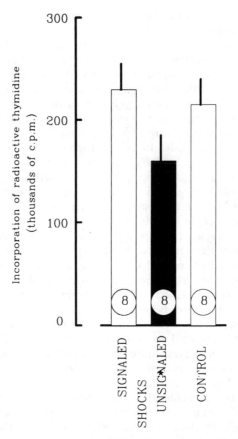

Figure 2. Influence of stressor predictability. Unavoidable and unsignalled shocks reduced the response of spleen lymphocytes to concanavalin A stimulation in vitro. This effect was completely reversed when a warning signal preceded every impending shock (from Mormède et al., 1988).

Diversity of the immune system.

Alongside the difficulties involved with the concept of stress, the complexity of the immune system represents a further challenge. The variety of approaches employed in the study of immune functions reflects not only the complexity of the immune system but also an uncertainty about which aspects of its function should be evaluated.

Early studies clearly demonstrated that the same experimental condition could be either beneficial or detrimental to the individual's resistance to disease. In the late 60's, Gross and Siegel (1965) showed that chickens maintained in a stressful social environment were more resistant to bacterial infections (such as E. coli), but more susceptible to viral infections (such as Marek's disease). Similarly, the selection of chickens for a high corticosteroid response to social stress was followed by an increased susceptibility to Marek's disease, but a reduced sensitivity to infection with E. coli (Gross and Colmano, 1971). In that case, the effects of

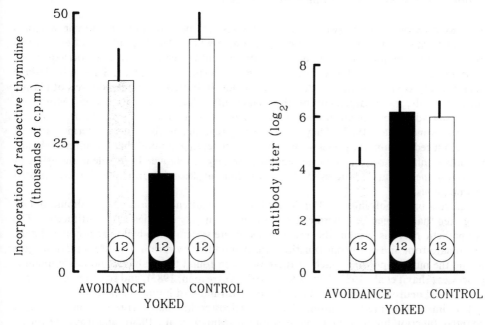

Figure 3. Influence of shock avoidance. Animals trained to avoid electric shocks according to a continuous avoidance schedule in a shuttle box did not show any change in the response of spleen lymphocytes to phytohemagglutinin stimulation in vitro, but the response was significantly depressed in yoked animals receiving the same shocks without behavioral control (left panel). The reverse was true when the circulating antibody response to sheep red blood cells was studied (right panel). In that case, the response was depresses in avoiding animals (from Mormède et al., 1988).

genetic and environmental factors were additive (Gross and Siegel, 1973).

The same holds true when selected immune functions are considered. For example, Blecha and coworkers (1982a) showed that a 2-hour restraint reduced the delayed-type hypersensitivity response to sheep erythrocytes in mice, but enhanced contact sensitivity reactions to dinitrofluorobenzene (DNFB), although both reactions are supposed to be mediated by T-cells. In the avoidance experiment described earlier, a group of animals was injected with sheep erythrocytes, and the levels of circulating agglutinins were measured five days later to evaluate their ability to establish a primary antibody response. As shown in figure 3, the agglutinin titer was significantly lower in the 'avoidance' group, as opposed to the proliferative response of spleen lymphocytes, which was significantly depressed in the 'yoked' animals (Mormède et al., 1988).

Although the effect of stress is frequently considered to be immunosuppressive, there is considerable evidence showing immunostimulating effects. For example, lymphoproliferative responses, or the capacity to generate a primary antibody response in vivo have been shown to be enhanced in stressful situations (see for instance Croiset et al., 1987). In order to understand how more complex immune functions are modulated by psychobiological influences, careful analytical studies will be required to identify the primary targets of the stress response within the immune system.

Multiplicity of the neuroendocrine stress responses.

The link between the brain (where stress responses are initiated) and the immune system is thought to involve the neuroendocrine system as a whole. The hypophyso-adrenocortical axis is clearly a major candidate, not only for conceptual reasons, but also, perhaps more importantly, for technical reasons. The stress concept as coined by Selye was defined operationally as an activation of the adrenal glands, indicated by increased levels of plasma corticosteroids, or enlargement of the glands after long-term stimulation. Moreover, corticosteroids have a potent suppressive action on many immune processes, both in vivo and in vitro (Munck et al., 1984). They are frequently regarded as the general intermediaries of stress-induced immunosuppression. It is also worth bearing in mind that the adrenal glands, the only significant source of corticosteroids, are easy to remove. Their role in the effect of stress on the immune process under study can, at least in principle, be investigated unambiguously.

Before a brief look into the Pandora's box of the adrenocortical axis, it should not be forgotten that the neuroendocrine stress response is highly complex and multifaceted. The sympathetic nervous system, the endogenous opiate systems, the hypophyso-gonadal axis, the somatotrope and lactotrope functions are all modified by various stressful situations, just to cite the most important. Each of these has been found to be able to modulate immune processes (Shavit et al., 1985; Felten et al., 1987; Grossman, 1985), and so can be considered as potential neuroendocrine intermediates of stress-induced immune changes.

The immunosuppressive properties of corticosteroids may represent an important adaptive function limiting the immune response (Munck et al., 1984), and it is noteworthy that corticosteroid secretion is activated by the immune response itself. Besedovski and Sorkin (1981) have even demonstrated the existence of a functional immuno-endocrine loop. However, the general immunosuppressive action of corticosteroids is far from being as general as is frequently assumed. It is increasingly clear that minimal concentrations of corticosteroids are necessary for a full expression of various immune processes such as the lymphoproliferation induced by plant lectins (Dunn, 1988). Even large doses of potent synthetic glucocorticoids can enhance immune processes such as interferon production in vitro (Adolf and Swetly, 1979).

Moreover, the stress-induced increase in circulating corticosteroids is not always followed by immunosuppression. In our experiments, both predictable and unpredictable shocks induced a maximal corticosteroid release, whereas only the unpredictable shocks affected the lymphoproliferative response. In the avoidance experiments described earlier, adrenocortical function was assessed both directly by the measure of circulating levels of corticosteroids at various steps of the procedure, and indirectly by measuring prolactin levels, which are a sensitive indicator of a sustained secretion of corticosteroids. Again the two shock conditions (avoidance and yoked) could not be distinguished on the basis of the neuroendocrine responses, despite the large changes observed in immune responses (Mormède et al., 1988).

On the other hand, suppression of the glucocorticoid response by removal of the adrenal glands does not always counteract the effect of stress. For instance, Blecha and coworkers (1982b) showed that adrenalectomy inhibited the effect of restraint stress on delayed-type hypersensitivity to sheep erythrocytes, but did not reverse the enhanced contact sensitivity response to DNFB. Adrenalectomy was similarly inefficient in counteracting the suppressive effects of footshock on the circulating lymphocyte proliferative response to mitogens (Keller et al., 1983). However, the apparent simplicity of adrenalectomy experiments is misleading.

Removal of endogenous corticosteroids has complex effects on the whole endocrine system. It is known that removal of corticosteroid feedback is followed by a surge of circulating levels of ACTH and other peptides originating from the precursor molecule, proopiomelanocortin (POMC). Among other mediators, circulating levels of beta-endorphin are increased, and their reactivity to stressors is considerably enhanced (De Souza and Van Loon, 1983). Interestingly, adrenalectomy is followed by a marked stimulation of the sympathetic nervous system, as evidenced by increased urinary excretion of noradrenaline and a faster turnover of this amine, these changes being corticosteroid-dependant (Dailey and Westfall, 1978).

The neuroendocrine mediation of the influence of stress on immune functions thus involves two levels of complexity. On the one hand, the neuroendocrine stress response is highly diverse, and on the other, any experimental manipulation of one system profoundly influences other systems, which are themselves potential candidates for an immunomodulatory function.

Conclusion

A better understanding of stress as a psychobiological response to environmental challenges is a prerequisite for the study of stress-induced changes in immune competence. More emphasis needs to be placed on the psychological appraisal of environmental stimuli, arising from specific characteristics of the situation (novelty, uncertainty, availability of behavioral control,...), as well as on individual factors resulting from the complex interaction of genetic and developmental influences. Alongside this dynamic concept of stress, the extreme complexity of the immune system should not be overlooked. This would avoid such oversimplifications as 'stress-induced immunosuppression'. In fact, certain immune functions may react to the same environmental influence in seemingly contradictory ways. Unitary processes underlying these divergent influences need to be elucidated. The neuroendocrine system is an obvious candidate for mediation of stress-induced changes in immune functions. However, there is a marked variation in the sensitivity of the various components of the neuroendocrine system to stressful stimuli, and any given hormone is capable of modulating many other parts of the immune system. Although corticosteroids are probably important modulators of the major immune functions, their general immunosuppressive effect in stress-induced changes is probably overestimated, and may well obscure the role of alternative pathways.

References

Adolf G.R. and Swetly P. (1979) Glucocorticoid hormones inhibit DNA synthesis and enhance interferon production in a human lymphoid cell line. *Nature*, 282, 736-738.

Besedovsky H.O. and Sorkin E. (1981) Immunologic-neuroendocrine circuits: physiological approaches. In 'Psychoneuroimmunology', Ader C. (Ed.). New York, Academic Press Inc., pp.545-574.

Blecha F., Barry R.A. and Kelley K.W. (1982a) Stress-induced alterations in delayed-type hypersensitivity to SRBC and contact sensitivity to DNFB in mice. *Proc. Soc. Exp. Biol. Med.*, 169, 239-246.

Blecha F., Kelley K.W. and Satterlee D.G. (1982b) Adrenal involvement in the expression of delayed-type hypersensitivity to SRBC and contact sensitivity to DNFB in mice. *Proc. Soc. Exp. Biol. Med.*, 169, 239-246.

Castagné V., Corder R., Gaillard R. and Mormède P. (1987) Stress-induced changes of circulating neuropeptide the rat: comparison with catecholamines. *Regul. Pept.*, 19, 55-63.

102

Coe C.L., Rosenberg L.T. and Levine S. (in press) Immunological consequences of psychological disturbances and maternal loss in infancy. In: 'Advances in Infancy Research' (Rovee-Collier C. and Lipsitt L.P. (Eds.). Norwood: Ablex Publication Corp..

Croiset G., Heijnen C., Veldhuis H.D., de Wied D. and Baillieux R.E. (1987) Modulation of the immune response by emotional stress. *Life Sci.*, 40, 775-782.

Dailey J.W. and Westfall T.C. (1978) Effects of adrenalectomy and adrenal steroids on norepinephrine synthesis and monoamine oxidase activity. *Eur. J. Pharmacol.*, 48, 383-391.

Dantzer R. and Mormède P. (1983) Stress in farm animals: a need for reevaluation. *J. Anim. Sci.*, 57, 6-18.

De Souza E.B. and Van Loon G.R. (1983) A triphasic pattern of parallel secretion of β-endorphin / β-lipotropin and ACTH after adrenalectomy in rats. *Amer. J. Physiol.*, 245, E60-E66.

Dorian B. and Garfinkel P.E. (1987) Stress, immunity and illness – a review. *Psychol. Med.*, 17, 393-407.

Dunn A. (1988) Nervous system-immune system interactions: an overview. *J. Recep. Res.*, 8, 589-607.

Felten D.L., Felten S.Y., Bellinger D.L., Carlson S.L., Ackerman K.D., Madden K.S., Olchowki J.A. and Livnat S. (1987) Noradrenergic sympathetic neural interactions with the immune system: structure and function. *Immunol. Rev.*, 100, 225-260.

Gross W.B. and Colmano G. (1971) Effect of infectious agents on chickens selected for plasma corticosterone response to social stress. *Poultry Sci.*, 50, 1213-1217.

Gross W.B. and Siegel P.B. (1973) Effect of social stress and steroids on antibody production. *Avian Diseases*, 17, 807-815.

Gross W.B. and Siegel P.B. (1965) The effect of social stress on resistance to infection with Escherichia coli and Mycoplasma gallisepticum. *Poultry Sci.*, 44, 998-1001.

Grossman C.J. (1985) Interactions between the gonadal steroids and the immune system. *Science*, 227, 257-261.

Hennessy J.W. and Levine S. (1979) Stress, arousal, and the pituitary-adrenal system: a psychoendocrine hypothesis. *Prog. Psychobiol. Physiol. Psychol.*, 8, 133-177.

Kelley K.W. (1985) Immunological consequences of changing environmental stimuli. In: 'Animal Stress', Moberg G.P. (Ed.). *American Physiological Society*, pp. 193-223.

Keller S.E., Weiss J.M., Schleifer S.J., Miller N.E. and Stein M. (1983) Stress-induced suppression of immunity in adrenalectomized rats. *Science*, 1983, 221, 1301-1304.

Laudenslager M.L., Ryan S.M., Drugan R.C., Hyson R.L. and Maier S.F. (1983) Coping and immunosuppression: inescapable but not escapable shock suppresses lymphocyte proliferation. *Science*, 221, 568-570.

Mason J.W. (1971) A re-evaluation of the concept of 'non specificity' in stress theory. *J. Psychiat. Res.*, 8, 323-333.

Mormède P., Dantzer R., Michaud B., Kelley K.W. and LeMoal M. (1988) Influence of stressor predictability and behavioral control on lymphocyte reactivity, antibody responses and neuroendocrine activation in rats. *Physiol. Behav.*, 43, 577-583.

Mormède P., Vincent J.-D. and Kerdelhué B. (1986) Vasopressin and oxytocin reduce plasma prolactin levels of conscious rats in basal and stress conditions. Study of the characteristics of the receptor involved. *Life Sci.*, 39, 1737-1743.

Mormède P. (1983) The vasopressin receptor antagonist dPTyr(Me)AVP does not prevent stress-induced ACTH and corticosterone release. *Nature*, 302, 345-346.

Munck A., Guyre P.M. and Holbrook N.J. (1984) Physiological functions of glucocorticoids in stress and their relation to pharmacological actions. *Endocrine Rev.*, 5, 25-44.

Raab A., Dantzer R., Michaud B., Mormède P., Taghzouti K., Simon H. and LeMoal M. (1986) Behavioural, physiological and immunological consequences of social status and aggression in chronically coexisting resident-intruder dyads of male rats. *Physiol. Behav.*, 36, 223-228.

Selye H. (1936) A syndrome produced by diverse nocuous agents. *Nature*, 1936, 138, 32-33.

Selye H. (1950) - Stress. Acta Inc., Medical Publications, Montreal.

Shavit Y., Terman G.W., Martin F.C., Lewis J.W., Liebeskind J.C. and Gale R.P. (1985) Stress, opioid peptides, the immune system, and cancer. *J. Immunol.*, 135, 834s-837s.

Sklar L.S. and Anisman H. (1979) - Stress and coping factors influence tumor growth. *Science*, 205, 513-515.

Weiss J.M. (1968) Effects of coping responses on stress. *J. Comp. Physiol. Psychol.*, 65, 251-260.

NEUROPEPTIDES AND BEHAVIOURAL AND PHYSIOLOGICAL STRESS RESPONSE: THE ROLE OF VASOPRESSIN AND RELATED PEPTIDES

B. Bohus, J.M. Koolhaas, C. Nyakas, P.G.M. Luiten, C.A.M. Versteeg,
S.M. Korte, D. Jaarsma, W. Timmerman and W. Eisenga
*Department of Animal Physiology, University of Groningen, Biological
Centre, P.O. Box 14, 9750 AA HAREN, The Netherlands*

Classical views on stress mechanisms maintain that the activation of the adrenal medulla and the cortex represent the major neuroendocrine aspects of stress. The functional significance of these hormones in both physiological and behavioural stress responses have been repeatedly emphasized (e.g. Mason, 1968; Selye, 1976; Bohus, 1984; Oliverio, 1987; etc.). Besides these stress hormones of the first generation a new class of neuroendocrine principles emerged that also serve stress functions (Bohus, 1984). These second generation consists of long known pituitary and their target hormones (e.g. prolactin, vasopressin, oxytocin, growth hormone, gonadotrophins, testosterone, thyroid-stimulating hormone, thyroxine; see Bohus, 1984; Oliverio, 1987) with a recently established stress function. In addition, newly discovered hormones such as the endorphins and enkephalins (e.g. Guillemin et al., 1977; Amir et al., 1980) and of hypothalamic neuroendocrine factors such as the Corticotropin-Releasing Hormone (CRH) from the hypothalamus (e.g. Koob, 1985) belongs to the family of stress hormones. For the majority of the pituitary and hypothalamic hormones neuronal cell bodies and axons with terminals can be localized in a number of brain regions outside the hypothalamus (see Nieuwenhuys, 1985). Accordingly, the brain should be added to the neuroendocrine compartment of the body: the brain is both a source and target of these hormones. These hormones with oligo- or polypeptide structures, are designated as neuropeptides both because of their action on nervous tissue (De Wied, 1987) and their neuronal origin (Bloom, 1987).

The present paper deals with one of these neuropeptides, namely vasopressin, an important modulator of physiological and behavioural stress responses via brain and peripheral mechanisms. Although Verney (1947) demonstrated that psychological stress results in antidiuresis in the dog, and Thompson and De Wied (1973) found that emotional stress increases biological antidiuretic activity in the blood of the rat, subsequent studies using the determination of radioimmunoassayable vasopressin failed to reinforce these findings (Van Wimersma Greidanus et al., 1979; Mens et al., 1982). Oxytocin (OXT), the structurally related hypothalamo-neurohypophyseal neurosecretory hormone is however released by divers stressors (Lang et al., 1983; Mens et al., 1982; Carter and Lightman, 1987; Gibbs, 1984; Williams et al., 1985). However, a number of studies showed relations between stress and vasopressin release into the cerebrospinal fluid or changes in brain vasopressin content following emotional stress (e.g. Laczi et al., 1983a,b; 1984). Early observations on the profound effect of vasopressin on stress-related behaviours – i.e. conditioned avoidance

S. Puglisi-Allegra and A. Oliverio (eds.), Psychobiology of Stress, 103–123.
© 1990 *Kluwer Academic Publishers. Printed in the Netherlands.*

responses (De Wied and Bohus, 1966; Bohus et al., 1973) have provided evidence of the stress function of the peptide. his stress function may be of primary importance at the level of the nervous system rather than at the periphery.

Vasopressin as a neuropeptide: localization in several systems in the brain

The nonapeptide arginine-8-vasopressin (AVP) is the major vasopressin- like peptide in mammalian brain. The hypothalamo-neurohypophyseal neurons represent the classical neuroendocrine pathway of AVP. The peptide is synthesized by the magnocellular neurosecretory cells of the supraoptic and paraventricular hypothalamic nuclei and transported to the posterior pituitary along the axons of the hypothalamo-neurohypophyseal tract. This neuroendocrine pathway ensures the 'classical' peripheral physiological effects of antidiuresis and/or blood pressure increase (see Cowley et al., 1980). Neurons originating from the parvocellular paraventricular nucleus and terminating in the median eminence on the portal vessels (e.g. Whitnall et al., 1985) represent another neuroendocrine limb. AVP released into the portal circulation regulates the anterior pituitary function. The corticotropin releasing potency of AVP is long known (e.g. De Wied, 1961).

Several networks in the brain use AVP as neurotransmitter (see Buijs, 1983; Weindl and Sofroniew, 1985). Vasopressinergic cells of the parvocellular hypothalamic paraventricular nuclei (PVN) project to lower brain stem and spinal autonomic nuclei such as the parabrachial nucleus (PB), n. tractus solitari (NTS), dorsal motor nucleus of the vagus nerve (PMV) and the intermedio-lateral (IML) column of the spinal cord. AVP- containing cells of the suprachiasmatic nuclei project to the habenular area, the organum vasculosum laminae terminalis (OVLT), the dorsomedial hypothalamic nucleus. Vasopressinergic cell bodies have been found in limbic areas such as the medial amygdaloid nuclei (AM), the bed nucleus of the stria terminalis (BNST) and the n. basalis (Meynert n.). The projections from the AM reach the ventral hippocampus and the lateral septum. The BNST projections terminate also in the lateral septal area. In addition, the habenular complex and the locus coeruleus are terminal areas of these fibers. Some scattered cell bodies have been described in the locus coeruleus and terminals of unknown origin in the ventral tegmental area. Although not much is known about the regulation of the diverse vasopressinergic networks, the sexual differentiation within the AM and BNST system is remarkable: there is a higher amount of immunostainable AVP in the neurons of male rats. This sexual difference is testosterone-dependent (De Vries et al., 1985).

The actions of AVP in the brain are mediated by cell membrane receptors. Two types of vasopressin receptors, designated as V-1 and V- 2 receptors, have been identified in various peripheral tissues. The V- 1 receptors mediate vascular/metabolic effects while the V-2 receptors are mainly responsible for the antidiuretic action. The vasopressin receptors in the brain are presumably of a subtype of the V-1 receptor, designated as V-1a type (see Jard et al., 1987). The V-1b subtype receptors have been localized in the anterior pituitary (see Jard et al., 1986). In addition to the V-1a receptors the involvement of the oxytocin receptor in the action of AVP in the brain has been suggested: AVP acts as agonist while oxytocin is an inverse agonist of this receptor (Elands, 1989).

Vasopressinergic modulation of a neurogenic pressor response

Behavioural action of vasopressin as observed in the rat upon peripheral injection (De Wied and Bohus, 1966; Bohus et al., 1973) raised the question whether it is caused by an increased arousal. Therefore, studies were aimed to investigate this hypothesis by using a physiological measure of arousal – i.e. a pressor response evoked by the electrical stimulation of the mesencephalic reticular formation at the level of the cuneiform nucleus. Classical studies by Moruzzi and Magoun (1949) showed that such a stimulation results in cerebral cortical and limbic arousal.

Administration of lysin-8-vasopressin (LVP) intravenously in urethane-anaesthetized rats appears to increase the current threshold to induce a minimal neurogenic pressor response, and also reduces the magnitude of the pressor response evoked by electrical stimulation of the posterior hypothalamus, a well-defined pressor area in the brain (Bohus, 1974). The relatively long latency of action – i.e. 20 min minimally – suggested an effect on brain mechanisms involved in the organization of the pressor response.

In order to justify the hypothesis of a central site of action, the influence of AVP administered into a lateral cerebral ventricle on the neurogenic pressor response was then studied in detail. The pressor response was induced by the electrical stimulation of the area of the cuneiform nucleus, using stimulation frequencies of 10, 30, 50, 70 and 90 Hz with a fixed current intensity (40-100 fA). These stimulations result in a frequency dependent increase of systolic blood pressure (SBP): the higher is the stimulation frequency, the larger is the magnitude of the SBP response (for methodological details see Versteeg et al., 1982a; Bohus et al., 1983).

AVP reduces the magnitude of the centrally evoked pressor response following lateral ventricular injection in urethane-anaesthetized rats by up to approximately 40% for at least 60 min. Significant changes in the pressor response can be seen after 20 min. The attenuation is dose- dependent in the range of 3-25 ng. The peptide fails to affect the baseline SBP at any time within the 60 min post-injection observation period. The heart rate change that accompanied the pressor response appears to be a frequency-dependent bradycardia. AVP reduces the magnitude of the cardiac response, but this effects fails to show a dose- and frequency-dependent character (Versteeg et al., 1982a).

These experiments suggested that vasopressin has an inhibitory action on brain mechanisms involved in the organization of a neurogenic pressor response, but fails to influence basal blood pressure control. Immunohistochemical studies have not demonstrated yet vasopressinergic innervation of the cuneiform area in the mesencephalon. It is therefore possible that the site of action is localized in regions other than the stimulation site. Therefore, the peptide was administered into the fourth cerebral ventricle – i.e. in the vicinity of brainstem cardiovascular control areas to receive vasopressinergic fibers from the paraventricular hypothalamic nuclei (see Buijs, 1983; Weindl and Sofroniew, 1985).

It appears that comparable doses of AVP administered into the fourth ventricle did not mimic the action of the peptide given into a lateral cerebral ventricle (Versteeg et al., 1983a). It seemed therefore, that AVP cannot easily get access to lower brainstem areas at some distance from the fourth ventricle. Alternatively, the site of action of AVP is localized in structures nearby the lateral ventricle. Limbic system areas that have definite or scattered vasopressinergic innervation from the BNST or ME have been implicated in behavioural and some neurochemical (noradrenergic) action of the peptide (Kovács et al., 1979; Bohus et al., 1982). Accordingly, AVP was administered into a lateral cerebral ventricle in rats with bilateral dorsal hippocampal damage. In addition, the peptide was also locally injected into the dentate gyri of the hippocampus. The inhibitory action on the neurogenic pressor response of AVP administered into a lateral cerebral ventricle is totally absent in rats with bilateral damage of the dorsal hippocampus. Microinjection of the peptide in the dentate gyri of the dorsal hippocampus in a dose that is ineffective intracerebroventricularly attenuates the amplitude of the pressor response (Versteeg et al., 1983b). These observations suggest that the dorsal hippocampus may be but one location in the forebrain where cardiovascular regulatory functions in relation to arousal are modulated by vasopressin.

Subsequent experiments led to the conclusion that OXT shares some of the effect on the neurogenic pressor response of AVP. Administration of OXT into a lateral cerebral ventricle also attenuates the neurogenic pressor response evoked by the electrical stimulation of the cuneiform areas in the brainstem. OXT is, however, less potent then AVP upon this route of administration (Versteeg et al., 1982b). Administration of OXT into the fourth cerebral ventricle results in a marked reduction of the amplitude of the neurogenic pressor response. The 7-9 fragment of OXT, often designated as MIF (MSH Inhibiting Factor) or PLG (Prolyl-leucyl-glycinamide) is even more potent than OXT upon fourth ventricular administration (Versteeg et al., 1983a).

Together, these and the foregoing data suggest that both AVP and OXT are able to modulate blood pressure response that is related to arousal. Immunohistochemical studies show vast innervation of PB complex, NTS, DMV and IML of the spinal cord by oxytocinergic neurons from the PVN (see Buijs, 1983). There are obviously more locations and multiple mechanisms in the brain through which the modulatory action of the peptides is effectuated. It was therefore an increasing need to explore the network through which the pressor response of cuneiform origin is organized.

Organization of a brainstem pressor response: physiological and neuroanatomical studies

The first indication of the organization principles of the pressor response as induced by the stimulation of the mesencephalic cuneiform area is provided by transection studies, placed cranially or caudally from the stimulation sites with the aid of a microknife (Bohus et al., 1983). Ipsilateral cuts caudal from the stimulation site which disconnect the descending pathways, reduce the magnitude of the pressor response to approximately 20% of the preoperative level. Cuts placed cranially reduce the response by about 50%. Since cranial cuts disconnect ascending connections, the findings suggest that a substantial part of the pressor response is mediated and/or modulated by higher brain structures.

Subsequent studies were then devoted (a) to exactly localize the cuneiform pressor area by series of stimulations; (b) to determine the afferent connections of the pressor area both cranially and caudally; and (c) to determine the functional significance of the efferentations for the pressor effects (see Bohus et al., 1989). Although electrical stimulation of the mesencephalic reticular region results in an SBP elevation within a relatively large area (approximately 800 by 500 μm), the maximal magnitude of the response – i.e. 75-100 mm Hg increase – can be evoked in or directly around the cuneiform nucleus.

The efferent connections of the effective stimulation sites were determined by the Phaseolus vulgaris Leucoagglutinin (PHA-L) immunohistochemical anterograde tracing technique (Gerfen and Sawchenko, 1984; Ter Horst et al., 1984). Plant lectins such as PHA-L are easily taken up by the cell bodies and transported along the axons to the terminal areas.

In most cases the presynaptic boutons can be detected even at a long distance from the cell body. The iontophoretically injected PHA-L can be visualized by a double immunohisto-chemical detection technique (for methodological details see Ter Horst et al., 1984).

Ascending fibers from the cuneiform nucleus ipsi- and contralaterally reach the colliculus superior, and a branch of axons travels in the direction of central gray and dorsal raphe nucleus. Few terminals can be detected in the substantia nigra and the ventral tegmental area. The ascending axons through the central gray and the reticular formation are directed towards the thalamic parafascicularis and centrum medianum nuclei. Another branch of the ascending axons runs along the 3rd ventricle. Their major termination areas are the PVN of the hypothalamus, the lateral hypothalamus, the central and ventromedial amygdaloid n., the BNST and the lateral preoptic area.

Most of the descending connections appear to remain ipsilateral. The major branch of the axons travels in the direction of the n. reticularis pontis. Many collaterals terminated in the locus coeruleus and the PB nuclei. A significant branch ends with many presynaptic boutons in the raphe magnus. Few axons reached caudal areas such as the ambiguus nucleus, n. reticularis parvocellularis, the NTs and the DMNV.

Physiological studies suggest that the descending connection to the PB nuclei is of importance for the organization of the pressor response. Cuts placed rostrally from this nucleus substantially reduce the magnitude of the pressor response. The average reduction amounts about 40%.

The most significant changes in the magnitude of the pressor response were found after PB deafferentation. Therefore, the attention was then focused on this area. In contrast to the cuneiform area, more is known about the physiology of the parabrachial nuclei, at least in the cat (Mraovitch et al., 1982; Darlington and Ward, 1985). The neuroanatomy of the efferent connections of the various subdivisions of the PB complex is well-documented (see Fulwiler and Saper, 1984). The physiological studies with electrical stimulation (5-100 HZ frequency and 40 ± 10 fA intensity) show a marked increase in the SBP (about 75-100 mm Hg) in medial and lateral PB sites. In addition, a large response is induced by the stimulation of the Kolliker-Fuse (KF) nucleus that is located ventrolaterally from the PB complex. Heart frequency changes at stimulation are not regular. In most cases mild tachycardia is present. An about 40% reduction in the magnitude of the pressor response occurs following caudal cuts destroying the medullary connections. Practically no changes can be found after rostral cuts. Neuroanatomical studies following PHA-L injection in the PB area show a widespread afferentation both cranially and caudally. The descending axons from the lateral and medial extern region of the PB-KF complex terminate in the n. facialis, n. rostroventrolateralis (RVLM) and caudoventrolateralis medullaris (CVLM), and in the IML of the spinal cord, mostly in the region of the thoracal 1-3 and 7-9 segments. Few terminals can also be found in the various regions of the NTS, n. raphe magnus (RM) and the n. ambiguus. Although the terminals are localized bilaterally, about 20% of the terminalia was found contralaterally.

Ascending axons terminate in the dorsal raphe, ventrolateral central gray, and the reticular formation of the mesencephalon. More rostral projections can be detected in the zona incerta, ventral parvocellular thalamic n., and in the lateral hypothalamus. Most of the terminals can be found in the central amygdala (CA). The projections are bilateral with an ipsilateral dominance.

These physiological and neuroanatomical studies reinforce the notion that the PB complex, particularly the external areas around and in the KF nucleus, is a sensitive pressor area of the lower brainstem with differential efferentation both in descending and ascending direction. The majority of the fibers appears to be descending, travelling to and terminating in major medullary cardiovascular control areas such as the RVLM, CVLM and the NTS (see Korner, 1979). In addition, fibers terminate in the IML. It is not clear yet whether these fibers contribute to the cardiovascular control.

Subsequent deafferentation experiments showed that microknife cuts placed 1 mm rostrally from the stimulation points in the PB area fail to alter the magnitude of the pressor response. In contrast, caudal cuts result in an average of 43 present reduction of the pressor response. It is most probably that destruction of the connections to the RVLM has the most significant effect on the pressor response. The functional significance of the RVLM is to regulate the basal vasomotor tone and the sympathetic innervation of the heart (Damphney et al., 1982; Ross et al., 1984). It is remarkable that the RVLM contains the adrenergic C-1 cells (Kalia et al., 1985a, b, c). The adrenergic cells are responsible for sympathetic effects (Ross et al., 1984). The other major projection area – i.e. the CVLM – contains cell bodies of the A-1 noradrenergic system. From a functional point of view is the CVLM, particularly via noradrenergic projection to the RVLM, sympatho-inhibitor in nature (Granata et al., 1986). It is therefore likely that the activation of the PB complex inhibits the activity of the CVLM. Concerning minor projections, the termination of fibers in the NTS and the RM may contribute to the final form of the pressor response. Chemical stimulation of the RM results in a pressor response (Pilowsky et al., 1986). The significance of the NTS in the baroreceptor regulation of cardiovascular function is excellently documented (see Korner, 1979).

Taken together, the physiological and neuroanatomical observations suggest the existence of a cardiovascular circuit that is related to a specific form of autonomic arousal: a mild pressor response with a genuine – i.e. baroreceptor independent bradycardia. Versteeg et al. (1982a) failed to find a correlation between the magnitude of the SBP rise and the heart rate decrease. This autonomic arousal reaction may accompany certain forms of behavioural arousal (see further in this paper). The proposed circuit takes its origin from the mesencephalic reticular formation, particularly from the cuneiform area. Direct descending pathways that induce the pressor response involve the medullary cardiovascular regulatory areas such as the RVLM and the spinal cord (IML) via the PB complex. These connections represent the sympathetic limb of the circuit. The bradycardiac effect is probably mediated by the direct activation of the vagal output via the CMNV: this is the parasympathetic limb of the circuit. The transection studies suggest that the direct descending projections form only a part of the complex response. The ascending connections reach the PVN, the CA, and the BNST. The autonomic (cardiovascular) function of the PVN and CA is well documented (e.g. Swanson and Sawchenko, 1983; Hilton and Zbrozyna, 1963; Kapp et al., 1982). Less is known about the BNST. In a recent study LeDoux et al. (1988) found that lesions of the BNST fail to affect behavioural and autonomic (blood pressure) correlates of conditioned fear. Extensive neuroanatomical data suggest the existence of monosynaptic connections from the PVN, CA and BNST to the PB, DMNV, NTS and IML (e.g. Swanson and Sawchenko, 1983; Venning et al., 1984; Gray and Magnuson, 1987). The PVN, CA and BNST are also interconnected with each other (e.g. Swanson and Cowan, 1979). In addition, the recent and other studies show that both the PB and the PB and the NTS project directly to these limbic nuclei (e.g. Fulwiler and Saper, 1984; etc.).

The ascending limb of the circuit may serve for coupling the behavioural and autonomic response, particularly through the CA and BNST. In turn, the descending connections of the CA, BNST and PVN may fine-tune or shape the autonomic responses through the monosynaptic projections to the lower brainstem, medullary and spinal sympathetic and parasympathetic outflow stations. It is noteworthy that the majority of these ascending and descending long neurons use neuropeptides as transmitters (e.g. Gray et al., 1986; Gray and Magnuson, 1987; Yamano et al.,1988).

The exact localization of action of AVP within this circuit awaits for further studies. Preliminary observations with local microinjection of AVP and OXT show the PB complex as one probably site of suppression of the pressor response. Inhibitory effects of neurohypophyseal peptides in the caudal brainstem (Morris et al., 1980) and in the intermediolateral cell column of the spinal cord – i.e. sympathetic preganglionic neurons (Gilbey et al., 1982) were also reported. The NTS may also serve as a relay of the suppressive effect (Brattström et al., 1986). Electrophysiological observations show the activation of the DMVN by locally administered OXT and AVP (Charpak et al., 1984). Schmid et al. (1984) also suggested that vasopressin activates parasympathetic pathways to the heart and activates the baroreflex function through a central action. This action may be responsible for the enforcement of vagal activation as observed in a number of behavioural paradigms in freely moving rats. However, it should be mentioned that some investigators (e.g. Pittman et al., 1982; Schmid et al., 1984; Vallejo et al., 1984; Unger et al., 1984; Rohmeiss et al., 1986; Undesser et al., 1986) report an increase of sympathetic outflow and increase in blood pressure due to vasopressin administered centrally.

Vasopressin and bradycardiac stress responses in the rat

A stressful condition in which sympathetic and parasympathetic outflow is parallelly activated is related to an inhibitory (passive) avoidance situation. Emotional stress of fear of a painful electric footshock as determined 24 h or later after a single learning trial results in an increase in blood pressure, and an initial relative decrease in heart rate. The bradycardiac component slowly disappears during the 5 min test but the blood pressure response is persisting. This autonomic pattern is independent of the nature of the behaviour as imposed on the rat. The pattern accompanies both passive (inhibitory) avoidance response or the forced exposure to this avoidance of the former shock compartment in a step-through type apparatus (Bohus, 1985; 1988). Because of the appearance of the pattern in the experimental room in which the apparatus is placed, one may suppose an autonomic conditioning to a complex set of stimuli that accompany the emotional stress. This autonomic response pattern is designated therefore as a preparatory, expectancy pattern. The bradycardiac response is a relative one and it is considered as the result of a baroreceptor independent increase of vagal outflow. The designation relative refers to the fact that the non-stressed, freely- behaving controls show signs of substantial sympathetic activation of their heart rate due to the somatic (exploratory) activity and/or alert behavioural state. The vagal inhibition of the cardiac rhythm is superimposed on this rather tachycardiac 'baseline". That the vagal activation is a genuine one is suggested by the finding that an almost complete blockade of blood pressure rise in the emotional stress situation as it is the case in neonatally sympathectomized rats, the bradycardiac response is intact (Bohus et al., 1976).

A number of evidences has been collected supporting the view that bradycardiac components of diverse emotional stress responses are enhanced by vasopressin. Peripheral administration of LVP enhances the bradycardiac heart rate changes during the intertrial intervals of a classically conditioned fear response (Bohus, 1973). The vasopressin analogue desglycineamide-9-LVP (DG-LVP) enhances the magnitude of the relative bradycardiac response that accompanies inhibitory avoidance behaviour of the male rat (Bohus, 1975). Analysis of the distribution of the interbeat intervals suggests that the vagal influence on the cardiac rhythm is intensified by the peptide (Bohus, 1977). Remarkably, the bradycardiac response during inhibitory avoidance behaviour is practically absent in rats suffering from hereditary hypothalamic diabetes insipidus. Administration of the vasopressin analogue restores the bradycardiac response in conjunction with the normalization of the avoidance behaviour (Bohus, 1977). Rats of the Brattleboro strain, homozygous for the disease lack the ability to synthesize vasopressin in the hypothalamus (Valtin and Schröder, 1964) due to a defect in the mRNA message (Richter, 1985). Profound behavioural deficits in the avoidance situation were found in the diabetes insipidus rats of the Dutch substrain (e.g. Bohus et al., 1975; De Wied et al., 1975). These deficits appeared to be absent in other substrains of the Brattleboro rats (see Sahgal, 1984).

Recent studies in old and senescent rats showed that the vagally mediated cardiac stress response is diminishing with the age (Nyakas et al., 1989). Peripheral administration of AVP results in the reappearance of the bradycardiac response (Nyakas et al., in preparation). Extensive evidence suggests that certain compartments of the vasopressinergic system in the rat brain display degenerative changes (Fliers et al., 1985; Roozendaal et al., 1987). This changes concern primarily the vasopressinergic cell bodies in the suprachiasmatic nuclei and their projection areas. Whether the brain areas that are involved in the vasopressinergic regulation of autonomic outflow are also affected by aging remains to be explored.

Table 1 shows an example of the effect of AVP on the cardiac response of aged (14 mo old) rats in comparison with its action in young adult (3 mo old) male animals.

Table 1. Effect of arginine vasopressin (AVP) on the cardiac response to emotional stress in young adult and aged male rats.

Age*	Treatment**	Cardiac rate difference[#]
3	Placebo	11.6 ± 4.8
	AVP	15.2 ± 3.2[##]
14	Placebo	1.2 ± 1.2
	AVP	30.0 ± 12.0[##]

* Expressed in months; ** Treatment was given subcutaneously 60 min before the test, the dose of AVP was 10 fg/kg; [#] Difference between non-stressed and stressed cardiac rate of each individuals expressed in mean interbeat interval (msec $\pm$ S.E.M.); [##] $p < 0.05$ (paired t-test); 7 observations per age.

The data indicate that the bradycardiac response is absent in the aged rat (low difference score in placebo treated-state) but the effect of AVP is more pronounced.

Vasopressin appears to influence a vagally mediated cardiac response under non-aversive, but stressful conditions. Orientation/attention to stimulus changes is accompanied by a decrease in heart rate in the rat (e.g. Graham and Clifton, 1979). Sudden silence superimposed on low intensity background noise (65 dB) is accompanied by immediate behavioural arrest and orientational movements. The stressful character of the response is suggested by a substantial rise in plasma corticosterone levels (Bohus and Balkan, in preparation). Peripheral administration of AVP results in a prolongation of the bradycardiac response in young adult (Hagan and Bohus, 1984) and aged male rats (Buwalda et al., in preparation).

Taken together, experiments in freely behaving rats in various stressful situations suggest that AVP serve as an important modulator of a vagally mediated cardiac response to emotional stressor.

Behavioural and cardiac effects of intraventricularly administered arginine-vasopressin: a comparison with corticotropin releasing hormone

The experiments described above suggest that vasopressin may be involved in a network that serves a parasympathetically ruled cardiac response to stress. The sympathetic limb of the output may under certain circumstances be inhibited by this neuropeptide or related principles. The obvious question that is raised as to whether AVP inhibits arousal per se leading to the described autonomic pattern or, alternatively AVP induces (or promotes) a certain form of arousal that results in a conservation rather than a discharge of bodily resources.

Former observations concerning AVP and arousal were not equivocally conclusive. For example, Dunn and his associates (see Rees et al., 1976; Delanoy et al., 1978) Meisenberg and Simmons (1981, 1982, 1983) reported behavioural excitation by centrally administered vasopressin in mice. Central administration of vasopressin may also cause pressor effects and increased sympathetic outflow in resting conditions (e.g. Pittman et al., 1982; Berecek et al., 1984; Rohmeiss et al., 1986; Unger et al., 1984). Accordingly, in order to decide the precise role of the peptide it was of interest to investigate the behavioural and cardiac consequences of centrally administered AVP in well habituated and stressed rats. For comparison the effects of CRH were also investigated (Bohus et al., in preparation).

The interest for comparison with CRH was deemed by a number of reasons. First of all, a complementary action of CRH and AVP on the stress-induced activation of the pituitary-adrenocortical system due to an effect on pituitary corticotrop cells is long known (e.g. Gillies and Lowry, 1979). Physiological importance of such complementary action is supported by observations showing that the two peptides are colocalized in the parvocellular cells of the hypothalamic paraventricular nucleus, and cotransported to the median eminence (e.g. Whitnall et al., 1985). Secondly, vasopressinergic and CRH-containing neurons frequently terminate in the same brain stem autonomic nuclei although the respective cell bodies are differentially localized (see e.g. Nieuwenhuys, 1985). Thirdly, the view is frequently advocated that the brain CRH neurons mediate both behavioural and autonomic stress response (Koob, 1985; Dunn and Berridge, 1987).

Slow infusion of AVP into a lateral cerebral ventricle leads to an immediate activation of the behaviour of the resting rat that is well- habituated to handling, the infusion and cardiac recording procedure (Table 2). This activation is characterized by waking up the animal, sniffing, exploring and grooming occur instead of sleep and immobility. Eating is also observed when food pellets are available. The dose- response curve shows an inverted U-shape in the dose range of 1 to 30 ng with a maximal effect of 10 ng. Termination of the infusion results in an almost immediate cessation of the behavioural activation. The cardiac reaction to the peptide infusion is variable between a slight bradycardia with the maximum changes at the dose of 10 ng. A slight tachycardia is observed after the termination of the infusion. Similar effects can be observed when the peptide was given intracerebroventricularly as a single injection, except that the bradycardiac effect is practically absent.

A completely different picture emerges during and after the infusion of CRH into a lateral cerebral ventricle of the well-habituated rat in resting state. The behavioural activation during the infusion is mostly characterized by sniffing and eating while an increase in the exploratory activity and grooming is observed after the termination of the infusion. There is a linear dose-response relationship in the range of 10 to 300 ng with the maximum at 300 ng. The infusion of CRH results in a tachycardiac response. This response lasts after the termination of the infusion (Table 2). Injection of the peptide results in similar alterations in behaviour and cardiac activity.

Table 2. The effects of arginine vasopressin (AVP) and corticotropin- releasing hormone (CRH) infusion into a lateral cerebral ventricle on the behaviour and cardiac rate of well-habituated rats.

Infusion		Behavioural activity* during-infusion-after [#]		Cardiac rate** during-infusion-after	
Placebo		6.0 ± .4	5.9 ± .4	1.1 ± .6	1.7 ± 1.5
AVP	1 ng	19.7 ± 2.4	2.8 ± .5	3.8 ± 2.5	.5 ± .4
	10 ng	52.4 ± 3.2	15. ± 2.4	7.6 ± 3.8	− 8. ± .9
	30 ng	12. ± .8	11.2 ± 1.5	.7 ± 2.2	− 4.9 ± 3.2
CRH	10 ng	6.6 ± .7	9.8 ± .9	1.3 ± .6	− .3 ± 1.2
	30 ng	8.1 ± .7	20.7 ± 2.2	− .8 ± 3.2	.1 ± 1.1
	100 ng	15.3 ± 1.4	24.3 ± 1.2	− 7. ± .9	−13.7 ± 2.1
	300 ng	21.2 + 1.2[##]	31.1 + 1.9[##]	− 12.1 + 1.9[##]	−13.5 + 2.8[##]

* Percentage of time spent with activity; ** Delta mean interbeat interval in msec in comparison to the preinfusion heart beat intervals; [#] Duration of infusion: 20 min; Infusion rate: .25 fl /min; post-infusion period: 45 min. The data are presented as means ± S.E.M.; 6-8 observations per dose; [##] ANOVA p < 0.01; The dose of the peptides refer to the total concentration per 20 min.

Endocrine experiments led to the suggestion that the ACTH-releasing effect of CRH is potentiated by AVP (see Gillies et al., 1982). Therefore, in one experiment the two peptides were simultaneously injected into a lateral cerebral ventricle. The concomitant administration of AVP and CRH fails to induce stronger behavioural and cardiac effects with larger magnitude than the maximal actions observed following single injections of AVP and CRH. Accordingly, potentiation seems to be absent. However, it should be mentioned that the maximal doses of the two peptides were used.

A short exposure to a novel environment of a complex maze is followed by a substantial increase in behavioural activities in the home cage of the animal. The activity is diminishing within 30 min. The cardiac response is a tachycardia that more-less parallels the behavioural activation. AVP infusion leads to a slight diminution of the stress-induced activation of the behaviour. The tachycardiac response is almost completely blocked by this peptide (Table 3). The effects of AVP disappears after the termination of the infusion.

Table 3. The effects of arginine vasopressin (AVP) and corticotropin- releasing hormone (CRH) infusion into a lateral cerebral ventricle on novelty-induced behavioural and cardiac response in the rat.

Infusion		Behavioural activity during-infusion-after		Cardiac rate during-infusion-after	
Placebo		34.9 ± 1.	14.7 ± 1.1	− 8.1 ± 1.3	− 6.7 ± 1.5
AVP	30 ng	30. ± 1.1	26.5 ± 2.1	− 2.5 ± 3.2	− 6.4 ± 1.6
CRH	300 ng	33. ± 1.7	37.5 ± 1.3	− 4.4 ± 1.1	−10.3 ± 1.6

The rats were exposed to a novel environment of a complex maze for a period of 2 min. Subsequently they were returned to their home cage where the 20 min infusion has started. For other details and abbreviations see Table 2.

Infusion of CRH fails to influence the magnitude of the stress-induced behavioural activation, but the rats remained active in the post-infusion period. The tachycardiac response to arousal is slightly diminished during CRH infusion, but marked tachycardia can be observed during the post-infusion period.

These experiments clearly demonstrate that the central behavioural and cardiac actions of AVP and CRH are of a differential nature. In well- habituated resting and the stressed animal. A potentiation of each others action as seen at the anterior pituitary level is absent. The major findings of our experiments is some aspects agree with former studies, but a number of novel properties of the effects of the two peptides emerged.

First of all, AVP causes, but not an undifferentiated behavioural activation. The activating effect is seen in the sleeping/resting animal, but the behavioural activation caused by a novel stressful environment is attenuated by the peptide. These effects may be explained by a single mechanism acting differentially at various activation level of the central nervous system: it is arousing at low level of activation while it reduces arousal at high level of activation. The activation level is low in the non-stressed, well-habituated animal, but high in the stressed one.

Yerkes and Dodson (1908) were the first to suggest that performance and arousal are related in an inverted-U manner: a poor performance occurs both at very low and an exceptionally high level of arousal. Sahgal (1984) argued that effects of vasopressin on learning and memory are due to an effect on arousal. He reported repeatedly disruptive effects of vasopressin on passive avoidance performance given after learning (see Sahgal, 1984). We have also seen that postlearning administration of a vasopressin analogue facilitate or inhibit the retention of a shuttle-box avoidance response in the rat in a dose-dependent way: low doses facilitate while high doses attenuate performance (Hagan et al., 1982). If Sahgal's (1984) hypothesis is correct than changes in spontaneous behaviour should be a monotonous function of the dose of the peptide. Sahgal (1984) refers to the studies that report on increased activity and grooming in mice upon vasopressin administration (Rees et al.,1976; Delanoy et al;., 1978; Meisenberg and Simmons, 1982; etc.). In our experiments we failed to find such a monotonous function. This suggests that the effect of AVP cannot be explained by a simple arousal mechanism. It is likely that the peptide is arousing by itself, but the behavioural manifestation appears only up to an optimum level. The effect of the peptide may be protective or reparative of a pre-stress level of activation if the nervous system is over-aroused by non-specific stressful stimuli. The mnemonic action of the peptide may fit into the hypothesis of a protective action: AVP may enhance or disrupt the performance of a behavioural stress response that is specific for a certain environment. We have argued that learning and memory may be interpreted in terms of specific a stress response (Bohus, 1984).

The effect of CRH on the behaviour and cardiac rate seems to be more straightforward and uncomplicated: behavioural activation is a monotonous function of the dose of the peptide. The behavioural findings in the well-habituated rats agree with observations Koob and his associates (see Koob, 1985). Sutton et al. (1982) reported first that intracerebroventricularly administered CRH produces behavioural activation in the rat. A remarkable property of the peptide is to cause a behavioural activation that outlasts the infusion period. Sutton et al. (1982) found an activating effect of a single injection even after 3 hours. In our experiment the behavioural activation appears to be somewhat more pronounced after the termination of the infusion. Activation of the sympathetic nervous output by CRH and thereby the cardiovascular system has been reported by Brown et al. (1982) and Fisher et al. (1982). In our study increased heart rate suggest sympathetic activation. This activation parallels behavioural arousal, but it is probably not the direct consequence of increased locomotor activity. Doses lower than 100 ng cause an increase in locomotor activity, but fail to induce a tachycardia. Our experiments also demonstrate that the cardiac response alike the behavioural one outlasts the infusion period. The neurobiological background of the long lasting behavioural and autonomic activation is not known yet. It is likely that the peptide affects two mechanisms: it activates an activational or arousal system, and inhibits a deactivating or dearousal system.

Based on experimental observations in open field and conflict tests (Sutton et al., 1982; Britton et al., 1982) Koob (1985) suggests that CRH is involved in the behavioural response to stress by potentiating the effects of novel, actually aversive environments on the behaviour of the rat. Similar findings were reported by Berridge and Dunn (1986) in mice. In addition, anxiogenic properties of CRH were also reported (Thatcher- Britton et al., 1985; Dunn and File, 1987). Our experiments investigating the effect of CRH on novelty-induced arousal show two interesting aspects. Firstly, CRH infusion fails to influence the behavioural activation in the home-cage induced by the stressfulness of an another environment. Taking into account the findings by Koob and his associates (see Koob, 1985) and of Berridge and Dunn (1986) it is likely that the peptide must act in the stressful situation itself. In order to affect the hypothetical arousal mechanisms. Secondly, the peptide appears to be active after termination of the infusion in the previously aroused animal. This can be interpreted as an inhibition of dearousing mechanisms because the rats' activity following placebo infusion is almost back to the resting level. Accordingly, the inhibition of dearousing mechanism is independent of the coupling between the presence of peptide and the stressor (arousing) event.

Vasopressin and other neuropeptides in stress: a synthesis

A well concerted action of behavioural and physiological systems is the essential requisite of coping with environmental challenge. Stress and adaptation, stress and coping, stress and neuroendocrine state are inseparable terms. In the present paper an attempt was made to place vasopressin in the proper site of the jigsaw puzzle of the physiology of stress. Vasopressin and related peptides induce a variety of behavioural and physiological action when injected peripherally or centrally. The following summarizes some of the vasopressin actions:

1. Hypothalamo – posterior pituitary axis
   a. antidiuresis
   b. blood pressure increase due to contraction of the vascular smooth muscle
   c. activation of glycogenolysis in the liver
2. Hypothalamo – anterior pituitary axis
   a. potentiation of the release of opiomelanocortins
3. Autonomic nervous system
   a. attenuation of a neurogenic pressor response
   b. enhancement of the cardiac vagal outflow related to emotional stress

    c. enhancement of the cardiac vagal outflow related to orientation/attention
    d. re-establishment of cardiac vagal activation in aging
    e. attenuation of arousal-induced sympathetic activation
4. Central nervous system
    a. effects on learned behaviours
    b. induction of behavioural activation
    c. attenuation of stress-induced behavioural activation

This list is obviously incomplete, and describes mostly actions that are relevant to the recent discussion. It is clear that vasopressin is involved in many aspects of fundamental mechanisms that serve coping with environmental challenge. The behavioural and cardio-vascular studies suggest that vasopressin is primarily serving a brain circuit that is involved in the organization of a specific stress pattern. This notion presupposes a central action of the peptide. The presence of well-defined vasopressinergic network(s) in the brain and of a complex receptor system supports the view that centrally released (and administered) AVP acts via these networks and receptors. However, many of the effects of AVP on behavioural and cardiac responses could also be induced by peripheral administration. It was proposed that from peripheral sources small amounts of the peptide may penetrate into the brain and thereby affects brain functions via interactions with classical neurotransmitters (e.g. De Wied, 1976; Bohus, 1982). This hypothesis was challenged by Le Moal et al. (1984). They suggested that peripherally administered AVP induces physiological-endocrinological effects and the viscerally induced signals alert and arouse the animal. Subsequently Koob et al. (1985) proposed the existence of separate but parallel central and peripheral AVP systems that may act on a similar behavioural substrate in an independent way. At the recent stage of our research the majority of the data suggest a central nervous action of AVP, but the exact mechanism(s) of signal transduction from peripheral sources remains to be elucidated. Interestingly, vasopressin-like peptides are present in the neurons of noradrenergic sympathetic ganglia and nerve fibers innervating peripheral tissues (Hanley et al., 1984). Vasopressin-like material may be a possible mediator of the non-adrenergic responses of peripheral organs to sympathetic activation.

    Theories of cardiovascular psychophysiology maintain that heart rate changes reflect somato-motor activity (Obrist, 1981). According to Obrist (1981) the concept of coping can be viewed along a dimension of activity. In this term somato-motor acts are expected in active and effective coping. As a consequence, cardiac rate is increasing in a metabolically relevant way. Bradycardia may also occur, but it reflects inhibition of the ongoing somatic activity. Obrist (1981) argues that if coping is viewed in terms of activity then passive coping means immobility, a case when action is not a good behavioural strategy. Vagal activation is, however, metabolically not relevant. We have also argued that passive coping, characterized by absence of action (probably due to inhibition) and ruling parasympathetic activation, is a successful coping strategy, and serves as an alternative of the active, sympathetically ruled coping (see e.g. Koolhaas et al., 1986; Bohus et al., 1987; Benus, 1988). However, the two strategies cannot be viewed simply along a somato-motor coupling. Data presented here and in other papers from the authors (e.g. Bohus, 1977; 1985) show that discoupling frequently occurs: vagal activation is not necessarily coupled with immobility, and immobility is not necessarily accompanied by bradycardia. Accordingly, behavioural and physiological characteristics of the two kind of coping strategies can be coupled, but this coupling is not necessarily a causal relationship. Recent findings also suggest that brain circuits serving the behavioural and cardiac component of emotional stress responses may be differentially organized. LeDoux et al., (1988) found that destruction of the lateral- central nuclei of the amygdala disrupt both autonomic and behavioural correlates of conditioned fear. Lesions of the lateral hypothalamus disrupt the autonomic response but not the immobility behaviour. Conversely, destruction of the caudal central gray abolishes the behavioural response, but the autonomic component remains intact.

In one of the former sections of this paper a neuronal circuit has been described that may serve a specific autonomic response pattern to stress: parallel activation of the sympathetic (pressor response) and parasympathetic limb (bradycardia). This response often accompanied by immobility behaviour. The cuneiform nucleus in the mesencephalic reticular formation is placed in the centre of this circuit. Korte et al. (in prep.) found that low intensity electrical stimulation of the cuneiform nucleus results in immobility by interrupting ongoing behavioural activities such as grooming, exploration, etc. In the meantime blood pressure is increasing. Freezing behaviour can also be evoked by the injection of glutamate into the cuneiform area (Mitchell et al., 1988). This finding reinforces the view that cell bodies rather than passing axons are involved. Neuroanatomical studies by Redgrave et al. (1988) suggest an important output from the superior colliculus. The multimodel input to the superior colliculus from sensory areas and the tectocuneiform connection may be involved in some particular subset of rats' defensive behaviour (Dean et al., 1988). Because of the obvious vagal component of the response, and its appearance (conditioning) in conjunction with a wide variety of stimuli accompanying the stressful character of the environment we proposed to designate this pattern as an expectancy/anticipatory pattern.

The proposed response pattern is one of the many that have been identified physiologically, neuroanatomically and sometimes behaviourally. The defense reaction as originally proposed by Hilton (1982) is characterized by a pattern of response in anticipation of severe muscular exertion. The cardiovascular component includes muscle vasodilatation, vasoconstriction in splanchnic and renal nerve beds and in skin, venoconstriction, increased cardiac activity, and inhibition of baroreceptor reflex (see Hilton and Spyer, 1980). The circuit involves the amygdala, hypothalamus and the central gray and the response is strongly sympathetic in nature (see Hilton and Spyer, 1980; Spyer, 1982;). Different circuits have been proposed by Smith and his associates (Smith et al., 1980; Smith and DeVito, 1984): activation of a hypothalamic area controlling emotional response (HACER) results in blood pressure, heart rate and terminal aortic flow increase and renal flow decrease. Loewy and McKellar (1980) have used a neuroanatomical approach to propose a circuit for autonomic activation. The autonomic, behavioural and anatomic studies on the organization of the correlates of conditioned fear in the rat by LeDoux et al. (1988) suggest differential mediation of the behaviour and the autonomic response. Although there are differences in the behavioural and autonomic patterns underlying the different responses (e.g. flight-freeze; sympathetic-parasympathetic activation), there are a few key areas in the limbic system (e.g. CA, PVN, BNST) that are involved in the organization in more than one pattern of responses.

How can one envisage the existence of different circuits? Are those simply experimental artifacts or true physiological mechanisms? Although a definite answer cannot be given yet, firm data suggest that the different circuits may exist parallely or the networks represent physiological mechanisms that function in different individuals. Henry and Stephens (1977) were the first to formulate a hypothesis that individuals of random populations differentiate in two kinds of response patterns in stressful social situations. The Cannonian fight-flight pattern is adopted by aggressive animals and it is characterized by high sympathetic neural and hormonal activity. The Selyean distress or conservation/withdrawal pattern that is adopted by non-aggressive individuals is characterized by high adrenal cortical activity as ruling stress response. On the basis of the large series of behavioural physiological experiments by Fokkema (1985) and Benus (1988) we have recently advocated the view of the existence of general behavioural and physiological active and passive strategies of rats and mice in functioning in their social or non-social environment (e.g. Koolhaas et al., 1986; Bohus et al., 1987). The active individuals being strongly sympathetically ruled may use an integratory mechanism that corresponds to either the defense or other emotional 'response patterns". The passive individuals show a ruling vagal, but with some sympathetic activation and seem to adopt the expectancy/anticipatory pattern.

Physiological and endocrine investigations suggest that the hormonal profile of rats adopting an active coping strategy is characterized by high sympatho-medullary activity, testosterone release, prolactin activity, etc. In contrast, passive animals show a strong activation of the pituitary-adrenal cortical axis and inhibition of testosterone secretion by psychosocial stressors (see Koolhaas et al., 1986; Bohus et al., 1987). Behavioural studies suggest that the output of these neuroendocrine principles is not only a marker of the activity-passivity of the rats, but the hormones create the neuroendocrine state of the brain thereby serving coping with the environmental challenges (Bohus et al., 1987). It seems that adrenomedullary hormones, epinephrine in particular, serves the active coping strategy, while adrenal corticosteroids are essential in a passive way of coping (Bohus, 1988).

In these terms, AVP should be considered as a neuroendocrine factor that principally serves passive coping activity both on central and peripheral mechanisms. The data presented here show that AVP has distinct effects on the expectancy/anticipatory cardiovascular pattern and on the stress-induced behavioural and cardiac activation. The enhancement of the vagal component is a primary action of AVP as discovered in various experiments. The expectancy/anticipatory pattern fits to a vagally ruled conservation/withdrawal type response as proposed by Engel (1977) rather than a Cannonian sympathetic massdischarge accompanying fight or flight. Of course, many questions remain to be answered, particularly concerning the various behavioural actions of AVP. Another question to be answered is as to whether AVP or OXT is the 'true' messenger in a passive coping. OXT shares a number of central nervous effects of AVP (e.g. Kovács et al., 1979; Versteeg et al., 1982b) probably due to the use of a common receptor for which AVP is an agonist and OXT an inverse antagonist (Elands, 1989).

Corticotropin-releasing hormone seems to be a counterpart of vasopressin in the peptidergic regulation of the stress responses. The behavioural and physiological profile of its action (see Koob, 1985) suggests that CRH has a role in a fundamental activating system that functions to mobilize resources to environmental challenge. In terms of coping strategies CRH may play a role in circuits that serve active coping.

Active and passive strategies can be viewed as extremes of a continuum. This means that certain individuals of a population can adopt only one of these strategies, primarily on genetic and ontogenetic background (Bohus et al., 1987), while others are able to select the proper strategy according to the environmental requirements as a consequence of learning (Bohus, 1988). The various neuropeptides are differentially involved in a higher organization of the response. There effects are brought about both via specific brain circuits and the peripherally released hormones. To elucidate their role in the stress- induced pathology remains a challenge for future research.

Acknowledgements

Part of the studies reported here were supported (in part) by the Dutch Heart Foundation (Grant nr. 84.002) and the Foundation for Medical and Health Research MEDIGON (Grant nr. 900-551-044). The editorial assistance by Mrs. Joke Poelstra is greatly acknowledged.

References

Amir, S., Z.W. Brown and Z. Amit (1980). The role of endorphins in stress: evidence and speculations. *Neurosci. Biobehav. Rev.* 4:77-86.

Benus, R.F. (1988). *Aggression and coping. Difference in behavioral strategies between aggressive and non-aggressive male mice.* Ph.D. Thesis, University of Groningen.

Berecek, K.H., H.R. Olpe, R.S.G. Jones and K.G. Hofbauer (1984). Microinjection of vasopressin into the locus coeruleus of conscious rat. *Am. J. Physiol.* 247:H675-681.

Berridge, C.W. and A.J. Dunn (1986). Corticotropin-releasing factor elicits naloxone sensitive stress-like alterations in exploratory behavior in mice. *Regul. Peptides* 16:83-93.

Bloom, F.E. (1987). Molecular diversity and cellular functions of neuropeptides. In: E.R. De Kloet, V.M. Wiegant and D. De Wied, Eds., *Progr. in Brain Res., Vol. 72*, pp 213-220, Amsterdam: Elsevier.

Bohus, B. (1973). Pituitary-adrenal influences on avoidance and approach behavior of the rat. In: W.H. Gispen, B.H. Marks and D. De Wied, Eds., *Progress in Brain Research, Vol. 39*, pp 407-420. Amsterdam: Elsevier.

Bohus, B. (1974). The influence of pituitary peptides on brain centers controlling autonomic responses. In: D.F. Swaab and J.P. Schadé, Eds., *Progress in Brain Research, Vol. 41*, pp 175-183, Amsterdam: Elsevier.

Bohus, B. (1975). Pituitary peptides and adaptive autonomic responses. In: W.H. Gispen, Tj.B. van Wimersma Greidanus and D. De Wied, Eds., *Progress in Brain Research, Vol 42*, pp 275-283, Amsterdam: Elsevier.

Bohus, B. (1977). Pituitary neuropeptides, emotional behaviour and cardiac responses. In: W. De Jong, A.P. Provoost and A.P. Shapiro, Eds., *Progress in Brain Research, Vol. 47*, pp 277-288, Amsterdam: Elsevier.

Bohus, B. (1982). Neuropeptides and memory. In: R.L. Isaacson and N.E. Spear, Eds., *The Expression of Knowledge*, pp 141-177, New York: Plenum.

Bohus, B. (1984). Neuroendocrine interactions with brain and behavior: a model for psychoneuroimmunology? In: R.E. Ballieux, J.F. Fielding and A. L'Abbate, Eds., *Breakdown in Human Adaptation to Stress: Towards a Multidisciplinary Approach, Vol. 2*, pp 638-652, "s Gravenhage: Nijhoff.

Bohus, B. (1985). Acute cardiac responses to emotional stressors in the rat; the involvement of neuroendocrine mechanisms. In: J.F. Orlebeke, G. Mulder and L.J.P. van Doornen, Eds., *Psychophysiology of Cardiovascular Control*, pp 131-150, New York: Plenum.

Bohus, B. (1988). Limbic-midbrain mechanisms and behavioral physiology of interactions with CRF, ACTH and adrenal hormones. In: D. Hellhammer, I. Florin and H. Weiner, Eds., *Neurobiological Approach to Human Disease*, pp 267-285, Toronto: Huber.

Bohus, B., W.H. Gispen and D. De Wied (1973). Effect of lysine- vasopressin and ACTH 4-10 on conditioned avoidance behavior of hypophysectomized rats. *Neuroendocrinol.* 11:137-143.

Bohus, B., Tj.B. Van Wimersma Greidanus and D. De Wied (1975). Behavioral and endocrine responses of rats with hereditary hypothalamic diabetes insipidus (Brattleboro strain). *Physiol. Behav.* 14:609-615.

Bohus, B., W. De Jong, A.P. Provoost and D. De Wied (1976). Emotionales Verhalten und Reaktionen des Kreislaufs und Endokriniums bei Ratten. In: E.W. Von Eiff, Ed., *Seelische und körperliche Störungen durch Stress*, pp 140-157, Stuttgart, Gustav Fischer.

Bohus, B., L. Conti, G.L. Kovàcs and D.H.G. Versteeg (1982). Modulation of memory processes by neuropeptides: interaction with neurotransmitter systems. In: C. Ajmone Marsan and H.-J. Matthies, Eds., *Neuronal Plasticity and Memory Formation*, pp. 75-87, New York: Raven.

Bohus, B., R.F. Benus, D.S. Fokkema, J.M. Koolhaas, C. Nyakas, G.Van Oortmerssen, A.J.A. Prins, A.J.H. de Ruiter, A.J.W. Scheurink and A.B. Steffens (1987). Neuroendocrine states and behavioral and physiological stress responses. In: E.R. De Kloet, V.M. Wiegant and D. De Wied, Eds., *Progress in Brain Research, Vol. 72*, pp 57-70, Amsterdam: Elsevier.

Bohus, B., C.A.M. Versteeg, W. De Jong, K. Cransberg and J.G. Kooy (1983). Neurohypophysial hormones and central cardiovascular control. In: B.A. Cross and G. Leng, Eds., *Progress in Brain Research, Vol. 69*, pp 445-457, Amsterdam: Elsevier.

Bohus, B., J.M. Koolhaas, P.G.M. Luiten, C.A.M. Versteeg, S.M. Korte and D. Jaarsma (1989). Vasopressin and related peptides: involvement in central cardiovascular regulation. In: F.P. Nijkamp and D. De Wied, Eds., *Neuropeptides, Brain and Hypertension*, pp 000, Amsterdam: Elsevier.

Brattström, A., W. De Jong and D. De Wied (1986). Barorezeptorischer Herz-Reflex während zentraler Vasopressin-wirkung. *Z. Klin. Med.* 41:1267- 1289.

Britton, D.R., G.F. Koob, J. Rivier and W. Vale (1982). Intraventricular corticotropin-releasing factor enhances behavioral effects of novelty. *Life Sci.* 31:363-367.

Brown, M.R., L.A. Fisher, J. Spiess, C. Rivier, J. Rivier and W. Vale (1982). Corticotropin-releasing factor: actions on the sympathetic nervous system and metabolism. *Endocrinology* 111:928-931.

Buijs, R.M. (1983). Vasopressin and oxytocin -- their role in neurotransmission. *Pharmacol. Ther.* 22:127-141.

Carter, D.A. and S.L. Lightman (1987). Oxytocin stress responses are dependent upon emotionality. *Psychoneuroendocrinology* 12:219-223.

Charpak, S., W.E. Armstrong, M. Muhlethaler and J.J. Dreifuss (1984). Stimulatory action of oxytocin on neurons of the dorsal motor nucleus of the vagus nerve. *Brain Res.* 300:83-89.

Cowley, A.W., Jr., S. Switzer and M.M. Guinn (1980). Evidence and quantification of the vasopressin arterial pressure control in the dog. *Circulat. Res.* 46:58-67.

Damphney, R.A.L., A.K. Goodchild and E. Tann (1982). Role of ventromateral medulla in vasomotor regulation: a correlative anatomical and physiological study. *Brain Res.* 249:223-235.

Darlington, D.A. and D.G. Ward (1985). Rostral pontine and caudal mesencephalic control of arterial pressure and iliac, celiac and renal vascular resistance. I: Anatomic regions. *Brain Res.* 361:284-300.

Dean, P., P. Redgrave and I.J. Mitchell (1988). Organization of efferent projections from superior colliculus to brainstem in rat: evidence for functional output channels. In: J.P. Hicks and G. Benedek, Eds., *Progr. in Brain Research, Vol. 75*, pp 27-36, Amsterdam: Elsevier.

Delanoy, R.L., A.J. Dunn and R. Tintner (1978). Behavioral responses to intracerebroventricularly administered neurohypophyseal peptides in mice. *Horm. Behav.* 11:348-362.

De Vries, G.J., R.M. Buijs, F.W. van Leeuwen, A.R. Caffe and D.F. Swaab (1985). The vasopressinergic innervation of the brain in normal and castrated rats. *J. Comp. Neurol.* 233:236-254.

De Wied, D. (1961). An assay of corticotrophin-releasing principles in hypothalamic lesioned rats. *Acta Endocrinol.* 37:288-297.

De Wied, D. (1976). Behavioral effects of intraventricularly administered vasopressin and vasopressin fragments. *Life Sci.* 19:685-690.

De Wied, D. (1987). The neuropeptide concept. In: E.R. De Kloet, V.M. Wiegant and D. De Wied, Eds., *Progr. in Brain Res., Vol. 72*, pp 93-108, Amsterdam: Elsevier.

De Wied, D. and B. Bohus (1986). Long term and short term effects on retention of a conditioned avoidance response in rats by treatment with long acting Pitresin and O-MSH. *Nature* 212:1484-1486.

De Wied, D., B. Bohus and Tj. B. Van Wimersma Greidanus (1975). Memory deficit in rats with hereditary diabetes insipidus. *Brain Res.* 85:152- 156.

Dunn, A.J. and C.W. Berridge (1987). Corticotropin-releasing factor adminsitration elicits a stress-like activation of cerebral catecholaminergic systems. *Pharmacol. Biochem. Behav.* 27:685-691.

Dunn, A.J. and S.E. File (1987). Corticotropin-releasing factor has anxiogenic action in the social interaction test. *Horm. Behav.* 21:193- 202.

Elands, J.P.M. (1989). *Neurohypophyseal hormone receptors*. University of Utrecht: Ph.D. Thesis.

Engel, G.L. (1977). Emotional stress and sudden death. *Psychol. Today* 11:114-118.

Fisher, L.A., J. Rivier, C. Rivier, J. Spiess, W. Vale and M.R. Brown (1982). Corticotropin-releasing factor (CRF): central effects on mean arterial pressure and heart rate in rats. *Endocrinology* 110: 2222-2234.

Fliers, E., G.J. De Vries and D.F. Swaab (1985). Changes with aging in the vasopressin and oxytocin innervation of the rat brain. *Brain Res.* 348:1-8.

Fokkema, D.S. (1985). *Social Behavior and Blood Pressure: A Study in the Rat*. Ph.D. Thesis. University of Groningen.

Fulwiler, C.E. and Saper, C.B., 1984, Subnuclear organisation of the efferent connections of the parabrachial nucleus in the rat, *Brain Res. Rev.* 7:229-259.

Gerfen, C.R. and P.E. Sawchenko (1984). An anterograde neuroanatomical tracing method that shows the detailed morphology of neurons, their axons and terminals: immunohistochemical localization of an axonally transported plant lectin, Phaseolus vulgaris Leucoagglutinin (PHA-L). *Brain Res.* 290:219-238.

Gibbs, D.M. (1984). Dissociation of oxytocin, vasopressin and corticotropin secretion during different types of stress. *Life Sci.* 35:487-491.

Gilbey, M.P., J.H. Coote, S. Fleetwood-Walker and D.F. Peterson (1982). The influence of the paraventricular-spinal pathway and oxytocin and vasopressin on sympathetic preganglionic neurones. *Brain Res.* 251:283- 296.

Gillies, G. and P. Lowry (1979). Corticotrophin releasing factor may be modulated by vasopressin. *Nature* 278:463-464.

Gillies, G., E. Linton and P. Lowry (1982). Corticotropin releasing activity of the new CRF is potentiated several times by vasopressin, *Nature* (London) 299:355-357.

Graham, F.K. and R.K. Clifton (1966). Heart rate changes as a component of the orienting response. *Psychol. Bull.* 65:305-320.

Granata, A.R., V. Numao, M. Kumada and D.J. Reis (1986). A1-noradrenergic neurons tonically inhibit sympathoexcitatory neurons of the C1 area in rat brainstem. *Brain Res.* 377:127-146.

Gray, T.S. and D.J. Magnuson (1987). Neuropeptide neuronal efferents from the bed nucleus of the stria terminalis and central amygdaloid nucleus to the dorsal vagal complex in the rat. *J. comp. Neurol.* 262:365-374.

Gray, T.S., T.L. O'Donohue and D.J. Magnuson (1986). Neuropeptide Y innervation of amygdaloid and hypothalamic neurons that project to the dorsal vagal complex in rat. *Peptides* 7:341-349.

Guillemin, R., T. Vargo, J. Rossier, S. Minick, N. Ling, J. Rivier, W. Vale and F. Bloom (1977). Beta-endorphin and adrenocorticotropin are secreted concomitantly by the pituitary. *Science* 197:1367-1369.

Hagan, J.J. and B. Bohus (1984). Vasopressin prolongs bradycardiac response during orientation. *Behav. Neural Biol.* 41:77-83.

Hagan, J.J., B. Bohus and D. De Wied (1982). Post-training vasopressin injections may facilitate or delay shuttle-box avoidance extinction. *Behav. Neur. Biol.* 36:211-228.

Hanley, M.R., H.P. Benton, S.L. Lightman, K. Todd, E.A. Bone, P. Fretten, S. Palmer, C.J. Kirk and R.H. Michell (1984). A vasopressin-like peptide in the mammalian sympathetic nervous system. *Nature* 309:259-261.

Henry, J.P. and P.M. Stephens (1977). *Stress, Health, and the Social Environment. A Sociobiological Approach to Medicine.* Berlin: Springer.

Hilton, S.M. (1982). The defence-arousal system and its relevance for circulatory and respiratory control. *J. exp. Biol.* 100:159-174.

Hilton, S.M. and K.M. Spyer (1980). Central nervous regulation of vascular resistance. *Ann. Rev. Physiol.* 42:399-411.

Hilton, S.M. and A.W. Zbrozyna (1963). Amygdaloid region for defense reactions and as efferent pathway to the brainstem. *J. Physiol.* (London) 165:160-173.

Jard, S., C. Barberis, S. Audigier and E. Tribollet (1987). Neurohypophyseal hormone receptor systems in brain and periphery. In: E.R. De Kloet, V.M. Wiegant and D. De Wied, Eds., *Neuropeptides and Brain Function, Progr. in Brain Res., Vol. 72,* pp 173-187, Amsterdam: Elsevier.

Jard, S., R.C. Gaillard, G. Guillon, J. Marie, P. Schoenenberg, A.F. Muller, M. Manning and W.H. Sawyer (1986). Vasopressin antagonists allow demonstration of a novel type of vasopressin receptor in the rat adenohypophysis. *Mol. Pharmacol.* 30:171-177.

Kalia, M. and K. Fuxe (1985a). Rat medulla oblongata. I: Cytoarchitectonic considerations. *J. Comp. Neurol.* 233:285-307.

Kalia, M., K. Fuxe and M. Goldstein (1985b). Rat medulla oblongata. II: Noradrenergic (A1 and A2) and adrenergic neurons, nerve fibers and presumptive terminal processes. *J. Comp. Neurol.* 233:308-332.

Kalia, M., K. Fuxe and M. Goldstein (1985c). Rat medulla oblongata. III: Adrenergic (C1 and C2) neurons, nerve fibers and presumptive terminal processes. *J. Comp. Neurol.* 233:333-349.

Kapp, B.S., M. Gallagher, M.D. Underwood, C.L. McNall and D. Whitehorn (1982). Cardiovascular responses elicited by electrical stimulation of the amygdala central nucleus in the rabbit. *Brain Res.* 234:251-262.

Koob, G.F. (1985). Stress, Corticotropin-releasing Factor and Behavior. *Perspectives on Behavioral Medicine* 2:39-52.

Koob, G.F., C. Lebrun, J.L. Martinez Jr., R. Dantzer, M. Le Moal and F.E. Bloom (1985). Arginine vasopressin, stress and memory. *Ann. N.Y. Acad. Sci.* 444:194-202.

Koolhaas, J.M., D.S. Fokkema, B. Bohus and G.A. Van Oortmerssen (1986). Individual differentation in blood pressure reactivity and behaviour of male rats. In: T.M. Dembrosky, T.H. Schmidt and G. Blümchen, Eds.,

Biobehavioral Bases of Coronary Heart Disease, Vol. 3, pp 517-526.

Korner, P.K. (1979). Central nervous control of autonomic cardiovascular function. In: F. Hamilton, Ed., *Handbook of Physiology, Cardiovascular Control*, pp 691-739, Washington: American Physiological Society.

Kovàcs, G.L., B. Bohus and D.H.G. Versteeg (1979). The effects of vasopressin on memory processes: the role of noradrenergic transmission. *Neuroscience* 4:1529-1537.

Laczi, F., O. Gaffori, E.R. De Kloet and D. De Wied (1983a). Differential regulation in immunoreactive arginine-vasopressin content of microdissected brain regions during passive avoidance behavior. *Brain Res.* 260:342-346.

Laczi, F., O. Gaffori, E.R. De Kloet and D. De Wied (1983b). Arginine- vasopressin of hippocampus and amygdala during passive avoidance behavior in rats. *Brain Res.* 280:309-315.

Laczi, F., O. Gaffori, M. Fekete, E.R. De Kloet and D. De Wied (1984). Levels of arginine-vasopressin in cerebrospinal fluid during passive avoidance behavior in rats. *Life Sci.* 34:2385-2391.

Lang, R.E., J.W.E. Heil, D. Ganten, K. Hermann, T. Unger and W. Rascher (1983). Oxytocin unlike vasopressin is a stress hormone in the rat. *Neuroendocrinology* 37:314-316.

LeDoux, J.E., J. Iwata, P. Cicchetti and D.J. Reis (1988). Different projections of the central amygdaloid nucleus mediate autonomic and behavioral correlates of conditioned fear. *J. Neurosci.* 8:2517-2529.

Le Moal, M., R. Dantzer, P. Mormede, A. Baduel, C. Lebrun, A. Ettenberg, D. Van der Kooy, J. Wenger, S. Deyo, G.F. Koob and F.E. Bloom (1984). Behavioral effects of peripheral administration of arginine vasopressin: a review of our search for a mode of action and a hypothesis. *Psychoneuroendocrinology* 9:319-339.

Loewy, A.D. and S. McKellar (1980). The neuroanatomical basis of central cardiovascular control. *Fed. Proc.* 39:2495-2503.

Mason, J.W. (1968). A review of psychoendocrine research on the sympathetic-adrenomedullary system. *Psychosomat. Med.*, 30:631-653.

Meisenberg, G. (1981). Short-term behavioral effects of neurohypophyseal peptides in mice. *Peptides* 2:1-8.

Meisenberg, G. and W.H. Simmons (1982). Behavioral effects of intracerebroventricularly administered neurohypophyseal hormone analogs in mice. *Pharmacol. Biochem. Behav.* 16:819-825.

Meisenberg, G. and W.H. Simmons (1983). Centrally mediated effects of neurohypophyseal hormones. *Neurosci. Biobehav. Rev.* 7:263-280.

Mens, W.B.J., M.A.H. Van Egmond, A.A. De Rotte and Tj. B. Van Wimersma Greidanus (1982). Neurohypophyseal peptide levels in CSF and plasma during passive avoidance behavior in rats. *Horm. Behav.* 16:371-382.

Mitchell, I.J., P. Dean and P. Redgrave (1988). The projection from superior colliculus to cuneiform area in the rat. II. Defence-like responses to stimulation with glutamate in cuneiform nucleus and surrounding structures. *Exp. Brain Res.* 72:626-639.

Morris, R., T.E. Salt, M.V. Sofroniew and R.G. Hill (1980). Actions of microiontophoretically applied oxytocin, and immunohistochemical localisation of oxytocin, vasopressin and neurophysin in the rat caudal medulla. *Neurosci. Lett.* 18:163-168.

Moruzzi, G. and H.W. Magoun (1949). Brain stem reticular formation and activation of the EEG. *Electroenceph. Clin. Neurophysiol.* 1:455-473.

Mraovitch, S., M. Kamuda and D.J. Reis (1982). Role of the nucleus parabrachialis in cardiovascular regulation in cat. *Brain Res.* 232:57- 75.

Nieuwenhuys, R. (1985). *Chemoarchitecture of the Brain*, Berlin: Springer.

Oliverio, A. (1987). Endocrine aspects of stress: central and peripheral mechanisms. In: P.R. Wiepkema and P.W.M. Van Adrichem, Eds., *Biology of Stress in Farm Animals*, pp 3-12, Dordrecht: Nijhoff.

Nyakas, C., A.J.A. Prins and B. Bohus (1989). Age-related alterations in cardiac response to emotional stress: relations to behavioural reactivity in the rat. *Physiol. Behav.* 00:000.

Obrist, P.A. (1981). *Cardiovascular Psycho-physiology: A Perspective.* New York: Plenum.

Pilowsky, P.M., V. Kapoor, J.B. Minson, M.J. West and J.P. Chalmers (1984). Spinal cord serotonin release and raised blood pressure after brainstem kainic acid injection. *Brain Res.* 366:354-352.

Pittman, Q. and L.G. Franklin (1985). Vasopressin antagonist in nucleus tractus solitarius/vagal area reduces pressor response to paraventricular nucleus stimulation in rats. *Neurosci. Lett.* 56:155-160.

Pittman, Q., T.D. Lawrence and L. McLean (1982). Central effects of arg- vasopressin on blood pressure in rats. *Endocrinology* 110:1058-1060.

Redgrave, P., P. Dean, I.J. Mitchell, A. Odekunle and A. Clark (1988). The projection from superior colliculus to cuneiform area in the rat. I. Anatomical studies. *Exp. Brain Res.* 72:611-625.

Rees, H.D., A.J. Dunn, and P.M. Iuvone (1976). Behavioral and biochemical response of mice to the intraventricular administration of ACTH analogs and lysine vasopressin. *Life Sci.* 18:1333-1340.

Richter, D. (1985). Biosynthessis of vasopressin. In: D. Ganten and D. Pfaff, Eds., *Current Topics in Neuroendocrinology, Vol. 4,* pp 1-16 Berlin: Springer.

Rohmeiss, P., H. Becker, R. Dietrich, F. Luft and T. Unger (1986). Vasopressin: mechanism of central cardiovascular action in conscious rats. *J. Cardiovasc. Pharmacol.* 8:689-696.

Roozendaal, B., W.A. Van Gool, D.F. Swaab, J.E. Hoogendijk and M. Mirmiran (1987). Changes in vasopressin cells of the rat suprachiasmatic nucleus with aging. *Brain Res.* 409:259-264.

Ross, C.A., D.A. Ruggiero, D.J. Park, T.H. Joh, A.F. Sved, J. Fernandez- Pardal and D.J. Reis (1984). Tonic vasomotor control by rostral ventrolateral medulla: effect of electrical or chemical stimulation of the area containing C1 adrenaline neurons on arterial pressue, heart rate and plasma catecholamines and vasopressin. *J. Neuroscience* 4:474-494.

Sahgal, A. (1984). A critique of the vasopressin-memory hypothesis. *Psychopharmacology* 83:215-228.

Seley, H. (1976). *Stress in Health and Disease.* Boston: Butterworths.

Schmid, P.G., F.M. Scharabi and G.B. Guo (1984). Vasopressin and oxytocin in the neuronal control of the circulation. *Fed. Proc.* 43:97-102.

Smith, O.A. and J.L. DeVito (1984). Central neural integration for the control of autonomic responses associated with emotion. *Ann. Rev. Neurosci.* 7:43-65.

Smith, O.A., C.A. Astley, J.L. DeVito, J.M. Stein and K.E. Walsh (1980). Functional analysis of hypothalamic control of the cardiovascular responses accompanying emotional behavior. *Fed. Proc.* 39:2487-2494.

Spyer, K.M. (1982). Central nervous integration of cardiovascular control. *J. Exp. Biol.* 100:109-128.

Sutton, R.E., G.F. Koob, M. Le Moal, J. Rivier and W. Vale (1982). Corticotropin releasing factor produces behavioural activation in rats. *Nature* 297:331-333.

Swanson, L.W. and W.M. Cowan (1979). The connections of the septal region in the rat. *J. Comp. Neurol.* 186:621-656.

Swanson, L.W. and P.E. Sawchenko (1983). Hypothalamic integration: organization of the paraventricular and supraoptic nuclei. *Ann. Rev. Neurosci.* 6:269-324.

Ter Horst, G.J., H.J. Groenewegen, H. Karst and P.G.M. Luiten (1984). Phaseolus vulgaris leucoagglutinin immunohistochemistry. A comparison between autoradiographic and lectin tracing of neuronal efferents. *Brain Res.* 307:379-383.

Thatcher-Britto, K., J. Morgan, J. Rivier, W. Vale and G.F. Koob (1985). Chlordiazepoxide attenuates response suppression induced by corticotropin-releasing factor in the conflict test. *Psychopharmacology* 86:170-174.

Thompson, E.A. and D. De Wied (1973). The relationship between the antidiuretic activity of rat eye plexus blood and passive avoidance behaviour. *Physiol. Behav.* 11:377-380.

Undesser, K.P., A.J. Trapani, W.W. Morgan and V.S. Bishop (1986). Role of central catecholamines on the potentiation of the baroreceptor reflex produced by vasopressin. *Circ. Res.* 58:882-889.

Unger, T., P. Rohmeiss, H. Becker, D. Ganten, R.E. Land and M. Petty (1984). Sympathetic activation following central vasopressin receptor stimulation in conscious rats. *J. Hypertension* 2 (suppl.3):25-27.

Vallejo, M., D.A. Carter and S.L. Lightman (1984). Haemodynamic effects of arginine-vasopressin microinjections into the nucleus tractus solitarius: a comparative study of vasopressin, a selective vasopressin agonist and antagonist, and oxytocin. *Neurosci. Lett.* 52:247-252.

Valtin, H. and H.A. Schroeder (1964). Familial hypothalamic diabetes insipidus in rats (Brattleboro strain). *Amer. J. Physiol.* 206:425-430.

Van Wimersma Greidanus, Tj.B., G. Croiset, H. Goedemans and J. Dogterom (1979). Vasopressin levels in peripheral blood and in cerebrospinal fluid during passive and active avoidance behavior in rats. *Horm. Behav.* 12:103-111.

Veening, J.G., L.W. Swanson and P.E. Sawchenko (1984). The organization of projections from the central nucleus of the amygdala to brainstem sites involved in central autonomic regulation: a combined retrograde transport-immunohistochemical study. *Brain Res.* 303:337-357.

Verney, E.B. (1947). The antidiuretic hormone and the factors which determine its release. *Proc. R. Soc. Lond., Ser. B* 135:25-105.

Versteeg, C.A.M., B. Bohus and W. De Jong (1982a). Attenuation by arginine- and desglycinamide-lysine-vasopressin of a centrally evoked pressor response. *J. Auton. Nerv. Syst.* 6:253-262.

Versteeg, C.A.M., B. Bohus and W. De Jong (1982b). Inhibition of centrally-evoked pressor responses by neurohypophyseal peptides and their fragments. *Neuropharmacology* 21:1939-1964.

Versteeg, C.A.M., K. Cransberg, W. De Jong and B. Bohus (1983a). Reduction of a centrally induced presssor response by neurohypophyseal peptides: the involvement of lower brainstem mechanisms. *Eur. J. Pharmacol.* 94:133-140.

Versteeg, C.A.M., W. De Jong and B. Bohus (1983b). Arginine-8-vasopressin inhibits centrally induced pressor response by involving hippocampal mechanisms. *Brain Res.* 292:317-326.

Weindl, A. and M. Sofroniew (1985). Neuroanatomical pathways related to vasopressin In: D. Ganten and D. Pfaff, Eds., *Current Topics in Neuroendocrinology, Vol. 4*, pp 137-196, Berlin: Springer.

Williams, T.D.M., D.A. Carter and S.L. Lightman (1985). Sexual dimorphism in the posterior pituitary response to stress in the rat. *Endocrinology* 116:738-740.

Whitnall, M.H., E. Mezey and H. Gainer (1985). Co-localization of corticotropin-releasing factor and vasopressin in median eminence neurosecretory vesicles. *Nature*: 317:248-250.

Yamano, M., C.J. Hillyard, S. Girgis, I. MacIntyre, P.C. Emson and M. Tohyama (1988). Presence of a substance-P-like immunoreactive neuron system from the parabrachial area to the central amygdaloid nucleus of the rat with reference to coexistence with calcitonin gene-related peptide. *Brain Res.* 451:179-188.

Yerkes, R.M. and J.D. Dodson (1908). The relation of strentgh of stimulus to rapidity of habit-formation. *J. Comp. Neurol. Psychol.* 18:459-482.

ROLE OF PROLACTIN IN STRESS-INDUCED BIOLOGICAL MODIFICATIONS IN ANIMALS

Filippo Drago, Francesco Spadaro, Velia D'Agata, Teresa Iacona, Carmela Valerio, Rocco Raffaele*, Carmelo Astuto, Nunzio Lauria and Matteo Vitetta**
Institute of Pharmacology and Neurological Clinic*, University of Catania Medical School, Catania, Italy
and
**Chair of Mental Hygiene, University of Messina Medical School, Messina, Italy.

Introduction

It is well established that the adenohypophyseal hormone prolactin (PRL) is released in high quantity by animals subjected to stress of physical and psychological nature. Swingle et al. (1951) first described the occurrence of pseudopregnancy in female rats subjected to different types of stressor stimuli. Later, it was found that stress promotes milk secretion, suggesting the possible involvement of PRL (Nicoll et al., 1960). In 1965, Grosvenor et al. described in detail the stress-induced depletion of PRL from the adenohypophysis. However, it is not yet clear whether this hyperprolactinemia is just the consequence of a general hypothalamic activation induced by stress or it plays any role in the biological phenomena caused by the application of stress.

In parallel with the development of psychoneuroendocrine research concerned with the effect of stress on PRL secretion, the brain as a target organ for this hormone received much attention. Animal experiments and further human observations led to the notion that PRL influences behavioral adaptation to environmental changes (Drago, 1982). In addition, evidence has been presented that PRL may also affect stress-induced autonomic responses, e.g. changes in gastric acid secretion and in thermoregulation (Drago and Amir, 1984).

The teleologic significance of the release of other pituitary hormones in stress responses has been recognized and is bound to the maintenance of physiological standards or to the stimulation of biological mechanisms necessary to restore body homeostasis. The adaptation to stress, however, is often accompanied by an 'endogenous' damage (e.g., hypertension or gastric ulcers) that is caused by the overall activation of nervous, endocrine and autonomic responses.

Stress-induced diseases are not only the result of a relative incapacity of the body to cope with stress under chronic situatios, but also of intrisic features of stress mediators. Thus, it seems likely that some pituitary hormones acting as stress mediators, besides positive effects on the neutralization of stress, exert negative effects resulting in an 'endogenous' damage. An example is given by the ACTH-corticosterone system that, besides responses elicited in

125

S. Puglisi-Allegra and A. Oliverio (eds.), Psychobiology of Stress, 125–133.

order to cope with stress, produces pathological alterations, such as hypertension, gastric ulcers and immune suppression.

Prolactin, which does not recognize any specific target organ in the periphery, exerts positive effects in the mechanisms of coping with stress. The present paper shows that this hormone may exert multiple effects under stress conditions. In fact, PRL seems to affect behavioral, neuroendocrine and autonomic responses to stress. There is now sufficient evidence indicating that this hormone also exerts a protective role against the 'endogenous' damage caused by stress. In some cases, this is due to an antagonism exerted by PRL on ACTH-induced responses.

The influence of prolactin on stress-induced behavioral changes

Hormones of pituitary origin, which are secreted in response to stress, exert central effects resulting in modifications of adaptive behavior, like avoidance, approach, or even aggressive and sexual behavior. Analgesia is also present in many stress conditions, in that enables the animal to be more resistant to stress stimuli.

The application of stress in rodents is often followed by the enhancement of grooming behavior. Novelty ('psychological stress') and mild physical stressors are fully capable to induce grooming behavior (Spruijt and Gispen, 1983). The same phenomenon is observed when the animal is in conflict or is frustrated. For example, rats placed on extinction in an alley or on a partial reinforcement schedule display an increase in grooming (Miller and Stevenson, 1936; Bindra, 1963). This stress- or conflict-related grooming, which seems to appear as irrelevant behavior ('out of context') is described as a displacemet activity (Borchelt, 1980). Considering that grooming may occur after extremely arousing conditions, it has been suggested that this behavior may play a deactivating role in restoring psycho-physical homeostasis (Delius et al., 1976; Gispen and Isaacson, 1981). Interestingly, analysis of the temporal aspects of novelty-related grooming in the rat revealed a difference in the effects of environmental stimuli. Hence, it was argued that the displayed behavior was primarily related to the arousing influence of the stimuli (Jolles et al., 1979).

The influence of PRL on stress-induced behavioral changes has been studied in adaptive and non-adaptive situations. Rats bearing a pituitary homograft under the kidney capsule that makes them hyperprolactinemic, exhibit a facilitated acquisition of active avoidance responses in the shuttle-box and in the pole-jumping test situations (Drago et al., 1982a). This effect is reduced either by dopamine or opioid receptor antagonists (Drago et al., 1981a, 1981b) and does not decrease in long-term hyperprolactinemic rats (Drago et al., 1982b).

Hyperprolactinemic rats exhibit a reduced pain sensitivity when they are tested for behavioral responsiveness to electrical footshock (Drago et al., 1982a). This effect has been confirmed in other tests for analgesia measurement (Ramaswamy et al., 1983) and is under control of opioid neurotransmission, in that the administration of the opioid receptor antagonist, naltrexone, is followed by a normalization of behavioral responsiveness to electrical footshock in hyperprolactinemic rats (Drago et al., 1981a).

Novelty-induced grooming behavior is particularly sensitive to PRL. Both endogenous hyperprolactinemia induced by pituitary homografts and intracerebroventricular (i.c.v.) administration of rat PRL enhance novelty-induced grooming in the rat (Drago et al., 1980). This effect disappears in long-term hyperprolactinemic animals (Drago et al., 1982b) and is diminished after the injection of dopamine or opioid receptor antagonists (Drago et al.,

1981a, 1981b). It is worth to mentioning that the enhanced grooming of hyperprolactinemic rats depends on the actual presence of this hormone (or of its behaviorally-active fragments) in the brain. In fact, i.c.v. administration of anti-PRL serum suppresses the excessive grooming of animals with high plasma levels of prolactin (Drago et al., 1986).

Other pituitary hormones are released during stress and share some aspects of the behavioral profile of PRL (Table 1). In fact, hormones such as ACTH, MSH, vasopressin and beta-endorphin may facilitate avoidance behavior, exert analgesic activity and enhance grooming (de Wied, 1969; de Wied et al., 1974; Olson et al., 1979). In particular, ACTH is most potent in enhancing novelty-induced grooming (Gispen and Isaacson, 1981).

Table 1. Profile of the effects of pituitary hormones in the stress

	PROLACTIN	ACTH	ENDORPHINS	VASOPRESSIN
Adaptive behavior	+	+	+ /−	+
Grooming	+	+	+	+
Analgesia	+	+ /−	+	+
Hypothermia	+	+	+	+
Corticosterone secretion	−	+	?	?
Gastric ulcers	−	+	+ /?	?
Immune response	+	−	−/ +	+

The interaction between PRL and ACTH in enhancing grooming behavior after i.c.v. administration has been studied in intact and hyperprolactinemic rats. In intact animals, it was first found that no cross-tolerance exist between PRL and ACTH (Drago et al., 1983). Furthermore, the i.c.v. injection of rat PRL into hyperprolactinemic rats failed to enhance further the grooming activity of these animals, whereas this behavior was substantially enhanced by i.c.v. injection of ACTH. In long-term hyperprolactinemc rats, which exhibit no facilitation of grooming, i.c.v. injection of rat PRL was unable to enhance this behavior. In contrast, at this time i.c.v. administration of ACTH again induced excessive grooming. These findings support the hypothesis that behaviorally-competent central mechanisms become hyposensitive to PRL under conditions of long-term hyperprolactinemia, while the responsiveness to ACTH remains preserved. These results also suggest that independent neural mechanisms are involved in PRL- and ACTH-induced grooming.

The influence of prolactin on ACTH-corticosterone system

Basal levels of plasma corticosterone in hyperprolactinemic rats bearing pituitary homografts under the kidney capsule have been found to be higher than those of animals with normal levels of PRL (Drago and Scapagnini, 1984). This finding is in agreement with other studies showing an increase in the secretory activity of adrenal glands in rats with hyperprolactinemia of various origin (Voogt et al., 1969; Doherty et al., 1980). The endocrine change appears to be accompanied by hypertrophy of the adrenal glands, as observed in post-mortem examination of hyperprolactinemic rats. The question rose as to whether the increased secretory activity of adrenal glands in hyperpolactinemic rats is due to a direct effect of PRL or to an activation of adenohypophyseal ACTH. In fact, hyper-

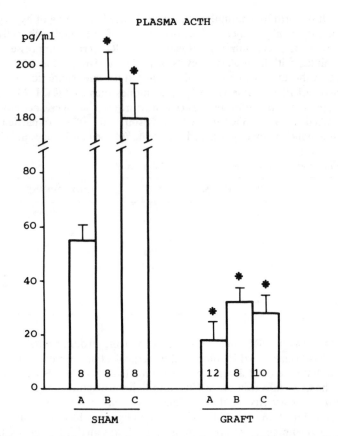

Fig. 1. ACTH plasma levels in hyperprolactinemic male rats bearing a pituitary homograft (graft) under the kidney capsule. In brackets the number of animals per each group. A = basal conditions; B = physical stress; C = after intravenous injection of vasopressin (5 mgkg)
* Significant difference as compared to control group (p < 0.05, Dunnett's test for multiple comparisons).

prolactinemic rats subject to hypophysectomy also show high levels of plasma corticosterone suggesting that a direct effect of PRL promotes the activation of the adrenal glands (Drago and Scapagnini, 1984).

When a physical stress is applied (e.g. forced swim in colf water), hyperprolactinemic rats show a suppression of the three-fold increase in plasma corticosterone levels observed in control animals given the same type of stress stimulus (Drago et al., 1986).

A possible explanation is that the increase in plasma cortisterone levels due to hyperprolactinemia causes, in turn, a reduction in ACTH secretion from the adenohypophysis. This can account for the suppression of the ACTH-corticosterone system under stress conditions. This hypothesis is confirmed by the finding that hyperprolactinemic rats show a decrease in basal plasma ACTH levels as compared to controls and no increase after either application of physical stress or injection of the ACTH releaser, vasopressin (Fig. 1).

It is interesting to noting that hyperprolactinemic lactating female rats show a marked suppression of plasma corticosterone levels after application of different types of stressor stimuli, accompanied by a decrease in the occurrence of stress-induced aggressive behavior (Kamoun, 1970; Endroczi and Nyakas, 1974).

Influence of prolactin on stress-induced hyperthermia

The effect of endogenous hyperprolactinemia on core temperature has been studied in rats before and after the application of restraint stress (Drago and Amir, 1984). Hyperprolactinemia is accompanied by a small but significant decrease in core temperature of freely moving rats, and this effect can be totally reversed by the administration of the opioid receptor antagonist, naloxone. In animals with normal levels of plasma PRL, the injection of naloxone is also followed by a hypothermic effect. Both animals with hyperprolactinemia or injected with naloxone fail to show any increase in normal core temperature after the application of physical stress.

Different types of stressor stimuli can induce a rise in body temperature (Stewart and Eikelboom, 1979). However, it is of great interest that most of the pituitary hormones released during stress cause hypothermia when administered exogenously. This is true for ACTH and MSH (Glyn and Lipton, 1981), vasopressin (Crine et al., 1981), and beta-endorphin (Martin and Bacino, 1979). Thus, PRL resembles other pituitary hormones in inducing hypothermia in animals under stress conditions.

Prolactin and stress-induced gastric ulcers

A number of findings have recently demonstrated that PRL may exert a protective action against a typical 'endogenous' damage caused by stress, i.e. gastric ulcers. In a preliminary study, it was found that hyperprolactinemia, as induced by pituitary homografts under the kidney capsule, is accompanied by an inhibition of development of gastric ulcers by cytoprotection in the model of cold-plus-restraint stress (Drago et al., 1986). This effect can be mimicked in animals injected i.c.v. with rat PRL or treated with the hyperprolactinemic agent, domperidone. A possible involvement of the dopamine transmission in this effect is suggested by the suppression of PRL cytoprotection after the injection of the dopamine receptor antagonist, haloperidol. The same suppression is observed when the inhibitor of prostaglandin synthesis, indomethacin, is injected. This suggests that the cytoprotective action of PRL on stress-induced gastic ulcers requires an intact prostaglandin synthesis pathway.

Recently, the role of central dopamine transmission in PRL cytoprotection has been studied in more detail. It was found that the injection of microdoses of 6-OHDA into the corpus striatum, leading to the distruction of the dopamine nerve endings in this area, is followed by a total abolition of PRL cytoprotective action on stress-induced gastric ulcers (Table 2). The same has been found after injection of 6-OHDA into the nucleus accumbens. These findings suggest that this effect of PRL is of central origin and involves both the nigrostrial and mesolimbic dopaminergic systems.

When the application of a physical stress of subliminal intensity is not followed by sustained gastric lesions, the contemporary injection of ACTH may substantially increase the occurrence of ulcers (Table 3). Thus, although ACTH does not show any ulcerogenic potency when injected in non-stressed animals, it potentiates the effects of physical stress in the development of gastric ulcers. Interestingly, this phenomenon is totally suppressed in animals with high levels of plasma PRL. Since the effect of ACTH may be mediated by peripheral corticosteroids, in this case it is likely that the antagonism exerted by PRL takes place at a peripheral level.

Table 2. Effects of 6-OHDA bilateral microinjections into the corpus striatum on gastric lesions following the application of cold-plus-restraint stress in hyperprolactinemic male rats bearing a pituitary homograft under the kidney capsule

Experimental conditions	No lesions	Petechiae	Ulcers
1) Intrastriatal saline + Sham-operation (8)	0.0	25.0	75.0
2) Intrastriatal saline + Homograft (8)	87.5*	2.5	0.0*
3) Intrastriatal 6-OHDA + Sham-operation (8)	0.0	12.5	87.5
4) Intrastriatal 6-OHDA + Homograft (8)	12.5	25.0	62.5

Values are expressed as percent of animals exhibiting no lesions, petechiae or ulcers. In brackets the number of animals per each group.
* Significantly different as compared to group 1 (p $<$ 0.05, Kruskal-Wallis test for multiple comparisons).

Prolactin and the immune system

The evidence that different types of stress can influence the immune system at various levels is now well established. On principle, a suppression of immune response occurs under conditions of chronic stress (Solomon, 1969; Hanson and Flaherty, 1981; Teshima et al., 1987).

Recent studies have shown that pitutary hormones which are released during stress may affect the functioning of the immune system. In particular, receptors for PRL have been found on the membrane of T and B lymphocytes and of monocytes (Russel et al., 1984). The immunostimulant properties of PRL have been demonstrated under different circumstances. The hormone may restore a normal immune response in hypophysectomized animals and stimulates antibody production (Berczi et al., 1981). Furthermore, PRL increases thymus and spleen weight in normal and hypophysectomized animals (Houston, 1984).

In contrast, ACTH is a potent inhibitor of antibody production (Johnson et al., 1982). Thus, PRL may exert a physiological antagonism against ACTH also with respect to stress-induced modifications in the immune response. Again, this antagonism seems to be exerted in order to protect the body against an 'endogenous' damage caused by stress.

Conclusive remarks

The physiological significance of PRL protective action against stress-induced biological modifications may reside, in a phylogenetic view, in the role played by this hormone in lactation and infant care. In fact, a number of authors have described a protective effect of

Table 3. Antagonism of prolactin on ACTH-induced gastric lesions following the application of restraint stress in male rats

Experimental conditions	No lesions	Petechiae	Ulcers
1) I.v. saline +			
Sham-operation, nonstressed (10)	100.0	0.0	0.0
2) I.v. saline +			
Sham-operation, stressed (10)	40.0*	40.0	20.0
3) I.v. ACTH +			
Sham-operation, nonstressed (10)	80.0	20.0	0.0
4) I.v. ACTH +			
Sham-operation, stressed (10)	0.0§	20.0	80.0*§
5) I.v. ACTH +			
Homograft, stressed (10)	100.0*	0.0*	0.0*

Values are expressed as percent of animals exhibiting no lesions, petechiae or ulcers. In brackets the number of animals per each group.
* Significantly different as compared to group 1 (p < 0.05, Kruskal-Wallis test for multiple comparisons).
§ Significantly different as compared to group 2 (p <0.05, Kruskal-Wallis test for multiple comparisons).
ACTH was administered intravenously (i.v.) in a single bolus of 1 mgrat.

PRL against stressor stimuli in lactating animals. Thoman et al. (1968) have found that hyperprolactinemic lactating rats exhibit a marked resistance to stress-induced changes in body temperature and a diminished secretory activity of adrenal glands under stress conditions. Moreover, the differences between lactating and non-lactating animals are accentuated as lactation increases, and disappear following weaning of the young animals. Lactating rats also show a marked suppression of plasma corticosterone levels after application of different types of stressor stimuli (Thoman et al., 1970). Under the same conditions, these animals exhibit a reduction of aggressive behavior in a shock-inuced fighting procedures.

Besides, as PRL release normally occurs in different stress situations, it is possible that its protective role has the more specific meaning of hindering any 'endogenous' damage induced by stress.

References

Berczi I., Nagy E., Kovacs K., Horuath E. (1981): Regulation of humoral immunity in rats by pituitary hormones. *Acta Endocrinol.* 98:506-514

Bindra D. (1963): Temporal analysis of relevant and irrelevant behavior components in partial reinforcement and extinction. *Psychol. Rev.* 13:551-563

Borchelt P.L. (1980): Care of the body surface (COBS). In: *Comparative Psychology: An Evolutionary Analysis of Animal Behavior* (edited by D. Ray Denny). Willey Pu., New York, pp. 362-384

Crine A.F., Bredart S. and Legros J.J. (1981): Effects of exogenous arginine vasopressin on rectal temperature in the albino rat. *Horm. Behav.* 15:226-231

Delius J.D., Craig B. and Chaudoir C. (1976): Adrenocorticotropic hormone, glucose and displacement activities in pigeons. *Z. Tierpsychol.* 40: 183-193

de Wied D. (1969): Effects of peptide hormones on behavior. In: *Frontiers in Neuroendocrinology* (edited by W.F. Ganong and L. Martini), Oxford University Press, New York, pp. 97-40

de Wied D., Bohus B. and van Wimersma Greiganus Tj.B. (1974): The hypothalamo-neurohypophyseal system

132

and the preservation of conditioned avoidance behavior in rats. *Prog. Brain Res.* 41:417-428

Doherty P.C., Smith M.S. and Bartke A. (1980): Hyperprolactinemia and reproductive function in male rats. Effects of adrenalectomy and bromocriptine. *Endocr. Soc. Meet. abs.* 704

Drago F. (1982): Prolactin and Behavior. Ph.D. Thesis, University of Utrecht, The Netherlands

Drago F., Canonico P.L., Bitetti R. and Scapagnini U. (1980): Systemic and intraventricular prolactin induces excessive grooming. *Europ. J. Pharmacol.* 65:457-458

Drago F., Gispen W.H. and Bohus B. (1981a): Behavioral effects of prolactin: involvement of opioid receptors. In: *Advances in Endogenous and Exogenous Opioids* (edited by H. Takagi and E.J. Simon), Kodansha Ltsd., Tokyo, pp. 335-337

Drago F., Bohus B., Canonico P.L. and Scapagnini U. (1981b): Prolactin induces grooming in the rat: possible involvement of nigrostriatal dopaminergic system. *Pharmacol. Biochem. Behav.* 15:61-63

Drago F., Bohus B. and Mattheij J.A.M. (1982a): Endogenous hyperprolactinaemia and avoidance behaviors of the rat. *Physiol. Behav.* 28:1-4

Drago F., Bohus B., van Ree J.M., Scapagnini U. and de Wied D. (1982b): Behavioral responses of long-term hyperprolactinaemic rats. *Europ. J. Pharmacol.* 79:323-327

Drago F., Bohus B., Gispen W.H., Scapagnini U. and de Wied D. (1983): Prolactin-enhanced grooming behavior: interaction with ACTH. *Brain Res.* 263:277-282

Drago F. and Amir S. (1984): Effects of hyperprolactinaemia on core temperature of the rat. *Brain Res. Bull.* 12:355-358

Drago F. and Scapagnini U. (1984): Prolactin-ACTH interactions in the neuroendocrine responses to stress. In: *Stress: The Role of Catecholamines and Other Neurotransmitters* (edited by E. Usdin et al.), Gordon and Brench Science Publishers, New York, pp. 311-324

Drago F., Amir S., Continella G., Alloro M.C. and Scapagnini U. (1986): Effects of endogenous hypeprolactinemia on adaptive responses to stress. In: *Prolactin – Basic and Clinical Correlates* (edited by R.M. MacLeod, M. Thorner and U. Scapagnini), Liviana Press, Padova, pp. 609-614

Endroczi E. and Nyakas Cs. (1974): Pituitary-adrenal function in lactating rats. *Endokrinologie (Budapest)* 63:1-5

Gispen W.H. and Isaacson R.L. (1981): ACTH-induced excessive grooming in the rat. *Pharmacol. Ther.* 12:209-237

Glyn J.R. and Lipton J.M. (1981): Hypothermic and antipyretic effects of centrally administered ACTH (1-24) and a-melanotropin. *Peptides 2:177-187*

Grosvenor C.E. (1965) Effects of nursing and stress upon prolactin-inhibiting activity of the rat hypothalamus. *Endocrinology* 77:1037-1042

Hanson P.G. and Flaherty D.R. (1981): Immunological responses to training in conditioned runners. *Clin. Sci.* 60:225-234.

Houston G. (1984): Hormones induced changes in lymphoid organs. Proc. B. Sc. Presentations at Univ. of Manitoba, 43

Johnson H.M., Smith E.M., Torres B.A. and Blalock J.E. (1982): Neuroendocrine hormone regulation of in vitro antibody production. *Proc. Natl. Acad. Sci. USA* 79:4171-4179

Jolles J., Rompa-Barendregt I. and Gispen W.H. (1979): Novelty and grooming behavior in the rat. *Behav. Neur. Biol.* 25:563-572

Kamoun A. (1970): Activité cortico-surrenale au cours de la gestation, de la lactation et du development pre et post-natal chez la rat. I. Concentration et cinique de disparition de la corticosterone. *J. Physiol. (Paris)* 62:5-32

Martin G.E. and Bacino C.B. (1979): Action of intracerebrally injected b-endorphin on the rat core temperature. *Europ. J. Pharmacol.* 59:227-236

Miller N.E. and Stevenson S.S. (1936): Agitated behavior of rats during experimental extinction and a curve of spontaneous recovery. *J. Comp. Physiol.* 21:205-231

Nicoll C.S., Talwalker P.K. and Meites J. (1960): Initiation of lactation in rats by nonspecific stresses. *Amer. J. Physiol.* 198:1103-1106

Olson R.D., Kastin A.J., Olson G.A. and Coy D.H. (1979): Behavioral effects after systemic injection of opiate peptides. Psychoneuroendocrinology 5:47-52.

Ramaswamy S., Pillai N.P. and Bapna J.S. (1983): Analgesic effect of prolactin: possible mechanism of action. *Europ. J. Pharmacol.* 96:171-173

Russel D.H., Matrisian L., Kibler R., Larson D.F., Poglos B. and Magun B.E. (1984): Prolactin receptor on human lymphocytes and their modulation by cyclosporin. *Biochem. Biophys. Res. Comm.* 121:889-896

Solomon G.F. (1969): Stress and antibody response in the rats. Int. Arch. Allergy Appl. Immunol. 35:97-107

Spruijt B.M. and Gispen W.H. (1983): ACTH and grooming behaviour in the rat. In: Hormones and Behaviour in Higher Vertebrates (edited by J. Balthazart, E. Prove and R. Gilles), Springer-Verlag, Berlin, 119-136

Stewart J. and Eikelboom R. (1979): Stress mask the hypothermic effect of naloxone in rats. Life Sci. 25:1165-1172

Swingle W.W., Fedor E.J., Barlow G., Collins E.J. and Perlmutt J. (1951): Induction of pseudopregnancy in rat following adrenal removal. Amer. J. Physiol. 167:593-598

Teshima H., Sogawa H., Kihara H., Nagata S., Ago Y. and Nagakawa T. (1987): Changes in populations of T-cells subsets due to stress. In: Neuroimmine Interactions: Proceedings of the 2nd Intern. Workshop of Neuroimmunomodulation (edited by B.D. Jankovic, B.M. Markovic and N.H. Spector), New York Academy of Sciences, New York, pp. 459-471

Thoman E.B., Wetzel A. and Levine S. (1968): Lactation prevents disruption of temperature regulation and suppresses adrenocortical activity in rats. Comm. Behav. Biol. 2:165-171

Thoman E.B., Conner R.L. and Levine S. (1970): Lactation suppresses adrenal corticosteroid activity and aggressiveness in rats. J. Comp. Physiol. Psychol. 70:364-369

Voogt J.L., Sar M. and Meites J. (1969): Influence of cycling pregnancy, labor and suckling on corticosterone-ACTH levels. Amer. J. Physiol. 216:655-658

PERINATAL DETERMINANT OF THE PITUITARY-ADRENAL ACTIVITY IN THE ADULT RAT

F.R. Patacchioli and L. Angelucci
Istituto di Farmacologia 2a, Medical Faculty, University of Rome
"La Sapienza", Rome, Italy

This paper reviews our work to date on the mother-conceptus pituitary- adrenal relationship in the rat, that we started to study some years ago, as determinant of the functional ontogenesis of the glucocorticoid receptor system in the rat hippocampus.

As summarized in table 1, there is strong evidence that contact with high levels of corticosterone in perinatal life may consistently affect pituitary-adrenal activity in later life in the rat and mouse.

Table 1. Some findings from the literature concerning the effects of neonatal corticosterone manipulation on postnatal development and adult life.

Adrenal function in the rat
- 100 mg/kg s.c. on day 3: suppression of circadian rhythm (Krieger, 1972)
- 250 mg/kg i.p. in the first week: transitory, or permanent reduction in basal levels of corticosterone in males and females, respectively; absence of circadian rhythm in males (Turner and Taylor, 1975)
- 160 microg/ml in drinking water from weaning: reduction of Adrenal weight at 32 days of age (Ramalay and Schwartz, 1980)

Behavior
- 1110-1160 mg/kg in cotton pellet on day 2: impairment in the acquisition of 2-way active avoidance in the rat (Olton et al., 1974)
- 250 mg/kg i.p. on day 3: reduced emotional activity in the rat (Turner and Taylor, 1975)
- 160 microg/ml in drinking water for six days at weaning: increased intraspecies aggressiveness in the mouse (Angelucci et al., 1980b)
- massive doses in early postnatal life: reduced emotional activity in the rat (Nyakas, 1983).

It is important to note that from the point of view of the pituitary- adrenocortical activity, mother and conceptus constitute a unit, both during pregnancy and lactation, by the way of reciprocal influences through the placenta and milk delivery. Disturbances of this relationship may be the common pathway through which perinatal exposure to different agents produces behavioral and concomitant endocrine consequences in adult life (Ward and Weiss, 1984). In fact, improper contacts of hormones with organisms in their early life may lead to temporally aberrant imprinting of the hormonal-tissue function relationship. Such consequences can be of great importance in the case of adrenocortical hormones, with regard to behavior and neuroendocrine activity of the hypothalamo-pituitary-adrenal axis in the adulthood.

S. Puglisi-Allegra and A. Oliverio (eds.), Psychobiology of Stress, 135–142.

In the rat, by the presence of a specific receptor system in the limbic structures, especially in the hippocampus (McEwen et al., 1969), glucocorticoid hormone regulates adaptive behaviors in their cognitive, such as arousal, learning, memory (for a review, see Bohus et al., 1982), and emotional-hormonal components, such as stress (Patacchioli et al., 1983; Sapolsky et al., 1984).

The studies we present today, concerning the mother-offspring relationship exclusively from the point of view of the pituitary- adrenocortical activity and the brain, started from a rather unusual observation: as shown in figure 1, binding capacity of the adrenocorticoid receptor system in the hippocampus, at variance with most biological parameters, has an exceptionally wide range of distribution in an apparently homogeneous population of rats (Angelucci et al., 1980a). Between extremes, from 150 to 500 fmoles of corticosterone bound per mg of cytosolic protein, there is more than three times variation. Since we have worked out these values from a highly inbred strain of the Wistar rat, it was difficult to think of this variation as determined by genetic factors, whereas environmental factors appeared a more plausible determinant. We attempted to elucidate the meaning of this phenomenon on the basis of the still partially unveiled physiological role of the adrenocorticoid receptor in the hippocampus. We were able to show (Angelucci et al., 1980a) that the distribution of binding capacity in the hippocampal adrenocorticoid system runs parallel to distribution in the same population of rats with adaptive ability in passive or active avoidance tasks.

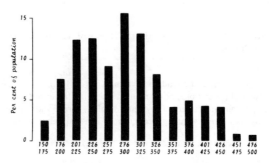

Fig. 1. Distribution of the hippocampal adrenocorticoid receptor system binding capacity, measured as 3H-corticosterone (40 nM) maximally bound in the cytosol of a population of 120 male Wistar rats, 24 hours after adrenalectomy.

The pituitary-adrenocortical mother-offspring unit in the rat is a useful model in perinatal pharmacology with regard to man. It allows the separate study of the effects on the newborn of drugs affecting the hormonal status of the lactating mother, during the stage of completion of brain ontogenesis and maturation of endocrine processes, with no substantial manipulation of the two distinct organisms.This stage in humans is mostly carried out in the last trimester of pregnancy but lasts through the first weeks of lactation.

In preliminary studies we have shown that plasma levels of corticosterone can be reliably increased in the lactating mother by drugs or stress and that a parallel increase is observable in her fetuses or sucklings (Angelucci et al.,1985). Cold stressing pregnant or lactating rats produced a remarkable increase of plasma corticosterone concentration in the mothers, as well as in their fetuses or sucklings, although to a lesser degree in the latter case (Angelucci et al., 1985). Since the lactating mothers were separately stressed for one hour and then returned to the home cage, the parallel increase in plasma concentrations of corticosterone in the sucklings could be due to various factors acting as stressors on the offspring: for instance, absence of the mother or her improper behavior after returning to the cage, although it is known that in the first days of postnatal life stress response is substantially absent (Turner and Taylor,1976).

These patterns show that drugs or environmental factors during pregnancy or lactation can produce maternal hypercorticosteronemia, raising the possibility of their indirect influence on the offspring through hormonal mediation. We have investigated the role of maternal adrenocorticoid hormones with regard to the development of regulation of the pituitary-adrenal axis and adaptive behaviors in the offspring, producing a deficiency (adrenalectomy) or an excess (exogenous administration) of the hormone in the last week of gestation or during lactation in the rat (Angelucci et al., 1983a; Angelucci et al., 1983b).

In the simplest terms, the question was: if corticosterone is present in the milk and penetrates the newborn's body, do variations in this passage disturb its maturational processes, with behavioral and endocrine effects observable in later life?

Corticosterone is indeed present in rat milk. Table 2 shows that corticosterone, at day 8 of lactation was present in rat milk, that it disappeared after adrenalectomy, and re-appeared when the hormone was exogenously replaced. Most remarkably, corticosterone rapidly equilibrates between plasma and milk: in fact, the hormone concentration in milk obtained from rats adrenalectomized immediately after a 30 min cold stress, fell in 30 min from 4.11 to 1.05 micrograms/100 ml. This indicates that intense but short activation of pituitary-adrenocortical secretion in the lactating mother could be reflected in increased glucocorticoid hormone passing to the sucklings.

Table 2. Morning milk and plasma corticosterone concentration (microg/100ml) in the rat.

	MILK	PLASMA
Intact rat	8.8 ± 1.04 (4)	40.1 ± 2.90 (4)
Intact rat, treated with corticosterone (200 microg/ml) in drinking water for 10 days	7.3 ± 1.23 (4)	24.4 ± 3.17 (4)
Rat ad.x at day 14 of pregnancy	0.4 ± 0.02 (2)	0.9 ± 0.03 (2)
Rat ad.x at day 14 of pregnancy and treated with corticosterone (200 microg/ml) in drinking water for 10 days	10.4 ± 1.57 (4)	4.2 ± 0.87 (4)

In parentheses: number of animals.

In order to answer the question of the importance of glucocorticoid hormone in milk, the following manipulations of the pituitary- adrenocortical mother-conceptus relationship were carried out: 'adrenocortical insufficiency' in the mother , by adrenalectomy at day 14 of pregnancy; 'adrenocortical hyperfunction' in the mother, by giving corticosterone 200 microgram/ml in drinking water, starting from day 2 post-partum.

Effects on the adult offspring of the lack of maternal corticosterone during pregnancy and/or lactation

Different types of offsprings were obtained by crossing over litters at birth. To avoid bias from the changing of the litter by itself, crossing over was also carried out within intact and adrenalectomized mothers. Four different groups were so obtained: offspring of an intact mother, lactated by an intact foster mother (i/i); offspring of an intact mother, lactated by an adrenalectomized foster mother (i/a); offsprings of an adrenalectomized mother, lactated by an intact foster mother (a/i); offsprings of an adrenalectomized mother, lactated by an adrenalectomized foster mother (a/a). Weaning of the offspring was carried out at 21-days of age, and groups of the same sex were randomly picked, each of 4 animals, to avoid litter bias.

The development of the regulation of the hypothalamo-pituitary- adrenal axis was apparently affected by lack of maternal corticosterone during pregnancy or lactation, or both. In fact, as shown in table 3, all endocrinally manipulated offspring groups had an increased basal level of corticosterone. However, the adrenocortical response to stress was attenuated in comparison with i/i, particularly in the a/a group of offspring.

Table 3. Morning plasma corticosterone concentration (microg/100 ml) in 90 day-old male rats with lack of maternal corticosterone during pregnancy or lactation, or both.

	BASAL	15' AFTER PSYCHIC STRESS	PERCENTAGE INCREASE
i/i *	1.69 ± 0.22 (20)	9.76 ± 1.73 (13)	477.84 ± 102.86
i/a	$3.49 \pm 0.44^{\circ\circ}$ (20)	7.23 ± 1.08 (8)	$107.09 \pm 31.13!$
a/i	$4.01 \pm 0.67^{\circ\circ}$ (28)	8.25 ± 0.96 (11)	$105.87 \pm 23.87^{\circ}$
a/a	$3.52 \pm 0.41^{\circ\circ}$ (24)	5.82 ± 1.26 (8)	$65.23 \pm 37.78^{\circ}$

Mean values $\pm$ S.E. *: for abbreviations, see the text. $^{\circ\circ}$: $p = 5\%$ vs i/i group, ANOVA. ! and $^{\circ}$: $p = 5$ and 1 % vs i/i group, respectively, Student "t" test. In parentheses: number of animals.

These 'in vivo' findings were accompanied by remarkable variations in the 'in vitro' activities of the three segments of the hypothalamo- pituitary-adrenal axis. In fact (data not shown), basal hypothalamic release of CRF was increased in the a/a group, whereas ACh-stimulated release was reduced or abolished in all the endocrine manipulated groups, in comparison with the i/i group. Basal pituitary release of ACTH in the a/a group was higher than in the i/i, and release stimulated by hypothalamic extract was increased, especially in the i/a and a/i groups of offspring. This increased sensitivity of the pituitary to CRF might be related to the impoverished CRF response of the hypothalamus to stimulation in these two groups. Basal adrenal production of corticosterone was reduced in all three endocrine manipulated groups of offspring, and also reduced, proportionately, was the responsivity to ACTH.

From these 'in vitro' results it appeared, on one hand, that perfect development of the pituitary-adrenal axis depends on a proper and strict sequence of events in the hormonal mother-conceptus relationship during perinatal life; on the other, that disturbances in such development following perinatal endocrine manipulations may be blunted in their consequences in the adult animal by homeostatic mechanisms, possibly through an increased pituitary response to hypothalamic CRH, at least with regard to adaptive responses of the hypothalamo-pituitary- adrenocortical axis.

As shown in table 4, lack of maternal corticosterone during perinatal life profoundly affected binding capacity for corticosterone in the brain. The amount of corticosterone normally present in milk was apparently able to regulate the development of the hypothalamo-pituitary- adrenal axis and the parallel development of the hippocampal glucocorticoid binding capacity in the offspring during completion of the ontogenesis in the first postnatal weeks.

Table 4. The effect of lack of maternal corticosterone during pregnancy or lactation, or both on the hippocampal glucocorticoid receptor system of the adult male rat offspring.

	Bmax fmoles/mg p	Kd nM
i/i*	460 (423-497)	5.9 (5.1-6.7)
i/a	336° (302-371)	5.2 (4.3-6.1)
a/i	453 (428-478)	5.4 (4.9-5.9)
a/a	350° (318-383)	5.3 (4.4-6.1)

Scatchard analysis of 3H-corticosterone binding on hippocampal cytosol from 6 rats; *: for abbreviations, see the text. °: Student 't' test, p = 5 % statistical difference vs i/i group. In parentheses: fiducial limits.

Besides the above endocrine and neurochemical effects, disturbances in the pituitary-adrenocortical mother-conceptus relationship were able to produce sex-dependent behavioral modifications in adult life, most probably through their action on perinatal development of the hippocampus with its glucocorticoid receptor system.

As shown in table 5, passive avoidance tended to be impoverished in the a/a group. It was not affected by the absence of glucocorticoid hormone in milk, but impoverished in the presence of the mother's 'adrenocortical insufficiency' during pregnancy. This indicates that this type of behavior in adulthood may be disturbed by alterations of the pituitary-adrenocortical mother-offspring relationship during prenatal life. Active avoidance behavior, too, in adulthood was affected by endocrine manipulations during perinatal life (Angelucci et al., 1983b).

The above findings are considered to be the consequences on the offspring of the mother's 'adrenocortical insufficiency' during pregnancy or lactation, a condition rather coarsely induced by adrenalectomy. At any rate, the findings point to the importance of glucocorticoid hormone physiologically present in milk in the regulation of maturational processes of the hippocampus, with its hormonal receptor, and of the hypothalamo-pituitary- adrenocortical axis, which become endocrinally and behaviorally manifested in later life.

Table 5. Passive avoidance behavior in 90 day-old rats, with lack of maternal corticosterone during pregnancy or lactation or both.

	MALES	FEMALES
i/i *	50.5	71
	29-110	23-226
	(14)	(13)
i/a	28	90
	6-45	34-179.5
	(11)	(9)
a/i	28.5	27°
	9.5-50	13.5-44.5
	(12)	(13)
a/a	15°	16°
	9.5-42.5	5-45
	(13)	(11)

Median latencies in seconds, with interquartile values. *: for abbreviations, see the text. °: $p = 5\%$ vs i/i group, Mann-Whitney 'U' test. In parentheses: number of animals.

Effects on the adult offspring of the excess of maternal corticosterone during lactation

The same indication was given by 'adrenocortical hyperfunction' in the lactating mother. In fact this condition was able to produce endocrine and behavioral effects on adult offspring.

Exogenous hormone administration to simulate such a hyperfunction did not substantially alter the mother's pituitary-adrenocortical axis: in fact, the day after corticosterone withdrawal, basal plasma levels of the hormone were no different from controls, and were fully characterized by a circadian rhythm (Angelucci et al., 1983b).

In a similar condition during lactation, an increased plasma level of corticosterone was found both in mothers and their 10 day-old sucklings (Angelucci et al., 1983b). When adult, this type of offspring ('corticosterone nursed') showed some sex- dependent anomalies. At 90 days of age, body weight was reduced in males and unchanged in females. Plasma level of corticosterone was significantly reduced in males, but significantly increased in females. As shown in table 6, binding capacity in the hippocampus for glucocorticoid hormone was increased, due to an increase in number of binding sites, both in males and females.

Table 6. The effect of excess of maternal corticosterone during lactation on the corticosterone secretion in adult offspring.

	MALES	FEMALES
Control-nursed	0.44 ± 0.06	1.40 ± 0.30
	(11)	(7)
Corticosterone-nursed	0.23 ± 0.06°	3.30 ± 0.40°°
	(6)	(7)

Mean values ± S.E. of plasma corticosterone (microg/100 ml). ° and °°: $p = 5$ and 1%, respectively, Student 't' test. In parentheses: number of animals.

Table 7. The effect of excess of maternal corticosterone during lactation on the passive avoidance behavior in adult offspring.

	MALES	FEMALES
Control-nursed	58	70
	10-83	31-277
	(15)	(14)
Corticosterone-nursed	280°	250°
	23-300	195-300
	(15)	(14)

Median latencies in seconds, with interquartile values. °: p = 5%, Mann-Whitney 'U' test. In parentheses, number of animals.

Behavioral studies showed that the 90-day-old offspring of mothers with 'adrenocortical hyperfunction' during lactation was in some way anomalous. As shown in table 7, passive avoidance performance was higher than in controls, both for male and female offspring.

These results strongly support the possibility that the mother- conceptus pituitary-adrenal interrelationship in the rat extends beyond birth by way of corticosterone in milk. We have reached the following conclusions: first: – alterations in the pituitary-adrenocortical mother-offspring relationship during perinatal life in the rat may follow variations in the activity of the mother's hypothalamo-pituitary-adrenal axis, possibly in the direction of either an increase or a decrease, which are sensed by the conceptus through the placenta or the milk. Second: – alterations in the pituitary-adrenocortical mother-offspring relationship can lead to remarkable, partially sex-dependent diversification in endocrine activity and behavior of the conceptus at puberty and adulthood. Third: – endocrine and behavioral diversification of the offspring resulting from the loss of physiological attenuation of stress response in the lactating rat, or from increased stress burden during pregnancy, may be one of the factors causing the ample distribution of hippocampal glucocorticoid binding capacity and ability in some adaptive behaviours, found in an otherwise apparently homogeneous population of rats.

References

Angelucci L, Valeri P, Grossi E, Veldhuis HD, Bohus B and De Kloet ER (1980a) Involvement of hippocampal corticosterone receptors in behavioral phenomena. In Brambilla F, Racagni G, De Wied D eds., Progress in Psychoneuroendocrinology. Amsterdam: Elsevier. pp 177-185.

Angelucci L, Valeri P, Palmery M, Patacchioli FR, Catalani A, (1980b) Brain glucocorticoid receptors: correlation of in vivo uptake of corticosterone with behavioral, endocrine and neuropharmacological events. In Pepeu G, Kuhar KJ, Enna SJ eds., Receptor for neurotransmitters and peptide hormones. New York: Raven Press. pp 391- 406.

Angelucci L, Patacchioli FR, Chierichetti C, Laureti S (1983a) Changes in behavior and brain glucocorticoid receptors detected in adult life of corticosterone-nursed rats. In Zbinden G, Cuomo V, Racagni G, Weiss B eds., Application of behavioral pharmacology in toxicology, Raven Press: New York, pp 277-291.

Angelucci L, Patacchioli FR, Chierichetti C, Laureti S (1983b) Perinatal mother-offspring pituitary-adrenal interrelationship in rats: corticosterone in milk may affect in adult life. *Endocrinol Exp* 17: 191- 205.

Angelucci L, Patacchioli FR, Scaccianoce S, Di Sciullo A, Cardillo A, Maccari S (1985) A model for later-life effects of perinatal drug exposure: maternal hormone mediation. *Neurobehav Toxicol Teratol* 7: 511- 517.

Bohus B, De Kloet ER, Veldhuis HD (1982) Adrenal steroids and behavioral adaptation: relationship to brain corticoid receptors. In Ganten D, Pfaff D eds., Adrenal action on the brain. Berlin: Springer-Verlag. pp 108-148.

Krieger DT (1972) Circadian corticosteroid periodicity: critical period for abolition by neonatal injection of corticosteroid. *Science* 178: 1205- 1206.

McEwen BS, Weiss JM, Schwartz LS (1969) Uptake of corticosterone by rat brain and its concentration by certain limbic structures. *Brain Res* 16: 227-241.

Nyakas C (1983) Behavioral effects of infantile administration of glucocorticoid hormones. In Zbinden G, Cuomo V, Racagni G, Weiss B eds., Application of behavioral Pharmacology in Toxicology. New York: Raven Press. pp 265-275.

Olton DS, Johnson CT, Howard E (1974) Impairment of conditioned avoidance in adult rats given corticosterone in infancy. *Dev Psychobiol* 8: 55-61.

Patacchioli FR, Capasso M, Chiappini P, Chierichetti C, Scaccianoce S, Tozzi W, Angelucci L (1983) Variations of hippocampal cytosol glucocorticoid binding capacity as an after effect of physiological increase in brain and blood corticosterone concentrations. In Endroczi E, De Wied D, Angelucci L, Scapagnini U, eds., Integrative neurohumoral Mechanisms, Amsterdam: Elsevier. pp 165-172.

Ramalay JA, Schwartz NB (1980) The pubertal process in the rat: effect of chronic corticosterone treatment. *Neuroendocrinology* 30: 213-219.

Sapolsky RM, Krey LC, McEwen BS (1984) Glucocorticoid-sensitive hippocampal neurones are involved in terminating the adrenocortical stress response. *Proc Natl Acad Sci USA* 81:6174-6177.

Turner BB, Taylor AN (1975) Postnatal corticosterone treatment: effect on reproductive development and open field behavior. *Fed Proc Fedn Am Soc Exp Biol* 34: 302-310.

Turner BB, Taylor AN (1976) Persistent alteration of pituitary-adrenal function in the rat by prepubertal corticosterone treatment. *Endocrinology* 98: 1-9.

Ward IL, Weiss J (1984) Differential effects of maternal stress on circulating levels of corticosterone, progesterone, and testosterone in male and female rats fetuses and their mothers. *Endocrinology* 114: 1635-1644.

COUNTERREGULATION OF STRESS-INDUCED HYPERGLYCEMIA BY THYROTROPIN-RELEASING HORMONE

Shimon Amir and Robin Pollock
Center for Studies in Behavioral Neurobiology, Department of Psychology, 1455 de Maisonneuve Blvd. West, Montreal, Quebec H3H 1M8, Canada,

Hyperglycemia is a frequent metabolic accompaniment of stress, and it is produced by a synergistic stimulatory action of glucagon and catecholamines on glucose production and an inhibitory action of catecholamines on insulin secretion and peripheral glucose utilization. Stress-induced hyperglycemia is most likely adaptive in function for, because the brain receives its energy from circulating glucose, increased metabolic fuel is required during times of stress for normal brain activity. However, if the rise in glucose levels is excessive, deleterious consequences are likely to ensue; glucoregulatory mechanisms most probably also evolved in order to reduce these pathologically elevated sugar concentrations and ensure survival of the organism.

Normoglycemia is maintained through the complex interplay of numerous agents, including the glucoregulatory hormones, glucagon, epinephrine, the glucocorticoids and insulin (Kraus-Friedman, 1984; Vranic et al., 1984), and it is under the controlling influence of various neurotransmitters and peptides that act either upon the endocrine pancreas and liver or at the level of the central nervous system (CNS) (Brown, 1981a; Frohman, 1983; Iguchi et al., 1983; Wood et al., 1983). Whereas in the advent of hypoglycemia most of the hormones and neural factors implicated in glucoregulation are called into play, only one peptide hormone, insulin, serves to counteract hyperglycemia. Insulin is released from pancreatic beta-cells in response to high circulating glucose levels; systemically, it counteracts these excess concentrations by both stimulating glucose utilization in muscle and adipose tissues and by inhibiting glucose output by the liver (Fain, 1984; Grodsky, 1970). The effect of insulin is also achieved centrally through action in the CNS, resulting in diminished glucose levels through the indirect modulation of sugar production and utilization (Amir & Shechter, 1987; Szabo & Szabo, 1972, 1975, 1983). However, during times of stress, the ability of insulin to lower circulating glucose concentrations is compromised by the inhibitory action of the catecholamines.

Thyrotropin-releasing hormone (TRH) is a hypothalamic tripeptide (pGlu-His-Pro-NH2) whose primary physiological function is the regulation of anterior pituitary thyrotropin and prolactin synthesis and secretion (Harris, Christianson, Smith, Fang, Braverman & Vegenakis, 1978; Jackson, 1982). TRH also acts outside the pituitary, being most notably involved in the influence of various autonomic functions such as cardiorespiratory activity, thermoregulation, gastric acid secretion, motor and appetitive functions, and release of pancreatic hormones (Griffiths, 1985; Jackson, 1982; Yarbrough, 1979). Furthermore, TRH has been the target of recent investigations concerning its possible role in glucoregulation

S. Puglisi-Allegra and A. Oliverio (eds.), Psychobiology of Stress, 143–149.
© 1990 *Kluwer Academic Publishers. Printed in the Netherlands.*

(Brown, 1981b; Dolva et al., 1983; Kabayama et al., 1985; Kato & Kanno, 1983; Knudtzon, 1981, 1985; Shen et al., 1985; Usman & Istanbullu, 1986; Usman & Koray, 1984). In our studies, we have observed that central injection of TRH consistently and dose-dependently results in rapid decreases in plasma glucose concentrations in non-diabetic, normoglycemic mice as well as in hyperglycemic mice (Amir et al., 1985a, 1985b).

These experiments involved microinjection of 0.1-$10\mu g$ of TRH into the lateral cerebroventricular system of normally-fed mice; powerful, sustained hypoglycemia that resembled, both in magnitude and in duration, the decrease in plasma glucose observed after systemic insulin challenge, was produced. In order to determine whether this effect of TRH was due exclusively to a local action of the peptide and not to a peripheral effect consequent to escape from the CNS into the systemic circulation, similar doses of TRH were injected systemically; no effect was observed, indicating that the hypoglycemia noted after central injection was indeed a local phenomenon.

The central effect of TRH on hyperglycemia was subsequently studied in a number of experiments. Hyperglycemia was induced in mice by treatment with one of the following agents: glucose, glucagon, morphine, clonidine, 2-deoxyglucose, compound 48/80 and lipopolysaccharide endotoxin. TRH was centrally injected immediately before or at various times after induction of hyperglycemia. It was observed that TRH potently and dose-dependently prevented or reversed, respectively, the rise in plasma glucose concentrations (Amir, 1985/1986; Amir, 1986a; Amir & Butler, 1988; Amir & Rivkind, 1988). Other experiments studied the effects of cerebroventricularly- injected TRH in mice administered prior and consequent to treatment with the stress-inducing hyperglycemic hormones epinephrine, glucagon and corticosterone (Amir, 1986b, 1988a; Amir et al., 1987; Amir et al., 1985b; Butler & Amir, 1986). Here, too, TRH was able to both prevent a rise in plasma glucose levels and to rapidly reverse the hyperglycemia that results from such treatments. Moreover, central TRH prevented the development of hyperglycemia due to electric foot- and anaphylactic shock, immobilization, and endotoxin administration. Systemic injection of TRH at doses similar to those given centrally (0.1-$10\mu g$) had either no effect (0.1-$1\mu g$) or only a mild effect ($10\mu g$), while doses greater than $100\mu g$ completely prevented the hyperglycemic response. All other neuroactive peptides tested in our studies, including somatostatin, ACTH, substance P, SRIF, CRF, MIF-1, enkephalin, alpha-MSH and met-enkephalin failed to block the development of or reverse drug- or stress-induced hyperglycemia, as TRH had (Amir et al., 1987). The ability of TRH to exert these actions was independent of the stimulus used to increase glucose levels: TRH was consistently able to prevent hyperglycemia, whether induced by peripherally or centrally acting agents or by stress, suggesting that the antihyperglycemic effect of TRH does not result from an interaction at the site of action of the hyperglycemic agent but, rather, is the result of an action on a common pathway residing at a level distal to the site(s) mediating the hyperglycemic effect. It was also found that strain, sex, age, nutritional status, diurnal rhythm and environmental temperature had no influence on the ability of TRH to exert its antihyperglycemic action (Amir et al., 1987).

We subsequently studied the role of endogenous insulin in its mediation of the central hypoglycemic effect of TRH. In one set of experiments, TRH was centrally administered to normoglycemic mice, and any changes in circulating insulin levels were recorded. It was found that central TRH elicited a significant, transient increase in plasma insulin concentrations, which paralleled decreases in circulating glucose levels (Amir et al., 1985a). Moreover, mice treated with the adrenergic agonist clonidine demonstrated reversal of drug-induced plasma insulin suppression consequent to central TRH administration, accompanied by an increase in systemic circulatory insulin levels (Amir & Rivkind, 1988). A second set of experiments considered the effects of TRH in insulin-depleted mice, brought about by treatment with alloxan or streptozotocin. Mice treated with a single large dose of the beta-cell toxin alloxan exhibited powerful hyperglycemia, and central TRH at all doses had no effect on plasma glucose levels (Amir et al., 1985b). A similar result was observed in streptozotocin- treated, epinephrine-challenged hyperglycemic mice (Amir et al., 1987). Thus the hyperglycemic effect of TRH apparently depends on the structural and functional integrity of pancreatic beta-cells, and is accompanied by an increase in circulating insulin levels, suggesting insulin as the peripheral agent mediating this effect.

Consequently, we investigated the central mechanisms mediating the antihyperglycemic effect of TRH. It was hypothesized that TRH initiates this effect by activating specific CNS receptors; to test this, we compared the central effect of native TRH on plasma glucose to that of many TRH structural analogs and fragments exhibiting diminished or absent binding affinity to TRH receptors. These experiments revealed that peptides differing from TRH by a single amino acid residue (e.g., pGlu-His, His-Pro, Glu-His-Pro-NH$_2$, pGlu-Phe-Pro-NH$_2$ and pGlu-His-Gly-NH$_2$) are unable to bring about lowered plasma glucose levels in epinephrine-treated, hyperglycemic mice. As well, peptides resembling TRH in amino acid composition but devoid of any substantial binding affinity to the receptor as compared to the native peptide are as active as TRH (i.e., pGlu-His-Pro-OH) or even more so (i.e., pGlu-His-Pro-Gly-NH$_2$) in reversing hyperglycemia (Amir & Butler, 1988; Amir & Rivkind, 1988; Amir et al., 1988). Thus the structural integrity of the peptide is essential for the antihyperglycemic action of TRH, although it is dissociated from interaction with the 'classic' TRH receptor.

The involvement of the calcium- and phospholipid-dependent second messenger, protein kinase C, was also investigated for its influence on the initiation of TRH action in hyperglycemia. Previously, this agent was shown to be involved in the mechanism of TRH stimulation of pituitary prolactin release (Albert, Wolfson & Tashjian, 1985; Drust & Martin, 1984; Dufy, Jaken & Barker, 1987; Kolesnick, 1987). Experiments were conducted in which the effect of direct activation of protein kinase C on the hypoglycemic effect of TRH was evaluated. It was found that direct activation of protein kinase C by central injection of the phorbol ester 12-O-tetradecanoylphorbol-13-acetate (TPA, 0.1-1.0μg) strongly and specifically enhanced the hypoglycemic effect of TRH; central injection of the inactive phorbol ester 4-alpha-phorbol, conversely, had no effect on central TRH action (Amir, 1988b). These results suggest that a protein kinase C-regulated mechanism is involved in modulating the effect of TRH, whether occurring at the locus of TRH binding or further downstream, at the level of the neural substrate activated by TRH after it binds to the receptor. Interestingly, some studies have shown that insulin may be capable of activating protein kinase C in the CNS (Akers & Routtenberg, 1984; Amir & Shechter, 1988); thus a potentially fruitful area of investigation would concern an examination of the possibility that the effect of insulin in enhancing the central glucoregulatory action of TRH, like that of TPA, occurs via a protein kinase C-dependent modulatory mechanism.

A further source of inquiry concerning the mediation of the central effect of TRH in hyperglycemia was whether pituitary factors might be involved. As is well known, an important function of TRH is to stimulate hormone release from the pituitary; additionally, a number of pituitary hormones purportedly responsive to TRH, such as ACTH, have been shown to stimulate insulin secretion (Beloff-Chain, Edwardson & Hawthorn, 1975; Genuth

& Leibovits, 1965; Kuzuya, 1962). In a series of experiments designed to ascertain whether the hypoglycemic action of TRH is dissociated from the pituitary, the central effects of native TRH were compared to those of several TRH analogs having TRH-like CNS activity profiles, but devoid of hypophysiotropic influences. As well, TRH was administered centrally to hypophysectomized mice. It was found that the TRH analogs pGlu-N-Val-Pro-NH_2, CG 3509 and DN 1417 are at least as, or even more, effective than is the native peptide at blocking glucagon-induced hyperglycemia; moreover, hypophysectomy did not impair TRH from its hypoglycemic action (Amir, 1988a). These results may be interpreted as meaning either that the central effect of TRH in blocking hyperglycemia is dissociated from its action on the pituitary, or that pituitary factors do not play a role in mediating this effect of TRH.

The autonomic system has also been studied as a possible mediator of central TRH-induced hypoglycemia. When TRH is administered centrally in experimental animals, stimulation of the sympathetic and parasympathetic pathways ensues. Moreover, autonomic mechanisms play a central role in the regulation of pancreatic hormone secretion (Miller, 1981; Rohner-Jeanrenaud et al., 1983), and central TRH has been shown to activate the autonomic efferent pathways, triggering the release of pancreatic hormones, including insulin (Kato & Kanno, 1983; Knudtzon, 1981, 1985). We subsequently studied the effects of pharmacological treatments that interfere with central sympathetic and parasympathetic outflow on the antihyperglycemic action of TRH. In one series of experiments, the effect of the muscarinic cholinergic antagonist atropine methyl nitrate in blocking parasympathetic outflow served to diminish, although not prevent, the hypoglycemic action of central TRH in normoglycemic mice (Amir & Butler, 1988; Amir et al., 1988a). Treatment with atropine in hyperglycemic mice only partially blocked the central effect of the peptide. Furthermore, selective blockade of the sympathetic system with various agents or surgical denervation of the adrenal glands had no effect on TRH action in both normo- and hyperglycemic mice; even combined sympathetic/parasympathetic blockade failed to completely abolish the central effect of TRH in blocking hyperglycemia (Amir & Butler, 1988; Amir et al., 1988a; Amir & Rivkind, 1988; Amir et al., 1985a, 1985b). Thus autonomic mechanisms play a variable role in mediating this central effect of TRH. The data suggest that, in normoglycemic mice, parasympathetic mechanisms are responsible for TRH action, and that in hyperglycemic mice, both parasympathetic and sympathetic mechanisms are involved in mediating TRH activity. Nonetheless, it should be noted that even complete blockade of these two neural systems failed to completely inhibit TRH-induced hypoglycemia in hyperglycemic mice. This suggests that non-neuronal factors contribute as well to TRH action in hyperglycemia; for example, peptides capable of stimulating insulin release, such as those demonstrated in the hypothalamus (Bobbioni & Jeanrenaud, 1983; Brouwer et al., 1982; Hill et al., 1977; Idahl & Martin, 1971; Miller, 1981; Rohner-Jeanrenaud et al., 1983), may be implicated.

Earlier studies on the neuroendocrinology of TRH have indicated that stressful stimuli may increase CNS TRH content (Arancibia, 1983), and that experimental manipulations of the circulating glucose levels are capable of modifying TRH turnover rates in the hypothalamus (Gonzalez et al., 1980; Kardon et al., 1977; Leung et al., 1975; Mitsuma & Nogimori, 1982; Pastor & Jolin, 1983). These findings led us to study the possibility that the effect of endogenous TRH in lowering stress-induced glucose levels may represent a physiological function of central TRH neurons. Unlike insulin, TRH cannot directly stimulate glucose metabolism in muscle and adipose tissue; however, it can influence it indirectly by stimulating insulin release via action in the CNS. We postulated that, in order to begin this stimulatory process, TRH mechanisms in the CNS are activated by stress or by the hyperglycemia that accompanies it and, in turn, protect against the potential harmful effects of excessively elevated glucose concentrations by promoting insulin secretion secondary to reversing the insulin release-inhibiting action of the catecholamines. We experimentally induced hyperglycemia in mice by treatment with 2-deoxyglucose and studied the effects of immunoneutralization of TRH in the CNS by means of central anti-TRH serum injection upon recovery from the hyperglycemic state (Amir & Jackson, 1988). The idea was to disturb the CNS TRH system through antibody treatment, which was hypothesized to impair insulin responses to, and delay recovery from, experimental hyperglycemia. Results indicated that immunoneutralization of central TRH significantly prolongs hyperglycemia or, alternatively, retards the reversal of hyperglycemia following 2-deoxyglucose treatment ; glucose levels were unaffected in control animals. As well, central injection of the antiserum had no effect on the development of hyperglycemia consequent to 2-deoxyglucose treatment. Thus reversal of stress-induced hyperglycemia may depend to some degree on the functional integrity of TRH mechanisms in the CNS. In contrast, TRH may not be involved in the mechanism mediating the development of the hyperglycemic response to stress.

In conclusion, the findings reported here demonstrate that TRH exhibits the pharmacological effect of reversing drug- or stress-induced hyperglycemia. Moreover, this effect may reflect a physiological function of endogenous TRH, serving to protect against the excessive and potentially dangerous build-up of circulating glucose levels during stress.

References

Akers, R. F., & Routtenberg, A. (1984). Brain protein phosphorylation in vitro: Selective substrate action of insulin. *Life Science*, 35, 809-813.

Albert, P. R., Wolfson, G., & Tashjian, A. H. (1988). Diacylglycerol increases cytosolic free Ca^{2+} concentration in rat pituitary cells. Relationship to thyrotropin-releasing hormone action. *Journal of Biol. Chemistry*, 262, 6577-6581.

Amir, S. (1985/1986). Thyrotropin-releasing hormone reverses morphine-induced hyperglycemia in mice. *Alcohol and Drug Research*, 6, 147.

Amir, S. (1986a). In J. W. Holaday, H. Loh, A. Herz & P.-Y. Law (Eds.), Advances in endogenous and exogenous opioid peptides. *NIDA Research Monographs*, 75, 375-379.

Amir, S. (1986b). Endorphins contribute to loss of glucose homeostasis in anaphylactic shock. In J. W. Holaday, H. Loh, A. Herz & P.-Y. Law (Eds.), Advances in endogenous and exogenous opioid peptides. *NIDA Research Monographs*, 75, 371-374.

Amir, S. (1988a). Thyrotropin-releasing hormone (TRH) blocks glucagon-induced hyperglycemia in mice: Dissociation of the hyperglycemic and pituitary actions of TRH. *Brain Research*, 455, 201-203.

Amir, S. (1988b). TPA (12-O-tetradecanoylphorbol-13-acetate) enhances the central hypoglycemic action of thyrotropin-releasing hormone in mice. *Brain Research*, 450, 369-372.

Amir, S., & Butler, P. D. (1988). Thyrotropin-releasing hormone blocks neurally-mediated hyperglycemia through central action. *Peptides*, 9, 31-35.

Amir, S., Harel, M., & Rivkind, A. I. (1987). Thyrotropin-releasing hormone potently reverses epinephrine-induced hyperglycemia in mice. *Brain Research*, 435, 112-122.

Amir, S., & Jackson, I. M. D. (1988). Immunological blockade of endogenous thyrotropin-releasing hormone impairs recovery from hyperglycemia in mice. *Brain Research.*, 462, 160-162.

Amir, S., & Rivkind, A. I. (1988). Prevention of clonidine-stimulated hyperglycemia by thyrotropin-releasing hormone. *Peptides*, 9, 527-531.

Amir, S., Rivkind, A. I., & Harel, M. (1985a). Central thyrotropin-releasing hormone induced systemic hypoglycemia in mice. *Society for Neuroscience Abstracts*, 11, 885.

Amir, S., Rivkind, A. I., & Harel, M. (1985b). Central thyrotropin-releasing hormone elicits systemic hypoglycemia in mice. *Brain Research*, 344, 387-391.

Amir, S., & Shechter, Y. (1987). Centrally-mediated hypoglycemic effect of insulin: Apparent involvement of specific insulin receptors. *Brain Research*, 418, 152-156.

Amir, S., & Shechter, Y. (1988). Apparent involvement of protein kinase C in the central glucoregulatory action of insulin. *Brain Research*, 450, 272-279.

Arancibia, S., Tapia-Arancibia, L., Assenmacher, I., & Astier, H. (1983). *Neuroendocrinology*, 37, 225-228.

Beloff-Chain, A., Edwardson, J. A., & Hawthorn, J. (1975). Influence of the pituitary gland on insulin secretion in the genetically obese (ob/ob) mouse. *Journal of Endocrinology*, 65, 109-116.

Bobbioni, E., & Jeanrenaud, B. (1983). A rat hypothalamic extract enhances insulin secretion in vitro. *Endocrinology*, 113, 1958-1962.

Brouwer, G. H., Lamptey, M. S., & Martin, J. M. (1982). Isolation and partial characterization of insulin-glucagon liberation from bovine hypothalamus. *Life Science*, 30, 703-710.

Brown, M. (1981a). Neuropeptides: Central nervous system effects on nutrient metabolism. *Diabetologia*, 20, 299-304.

Brown, M. (1981b). Thyrotropin releasing factor: A putative CNS regulator of the autonomic nervous system. *Life Sciences*, 28, 1789-1795.

Butler, P. D., & Amir, S. (1986). Thyrotropin-releasing hormone blocks stress-induced hyperglycemia in mice. *Society of Neuroscience Abstracts*, 12, 219.

Dolva, L. O., Hanssen, K. F., Flaten, O., Hanssen, L. E., & von Schenk, H. (1983). The effect of thyrotropin-releasing hormone (TRH) on pancreatic hormone secretion in normal subjects. *Acta Endocrinol.*, 102, 224-230.

Drust, D. S., & Martin, F. J. (1984). Thyrotropin-releasing hormone rapidly activates protein phosphorylation in GH3 pituitary cells by a lipid-linked, protein kinase C-mediated pathway. *Journal of Biol. Chemistry*, 259, 14520-14530.

Dufy, B., Jaken, S., & Barker, J. L. (1987). Intracellular Ca^{2+}-dependent protein kinase C activation mimics delayed effects of thyrotropin-releasing hormone on clonal pituitary cell excitability. *Endocrinology*, 121, 793-802.

Fain, J. N. (1984). Insulin secretion and action. *Metabolism*, 33, 672-679.

Frohman, L. A. (1983). CNS peptides and glucoregulation. *Annual Review of Physiology*, 45, 95-107.

Genuth, S., & Leibovits, H. E. (1965). Stimulation of insulin release by corticotropin. *Endocrinology*, 65, 109-116.

Gonzalez, C., Montoya, E., & Jolin, T. (1980). Effect of streptozotocin on the hypothalamic-pituitary-thyroid axis in the rat. *Endocrinology*, 107, 2099-2103.

Griffiths, E. C. (1985). Thyrotropin-releasing hormone: Endocrine and central effects. *Psychoneuroendocrinology*, 10, 225-235.

Grodsky, G. M. (1970). Insulin and the pancreas. *Vitam. Horm.*, 28, 37-101.

Harris, A. R. C., Christianson, D., Smith, M. S., Fang, S. L., Braverman, L. E., & Vegenakis, A. G. (1978). The physiological role of thyrotropin- releasing hormone in the regulation of thyroid stimulating hormone and prolactin secretion in the rat. *Journal of Clinical Investigation*, 61, 441-448.

Hill, D. E., Mayes, S., DiBattista, D., Lockhart-Ewart, R., & Martin, J. M. (1977). Hypothalamic regulation of insulin release in Rhesus monkeys. *Diabetes*, 26, 726-731.

Idahl, L. A., & Martin, J. M. (1971). Stimulation of insulin release by a ventrolateral hypothalamic factor. *Journal of Endocrinology*, 51, 601-602.

Iguchi, A., Sakamoto, N., & Burleson, P. D. (1983). The effects of neuropeptides on glucoregulation. *Advances in Metabolic Disorders*, 10, 421-434.

Jackson, I. M. D. (1982). Thyrotropin-releasing hormone. *New England Journal of Medicine*, 306, 145-155.

Kabayama, Y., Kato, Y., Tojo, K., Shimatsu, A., Ohta, H., & Imura, N. (1985). Central effects of DN 1417, a novel TRH analog, on plasma glucose and catecholamines in conscious rats. *Life Sciences*, 36, 1287-1294.

Kardon, F., Marcus, R. J., Winokur, A., & Utiger, R. D. (1977). Thyrotropin-releasing hormone content of rat brain and hypothalamus: Results of endocrine and pharmacologic treatments. *Endocrinology*, 100, 1604-1609.

Kato, Y., & Kanno, T. (1983). Thyrotropin-releasing hormone injected intracerebroventricularly in the rat stimulates exocrine pancreatic secretion via the vagus nerve. *Regul. Peptides*, 7, 347-356.

Knudtzon, J. (1981). Thyrotropin-releasing factor: A putative CNS regulator of the autonomic nervous system. *Hormone Metabol. Research*, 13, 371-375.

Knudtzon, J. (1985). Involvement of the autonomic nervous system in the in vivo TRH-induced increases in the plasma levels of glucagon, insulin and glucose in rabbits. *Hormone Metabol. Research*, 17, 53-57.

Kolesnick, R. N. (1987). Thyrotropin-releasing hormone and phorbol esters induce phosphatidylcholine synthesis in GH3 pituitary cells. Evidence for stimulation via protein kinase C. *Journal Biol. Chem.*, 262, 14525-14530.

Kraus-Friedman, N. (1984). Hormonal regulation of hepatic gluconeogenesis. *Physiology Review*, 64, 170-258.

Kuzuya, T. (1962). Regulation of insulin secretion by the central nervous system. II. The role of the hypothalamus and the pituitary gland upon insulin secretion. *Journal of the Japanese Society of Internal Medicine*, 51, 65-74.

Leung, Y., Guansing, A. R., Ajlouni, K., Hagen, T. C., Rosenfeld, P. S., & Barboriak, J. J. (1975). The effect of hypoglycemia on hypothalamic thyrotropin-releasing hormone (TRH) in the rat. *Endocrinology*, 97, 380-384.

Miller, R. E. (1981). Pancreatic neuroendocrinology: Peripheral neural mechanisms in the regulation of the islets of Langerhans. *Endocrine Review*, 2, 471-494.

Mitsuma, T., & Nogimori, T. (1982). Effects of streptozotocin-induced diabetes mellitus on hypothalamic-pituitary-thyroid axis in rats. *Endocrinol. Japon.*, 29, 695-700.

Pastor, R. M., & Jolin, T. (1983). Peripheral metabolism and secretion rate of thyrotropin in streptozotocin-diabetic rats. *Endocrinology*, 112, 1454-1459.

Rohner-Jeanrenaud, F., Bobbioni, E., Ionescu, E., Sauter, J. F., & Jeanrenaud, B. (1983). Central nervous system regulation of insulin secretion. *Advances in Metabolic Disorders*, 10, 193-220.

Shen, D. C., Lin, M. T., & Shian, L. R. (1985). Thyrotropin-releasing hormone-induced hyperglycemia: Possible involvement of cholinergic receptors in the lateral hypothalamus. *Neuroendocrinology*, 41, 499-503.

Szabo, O., & Szabo, A. J. (1972). Evidence for an insulin-sensitive receptor in the central nervous system. *American Journal of Physiology*, 223, 1349-1353.

Szabo, A. J., & Szabo, O. (1975). Influence of the insulin sensitive central nervous system glucoregulator receptor on hepatic glucose metabolism. *Journal of Physiology*, 253, 121-133.

Szabo, A. J., & Szabo, O. (1983). Insulin injected into CNS structures or into the carotid artery: Effect on carbohydrate homeostasis of the intact animal. *Advances in Metabolic Disorders*, 10, 385-400.

Usman, A., & Koray, Z. (1984). The effect of thyrotropin-releasing hormone (TRH) on the peripheral plasma levels of pancreatic glucagon and insulin in man. *Hormone Metabol. Research*, 16, 263-264.

Usman, A., & Istanbullu, S. (1986). Lack of effect of thyrotropin- releasing hormone (TRH) on the peripheral plasma levels of pancreatic glucagon in man. *Hormone Metabol. Research*, 18, 256-259.

Vranic, M., Gauthier, C., Bilinski, D., Wasserman, D., El Tayeb, K., Hetenyi, G., & Lickley, H. L. A. (1984). Catecholamine responses and their interactions with other glucoregulatory hormones. *American Journal of Physiology*, 247, E145-E156.

Wood, S. M., Polak, J. M, & Bloom, S. R. (1983). Neuropeptides in the control of the islets of Langerhans. *Advances in Metabolic Disorders*, 10, 401-420.

Yarbrough, G. G. (1979). On the neuropharmacology of thyrotropin- releasing hormone (TRH). *Progress in Neurobiology*, 12, 291-312.

NORADRENERGIC RECEPTOR MECHANISMS IN STRESS ADAPTATION

Eric A. Stone
Department of Psychiatry
New York University School of Medicine
New York, NY 10016

The noradrenergic system is one of the key neuronal systems involved in the central response to stress. It is known to be activated by most forms of stress, both emotional and physical, and is thought to participate in a variety of stress-related functions such as arousal, vigilance and anxiety and the coordination of various neuroendocrine and autonomic responses (Aston-Jones, 1985; Rasmussen et al., 1986; Svensson, 1987).

Much prior research on the noradrenergic system in stress has concerned presynaptic mechanisms involved in the synthesis and release of the neurotranmsitter, norepinephrine (NE). Relatively few studies have examined postsynaptic processes under these conditions. To understand more about the effects of stress on the latter processes therefore our group has undertaken a number of studies of postsynaptic receptor function in the CNS of stressed rats. Although primarily directed at the processes underlying stress, these studies have also helped to clarify general mechanisms of the regulation of central noradrenergic receptor function.

Most noradrenergic receptors in the CNS are either directly coupled to adenylate cyclase ($\beta 1$, $\beta 2$ and $\alpha 2$) or are known to indirectly modulate the activity of the enzyme ($\alpha 1$). Therefore to study the effects of stress on postsynaptic noradrenergic receptor function we examined the production of cAMP in response to catecholamines in brain slices from stressed animals.

Receptor desensitization during stress

Our initial studies were designed to determine if stress, by virtue of its ability to release brain NE, caused a desensitization of central noradrenergic receptors. For this purpose we examined responses from rats subjected either to acute or repeated forms of stress (footshock or restraint stress). We found that acute stress had relatively little effect but repeated stress (7-10 days, 1-2 hr/day) did produce a significant desensitization of the cAMP response to NE (Stone, 1978; 1979; 1981; Stone et al., 1984). This desensitization occurred in widely separated regions of the forebrain (cortex and hypothalamus) and appeared to be the result of alterations in postsynaptic receptor function since it could not be accounted for by changes in the inactivation of NE.

To determine which noradrenergic receptor had been altered by stress we undertook studies using selective beta and alpha receptor agonists and blocking agents (Stone et al., 1985; 1986). These studies showed that beta receptor function was only slightly reduced by stress but that there was a marked desensitization of a population of alpha-1 receptors which are involved in the augmentation of the beta-cAMP responses (so called cAMP-linked

151

S. Puglisi-Allegra and A. Oliverio (eds.), Psychobiology of Stress, 151–160.
© 1990 *Kluwer Academic Publishers. Printed in the Netherlands.*

alpha-1 receptors).

The fact that alpha-1 receptors were desensitized by stress more than beta receptors was somewhat surprising in view of the fact that beta receptors are known to be much more sensitive to the desensitizing effects of excessive catecholamine release, which presumably was the mechanism by which stress was causing the desensitization. This prompted us therefore to undertake direct studies of this mechanism (Stone et al., 1987a,b). We first examined whether the alpha receptors would be affected by alterations in brain NE release as occurs during stress. It was found, however, that unlike beta receptors, the cAMP-linked alpha-1 receptors were not altered by changes in NE availability induced by chronic administration of a reuptake inhibitor (desmethylimipramine, DMI) or by central denervation with 6-OHDA. This indicated that brain NE release during stress was unlikely to be the factor causing the desensitization. Since previous studies had suggested that pituitary-adrenal stress hormones have regulatory effects on noradrenergic receptors (Mobley et al., 1983; Duman et al., 1985) we next examined the actions of the latter compounds. It was found that chronic administration of either ACTH or corticosterone produced the same selective desensitization of cortical alpha-1 receptors as found after stress and that prior adrenalectomy attenuated the stress-induced effect (Stone et al., 1986; 1987a,b). These findings showed therefore that the alpha-1 desensitization was due to corticosterone release during stress and further confirmed the difference in regulation between beta and alpha-1 receptors during stress.

The finding that corticosterone release was responsible for the alpha-1 desensitization also helped to clarify the distribution of the effect in the brain. The glucocorticoid receptors that are believed to be activated by corticosterone during stress (type II) are known to have a widespread distribution throughout the forebrain (McEwen et al., 1986). Since the alpha-1 receptors that are linked to cAMP responses also have a widespread forebrain distribution (Stone & Herrera, 1986) this suggests that the stress-induced desensitization may involve the entire forebrain. In support of this, as discussed above, we have found that the desensitization occurs both in the hypothalamus and cerebral cortex, two widely separated forebrain regions and also that a similar desensitization effect occurs in the hippocampus after ACTH administration (Stone et al., 1987b).

In addition to its widespread anatomical localization, the alpha-1 desensitization also appears to have a broad biochemical influence as well. The above cAMP-linked alpha-1 receptors are known to modulate cAMP responses not only to beta receptor stimulation but also to stimulation of purinergic, histaminergic and VIPergic receptors presumably mediating interactions between the noradrenergic and these other neuronal systems (Daly, 1977; Magistretti & Schoderet, 1985). This suggests that stress may cause a decreased alpha-potentiation of cAMP responses in a variety of central receptor systems. In support of this we have found that there is a reduced potentiation of cAMP responses to adenosine and histamine as well as beta adrenergic receptor stimulation in the cortices of stressed rats (Stone et al., 1986).

The widespread neuroanatomical and biochemical nature of the alpha-1 desensitization suggests that the latter may have important consequences for long term behavioral reactions or adjustments to stress. The nature of these consequences however is not presently known because very little is established regarding the behavioral functions of cAMP-linked alpha-1 receptors. Although some alpha-1 receptors in the forebrain have been shown to have an activating effect on behavior during stress (Pichler & Kobinger, 1985), it is not clear if the alpha-1 receptors that mediate this behavioral activation are identical to the ones that modulate cAMP responses. There is, in fact, some evidence that they are different since the behaviorally active alpha-1 receptors are apparently fully activated by certain alpha-1 agonists (phenylephrine and methoxamine) which are partial agonists at the cAMP-linked receptors (Johnson & Minneman, 1986). Further research therefore will be necessary to define the behavioral function of the cAMP-linked receptors and to establish the behavioral significance of their desensitization during chronic stress.

In concluding this section it should be noted that the above studies have also raised the question of why beta adrenergic receptors are not desensitized more strongly during chronic stress in view of the high release rate of brain NE during stress. Although some authors have reported substantial reductions in brain beta receptors after stress most investigators have found weak effects in agreement with our results (see Stone, 1983a for review). Previous research has suggested two factors that may be responsible for this lack of effect, insufficient NE release and the release of substances that protect beta receptors from desensitization. As far as the first factor is concerned it has been found by Duncan et al., 1985 that pretreatment of rats with the NE reuptake inhibitor, desmethylimipramine, prior to stress (forced swim) leads to a rapid and significant down regulation of cortical beta receptors in rats. This suggests that processes of inactivation of NE or inhibition of NE release may be, in part, responsible for this resistance. With regard to the second factor, it has been found that glucocorticoid administration results in a prevention or reversal in various peripheral tissues of the desensitization of beta receptors caused by prolonged catecholamine exposure (see Davies & Lefkowitz, 1984). We have confirmed that this protective effect of corticosterone also occurs in the brain (see next section). Therefore it is possible that the release of corticosterone during stress is also a factor contributing to the above resistance and that the same factor that is responsible for the desensitization of alpha-1 receptors during stress actually protects beta receptors from desensitization. This differential regulation of beta and alpha-1 receptors during stress and corticosterone treatment strongly suggests that beta and alpha-1 receptor populations are only partially overlapped in their interaction and that substantial portions of their populations are either differentially localized or receive different endogenous inputs. The precise cellular localization of these receptors and the identity of their endogenous agonist(s) would therefore appear to be fertile areas for future investigations.

Measurement of receptor function in vivo

A second aspect of our research on central noradrenergic receptor function has concerned the development of a new method to study the activity of these receptors in vivo. All of our previous studies on the effects of stress on brain receptor function have utilized an in vitro preparation to study receptor function, i.e., the brain slice preparation. While this preparation has provided valuable data on various aspects of receptor function in isolation it has not been capable of providing data on overall neurotransmission at these receptors as caused by the release of endogenous NE because it lacks intact, functionally active noradrenergic neurons. This is unfortunate because stress, in addition to its postsynaptic effects, is also known to produce changes in the synthesis and release of NE (Stone, 1975; Weiss et al., 1975; Anisman & Zacharko, 1986) which will, of course, alter neurotransmission at these receptors. In order to measure overall neurotransmission at central noradrenergic receptors during stress it is necessary to obtain measures of receptor activity in vivo in unanesthetized animals. This has not been feasible in the past however because available techniques for in vivo receptor measurement are not generally applicable to behavioral studies. The latter methods involve either single cell electrophysiological recording (Bickford-Wimer et al., 1987) or assays of the in vivo accumulation of cAMP in brain tissue (Kant et al., 1981) and require either general anesthesia or stressful microwave fixation of the brain. For obvious reasons neither anesthesia nor stress is compatible with behavioral investigations. Furthermore the electrophysiological measure is applicable to neurons only whereas large percentages of noradrenergic receptors are located on glial cells (Zahniser et al., 1979).

The above considerations prompted us to undertake the testing and development of new techniques for measuring the activity of noradrenergic receptors in the intact brain. One method that we have tried that has yielded promising results involves the collection of extracellular brain cAMP by microdialysis. cAMP, formed in response to the activation of noradrenergic and other receptors, is known to be released in small quantities in proportion to its intracellular concentration into the extracellular fluid of the brain (Barber & Butcher, 1983; Lazareno et al., 1985; Stoof & Kebabian, 1981; Egawa et al., 1988). A number of years

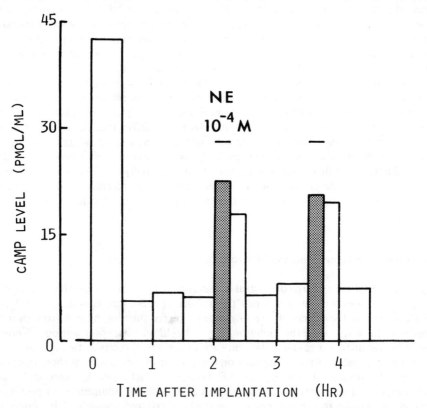

Fig. 1. Increases in extracellular cAMP levels in the frontal cortex after infusion of NE via a microdialysis probe. NE, 10^{-4} M, was infused for 15 min. at 2 and 3.5 hr after implantation. Collection periods were either 15 or 30 min. cAMP was assayed by RIA.

ago, Korf et al. (1976) and subsequently Schoener et al. (1979) showed that this extracellular cAMP was detectable using a push pull cannula and furthermore that it showed increases in level in response to the activation of brain dopaminergic and noradrenergic receptors by catecholamines. The push-pull cannula however has the disadvantage that it produces adverse physiological effects because it exposes tissues to direct fluid flow. In recent years, however, it has become possible to sample compounds in the brain's extracellular fluid without direct fluid flow by the use of microdialysis. We therefore attempted to determine if Korf et al's findings could be replicated with the microdialysis technique.

The results of our initial experiments with this technique have been highly promising (Egawa et al., 1988; Stone et al., 1988). We have shown using a sensitive RIA that cAMP is detectable in the extracellular fluid of the frontal cortex of urethane- anesthetized as well as unanesthetized rats and also that it shows a dose-dependent increase in response to infusion of NE via the probe (see Fig. 1). Furthermore, we have found that the latter response is blocked by infusion of a beta receptor blocking agent indicating that it results from activation of the latter receptors.

In preliminary studies we have also shown that the microdialysis technique can be used to demonstrate certain properties of brain noradrenergic receptors that have been observable hitherto only in vitro or ex vivo. For example, we have been able to show that central stimulation of alpha receptors with 6-fluoronorepinephrine produces a significant potentiation (2.5 fold) of the cAMP response to infusion of adenosine (Stone et al., In press). The latter effect had only been observed previously in brain slices and had not been reported to occur in cultured neurons or glial cells (McCarthy & DeVellis, 1979). Other effects that we have been able to observe in vivo include the desensitization of beta receptors in response to prolonged infusion of a beta agonist and the prevention of this desensitization by prior administration of corticosterone (see Fig. 2). Prior studies of the desensitization of brain beta receptors have been restricted to in vitro or ex vivo conditions. The present results indicate that the phenomenon can now be studied in the intact brain as well. The reversal of beta desensitization by glucocorticoid administration had been demonstrated previously only in peripheral tissues (Davies & Lefkowitz, 1984). The fact that it occurs in the brain as well suggests that it may be an important factor in the physiological maintenance of noradrenergic neurotransmission during times of chronic stress.

The above findings indicate therefore that the microdialysis-cAMP technique is a useful technique for studying receptor function in the intact brain and may eventually enable us to monitor ongoing neurotransmission in the central noradrenergic system during various behavioral states and stress.

Role of noradrenergic receptor activation in adaptation

Most of the studies on stress described above have been concerned with the role of receptor desensitization in adaptation. Another aspect of noradrenergic receptor function that may play an even more important role in this process is prolonged receptor activation. Considerable evidence has accumulated indicating that chronic activation of noradrenergic receptors (mainly beta which are protected from desensitization) during stress leads to a variety of adaptive changes in various tissues. In peripheral organs chronic activation produces various trophic and metabolic changes which generally enhance organ output and aid stress survival (Stone, 1983b) while in the CNS it may be involved in adaptive phenomena

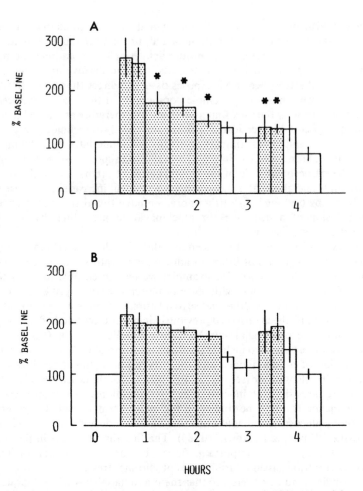

Fig. 2. Desensitization of the in vivo cAMP response to isoproterenol (ISO) (A) and its prevention by corticosterone administration (B) as measured by microdialysis probe in the frontal cortex. ISO, 10^{-4} M was infused via the probe for 2 hr. Corticosterone (10 mg/kg) was injected s.c. one hr prior to infusion. * differs from first 15 min response to ISO at $p < .01$.

involved in learning (Harley, 1987; Gervais et al., 1988), the recovery of function after injury (Feeney & Sutton, 1987), tolerance to drugs (Szabb et al., 1988) and the long term regulation of the responsiveness of other central neuronal systems (Green et al., 1980). These findings led us to propose a number of years ago that an important function of the central noradrenergic system, in addition to its acute behavioral functions, is the mediation of long-term changes in the brain that underlie the process of adaptation to stress (Stone, 1983b). Recently we performed an experiment to test this hypothesis in which we determining if enhancing the release of brain NE during stress by pharmacological means would facilitate adaptation to stress. In this experiment rats were given the NE reuptake inhibitor, DMI, to block the inactivation of brain NE shortly before they were subjected to restraint stress on each day of a 7 day period. The rate of adaptation to the stress was measured from the rate of recovery of exploratory behavior in an open field. We have shown that exploratory behavior, like a variety of other behavioral functions, is impaired initially by stress and recovers gradually as animals undergo adaptation (Trullas & Stone, unpublished findings; Weiss et al., 1975). It was found that administration of the reuptake inhibitor did cause a more rapid or more complete degree of behavioral recovery over the stress period (see Fig. 3). After 7 days of stress the DMI treated rats showed close to a 100% recovery of prestress exploratory levels whereas the vehicle treated rats showed only a 50% recovery. The greater recovery in the DMI treated group does not appear to be the result of an overall reduction in sensitivity to stress since the initial degree of impairment in these animals was actually greater than in vehicle treated controls. The latter enhancement of impairment after acute stress is in accord with a previous suggestion of ours that part of the initial behavioral deficit caused by stress is the result of excessive NE liberation (Stone, 1982).

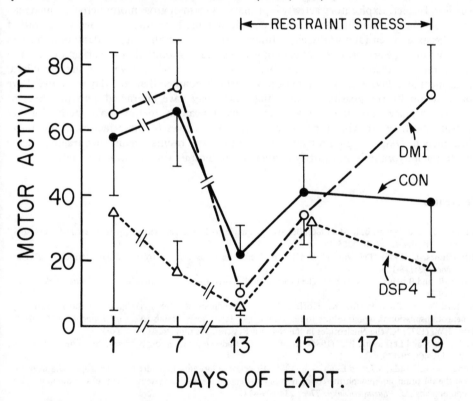

Fig. 3. Effect of treatment with DMI or DSP4 on the recovery of open field behavior during repeated restraint stress. Rats were restrained 2 hr/d on days 13-19. DMI, 10 mg/kg, was injected on days 7-19, 1 hr prior to stress. The rate of recovery of the DMI treated rats was significantly greater than that of vehicle injected controls.

The above results therefore suggest that chronic noradrenergic activity during stress may facilitate the development of long term adaptive processes. However it is not possible to conclude this with certainty at the present time since the above experiment is open to a number of alternative explanations. For one thing, the chronic administration of DMI is known to produce a desensitization of beta receptors (Sulser, 1982) which will add to the desensitization of alpha-1 receptors produced by stress and may eventually reduce rather than enhance noradrenergic receptor action in these animals. Recent studies on the net effects on neurotransmission of chronic reuptake blockade or chronic elevation of catecholamines however indicates that neurotransmission at peripheral and central noradrenergic receptors remains elevated during these conditions despite postsynaptic receptor desensitization (Bickford-Wimer et al., 1987; Hoffman, 1987). Therefore the desensitization is not likely to be a major confounding factor in this study. Other factors that may present more serious problems involve the specificity of DMI for the noradrenergic system, the use of exploratory behavior to measure adaptation to stress and the repeated measures design used in the above study. DMI is known to affect other neuronal systems in the brain besides the noradrenergic and it is possible that one of these non- noradrenergic systems is responsible for the above behavioral effect. Exploratory activity is primarily a form of gross motor activity which may have a unique relationship to the noradrenergic system. Therefore its recovery may reflect more the state of central noradrenergic function that the overall state of adaptation to stress in the organism. Furthermore, we cannot exclude the possibility that the effect of the drug on the rate of adaptation is an artifact of a greater ease of reversing the behavioral impairment at 7 days when the animals have partially adapted than at 1 day when they have not. And finally the possibility exists that DMI may have interacted with processes of habituation since the same animals were tested repeatedly in the above study. Further experiments therefore will be necessary to clarify the roles of these factors in the above phenomenon. Despite these problems however the latter results remain encouraging for the adaptational hypothesis and justify further research on its validity and generality.

References

Anisman, H. and Zacharko, R.M. (1986) 'Behavioral and neurochemical consequences associated with stressors.' *Ann. N.Y. Acad. Sci.* 467:205-25.

Aston-Jones, G.(1985) 'Behavioral functions of locus coeruleus derived from cellular attributes.' *Physiol. Psychol.* 13:118-126.

Barber, R. and Butcher, R.W. (1983) 'The egress of cyclic AMP from metazoan cells. *Adv. Cyclic Nucleot'. Res.* 15:119-38.

Bickford-Wimer, P.C., Parfitt, K., Hoffer, B.J. and Freedman, R. (1987) 'Desipramine and noradrenergic neurotransmission in aging: failure to respond in aged laboratory animals. *Neuropharmacology* 26:597-605.

Daly, J.W. (1977) 'Cyclic Nucleotides in the Nervous System'. Plenum Press, New York

Davies, A.O. and Lefkowitz, R.J. (1984) 'Regulation of B-adrenergic receptors by steroid hormones.' *Ann. Rev. Physiol.* 46:119-130.

Duman, R.S., Strada, S.J. and Enna, S.J. (1985) 'Effect of imipramine and adrenocorticotropin administration on the rat brain norepinephrine-coupled cyclic nucleotide generating system: alterations in alpha and beta components'. *J. Pharmacol. exp. Ther.* 234:409-414.

Duncan, G.E., Paul, I.A., Harden, T.K., Mueller, R.A., Stumpf, W.E. and Breese, G.R. (1985) 'Rapid down regulation of beta adrenergic receptors by combining antidepressant drugs with forced swim: a model of antidepressant induced neural adaptation. *J. Pharmacol. exp. Ther.* 234:402-408.

Egawa, M., Hoebel, B.G. and Stone, E.A. (1988) 'Use of microdialysis to measure brain noradrenergic receptor function in vivo.' *Brain Res.* 458:303-8.

Feeney, D.M. and Sutton, R.L. (1987) 'Pharmacotherapy for recovery of function after brain injury.' *CRC Rev. Neurobiol.* 3:135-197.

Gervais, R., Holley, A. and Keverne, B. (1988) 'The importance of central noradrenergic influences on the olfactory bulb in the processing of learned olfactory cues.' *Chem. Senses* 13:3-12.

Green, A.R., Costain, D.W. and Deakin, J.F. (1980) 'Enhanced 5- hydroxytryptamine and dopamine-mediated behavioral responses following convulsions. III. The effects of monoamine antagonists and synthesis inhibitors on the ability of electroconvulsive shock to enhance responses.' *Neuropharmacology* 19:907-914.

Harley, C.W. (1987) 'A role for norepinephrine in arousal, emotion and learning? Limbic modulation by norepinephrine and the Kety hypothesis.' *Prog. Neuro-Psychopharmacol. Biol. Psychiat.* 11:419-458.

Hoffman, B.B. (1987) 'Observations in New England Deaconess Hospital rats harboring pheochromocytoma.' *Clin. Invest. Med.* 10:555-560.

Kant, G.J., Bates, V.E., Lenox, R.H. and Meyerhoff, J.L. (1981) 'Increases in cyclic AMP levels in rat brain regions in vivo following isoproterenol.' *Biochem. Pharmacol.* 30:3377-3380.

Korf, J., Boer, P.H. and Fekkes, D. (1976) 'Release of cerebral cyclic AMP into push-pull perfusates in freely moving rats.' *Brain Res.* 113:551-561.

Johnson, R.D. and Minneman, K.P. (1986) ' Characterization of $\alpha 1$-adrenoceptors which increase cyclic AMP accumulation in rat cerebral cortex.' *Eur. J. Pharmacol.* 129:293-305.

Lazareno, S., Marriott, D.B. and Nahorski, S.R. (1985) 'Differential effects of selective and nonselective neuroleptics on intracellular and extracellular cyclic AMP accumulation in rat striatal slices.' *Brain Res.* 361:91-8.

Magistretti, P.J. and Schorderet, M. (1985) 'Norepinephrine and histamine potentiate the increases in cyclic adenosine 3':5'- monophosphate elicited by vasoactive intestinal polypeptide in mouse cerebral cortical slices: mediation by $\alpha 1$-adrenergic and H1-histaminergic receptors.' *J. Neurosci.* 5:362-368.

McCarthy K.D. and DeVellis, J. (1979) 'The regulation of adenosine 3':5'-cyclic monophosphate accumulation in glia by alpha adrenergic agonists.' *Life Sci.* 24:639-50.

McEwen, B.S., DeKloet, E.R. and Rostene, W. (1986) 'Adrenal steroid receptors and actions in the nervous system.' *Physiol. Rev.* 66:1121-1128.

Mobley, P.L., Manier, D.H. and Sulser, F. (1988) 'Norepinephrine- sensitive adenylate cyclase system in rat brain: role of adrenal corticosteroids.' *J. Pharmacol. exp. Ther.* 0:.

Pichler, L. and Kobinger, W. (1985) 'Possible function of $\alpha 1$-adrenoceptors in the CNS in anaesthetized and conscious animals.' *Eur. J. Pharmacol.* 107:305-311.

Rasmussen, K., Morilak, D.A. and Jacobs, B.L. (1986) 'Single unit activity of locus coeruleus neurons in the freely moving cat. I. During naturalistic behaviors and in response to simple and complex stimuli.' *Brain Res.* 371:324-334.

Schoener, E.P., Hager, P.J., Felt, B.T. and Schneider, D.R. (1979) 'Cyclic nucleotides in the rat neostriatum: push pull perfusion studies.' *Brain Res.* 179:111-119.

Stone, E.A. (1975) 'Catecholamines and Behavior' (Friedhoff, A.J., ed.), Plenum Press, New York in pp. 31-72.

Stone, E.A. (1978) 'Effect of stress on norepinephrine-stimulated accumulation of cyclic AMP in rat brain slices.' *Pharmacol. Biochem. Behav.* 8:583-591.

Stone, E.A. (1979) 'Reduction by stress of norepinephrine- stimulated accumulation of cyclic AMP in rat cerebral cortex.' *J. Neurochem.* 32:1335-1337.

Stone, E.A. (1981) 'Mechanism of stress-induced subsensitivity to norepinephrine. *Pharmacol. Biochem. Behav.* 14:719-723.

Stone, E.A. (1982) 'Noradrenergic function during stress and depression: an alternative view.' *Behav. Brain Sci.* 5:122.

Stone, E.A. (1983) 'Adaptation to stress and brain adrenergic receptors.' *Neurosci. Biobehav. Rev.* 7:503-9.

Stone, E.A. (1983b) 'Problems with current catecholamine hypotheses of antidepressant drugs.' *Behav. Brain Sci.* 6:535- 578.

Stone, E.A., Platt, J.E., Trullas, R. and Slucky, A.V. (1984) 'Reduction of the cAMP response to NE in rat cerebral cortex following repeated restraint stress.' *Psychopharmacology* 82:403- 405.

Stone, E.A., Slucky, A.V., Platt, J.E. and Trullas, R. (1985) 'Reduction of the cyclic adenosine 3',5'-monophosphate response to catecholamines in rat brain slices following repeated restraint stress.' *J. Pharmacol. exp. Ther.* 233:382-388.

Stone, E.A. and Herrera, A.S. (1986) 'Alpha adrenergic modulation of cyclic AMP formation in the rat CNS: highest levels in olfactory bulb. *Brain Res.* 384:401-403.

Stone, E.A., Platt, J.E., Herrera, A.S. and Kirk, K.L. (1986) 'The effect of repeated restraint stress, desmethylimipramine or adrenocorticotropin on the alpha and beta adrenergic components of the cyclic AMP response to norepinephrine in rat brain slices.' *J. Pharmacol. exp. Ther.* 230:702-707.

Stone, E.A., McEwen, B.M. and Herrera, A.S. (1987a) 'Role of corticosterone in stress-induced desensitization of cortical alpha-1 receptors.' *Abstr. Soc. Neurosci.* 13:897.

Stone, E.A., McEwen, B.S., Herrera, A.S. and Carr, K.D. (1987b) 'Regulation of alpha and beta components of noradrenergic cyclic AMP response in cortical slices.' *Eur. J. Pharmacol.* 141:347- 356.

Stone, E.A. and Egawa, M. 'Measurement of extracellular cAMP in brain by microdialysis: a method to study brain receptor function in vivo.' *Abstr. Soc. Neurosci.* In press.

Stone, E.A., Egawa, M. and Hoebel, B.G. (1988) 'In vivo studies of brain adrenoceptor function with microdialysis.' *FASEB J.* 2:A1800.

Stoof, J.C. and Kebabian, J.W. (1981) 'Opposing roles for D-1 and D-2 dopamine receptors in efflux of cyclic AMP from rat neostriatum.' *Nature* 294:366-8.

Sulser, F. 'Typical and Atypical Antidepressants: Molecular Mechanisms' (Costa, E. and Racagni, G., eds.), Raven Press, New York(1982) in pp. 1-20,

Svensson, T.H. (1987) 'Peripheral, autonomic regulation of locus coeruleus noradrenergic neurons in brain: putative implications for psychiatry and psychopharmacology.' *Psychopharmacology* 92:1- 7.

Szabb, G., Hoffman, P.L. and Tabakoff, B. (1988) 'Forskolin promotes the development of ethanol tolerance in 6- hydroxydopamine-treated rats.' *Life Sci.* 42:615-621.

Weiss, J.M., Glazer, H.I., Pohorecky, L.A., Brick, J. and Miller, N.E. (1975) 'Effects of chronic exposure to stressors on avoidance-escape behavior and on brain norepinephrine.' *Psychosom. Med.* 37:522-534.

Zahniser, N.R., Minneman, K.P. and Molinoff, P.B. (1979) 'Persistence of B-adrenergic receptors in rat striatum following kainic acid administration.' *Brain Res.* 178:589-595.

THE EFFECTS OF ACUTE EXPOSURE TO STRESSORS AND DRUG GROW WITH THE PASSAGE OF TIME

Seymour M. Antelman, Ph.D. and Anthony R. Caggiula, Ph.D.
University of Pittsburgh School of Medicine
Department of Psychiatry and University of Pittsburgh School of Medicine
Departments of Psychology and Behavioral Neuroscience

Sensitization refers to the ability of a strong stimulus to increase the subsequent response to that same, or a different agent. As such, sensitization represents one of the simplest and most common forms of learning. The sensitizing stimulus can be coincident with the augmented response, as is the case when a loud, background noise increases the startle response of rats to brief tone (Davis, 1974), or it can precede the behavioral measurement by days or weeks (Antelman, 1988a; Kandel, 1976). While it was once thought that most instances of sensitization were short lived (Domjan and Burkhard, 1986), more recent work indicates that sensitization is a more enduring phenomenon that, in many instances, actually grows with the passage of time (Antelman, 1988a).

The inducing stimulus can be exteroceptive or interoceptive. Effective exteroceptive stimuli include typical laboratory stressors (e.g., immobilization or electric shock) or environmental events, such as naturally occuring or man-made disasters (Antelman, 1988; Antelman, et al. 1988). Sensitization also can be induced by interoceptive stimuli, including pharmacological agents ranging from stimulants such as amphetamine and cocaine (Antelman and Chiodo, 1983; Robinson and Becker, 1986) to the anxiolytic, diazepam (Antelman, 1989b). Thus, sensitization can occur both within drugs – amphetamine augmenting the response to subsequent amphetamine – and across drugs, such as when diazepam increases an animals response to haloperidol (Antelman, 1988a). Moreover, bidirectional, cross-sensitization can be demonstrated between exteroceptive, stressful stimuli and drugs (Antelman and Chiodo, 1983). When one considers the fact that all drugs represent foreign substances, by virtue of their nonphysiological nature and/or their mode of introduction into the body, the results indicate that the sensitization seen after many drugs and nonpharmacological, environmental stimuli may be due to a common element, namely the nonspecific, stressful consequences of these agents.

In the present paper, data will be reviewed indicating that:
1. Stressful events have very long-term sequelae typically manifest as sensitized responses to subsequent presentation of the same or other stressful stimuli.
2. An earlier stressful event can similarly influence the actions of a diversity of drugs experienced much later.
3. Even a single experience with a drug can likewise have a long-term sensitizing influence if that drug is re-encountered.

S. Puglisi-Allegra and A. Oliverio (eds.), Psychobiology of Stress, 161–170.
© 1990 Kluwer Academic Publishers. Printed in the Netherlands.

In addition, we will suggest that the sensitizing effects of prior exposure to exteroceptive or interoceptive stimuli may be a special case of a more general phenomenon in which the effects of a strong stimulus grow with the passage of time: those effects may be manifest as increased responsiveness (sensitization) or decreased responsiveness (habituation or tolerance).

Stressor-Induced Sensitization to Subsequent Stressors

In an early experiment we examined the long-term influence of a stressor on electrical self-stimulation of the brain in rats. A number of studies have shown that self-stimulation – thought to be a prototype of reward behavior – induces the same hormonal concomitants as universally recognized stressful stimuli (McHugh, et al. 1966; Uretsky, et al. 1966; Terry & Martin, 1978) and it was therefore used as a recall stressor in this experiment. Animals with nucleus accumbens implants were trained to stable barpress rates and then food-deprived for 72 hours. Not only did this stressful event greatly enhance self-stimulation rates during its application but the enhancement continued to grow for the 23-day duration of the experiment following the resumption of feeding and normalization of intake and body weight (Antelman and Chiodo, 1983). This experiment presents the first of several examples indicating that the sensitizing influence of an earlier stressful event can grow with the passage of time.

In a very recent study we examined the influence of a single stressful event on later shock-induced analgesia, plasma corticosterone and dopamine (DA) activity in rat frontal cortex (Caggiula, et al. 1988). Animals exposed to a one hour footshock session (15, 0.5 sec. shocks) 10 days earlier showed significant attenuation of shock-induced analgesia, markedly elevated corticosterone levels and an increased DOPAC/DA ratio in the frontal cortex relative to that induced by shock immediately prior to analgesia testing.

All of these findings support the thesis of long-term effects of brief stressful events. However, while the corticosterone and DOPAC/DA data clearly indicate that prior stressful experience had a sensitizing influence, interpretation of the attenuating effect of preshock on shock-induced analgesia is less obvious, although it may have resulted from an enhancement (i.e., sensitization) of the effectiveness of the pain sensitivity test (a hot water bath).

The Effects of Acute Stressors on Later Drug Administration

Amphetamine

Most of the work involving the long-term effects of stressors on later drug treatments have utilized amphetamine (AM). This is – at least in part – the result of our suggestion that the sensitizing effects of AM are interchangeable with those of nonpharmacological stressors (Antelman, et al. 1980; Antelman and Chiodo, 1983).

In what was probably the first experiment to inquire whether stress has a long-term sensitizing effect on the actions of AM, rats were food-deprived for 72 hours and examined for AM-induced sniffing both one day before and 15 days following the deprivation period. Food deprivation 15 days earlier resulted in significant enhancement of the sniffing response to AM (4 mg/kg, i.p.), relative to that seen in undeprived controls (Antelman and Eichler,

1979). Body weights did not differ between experimentals and controls at the time of the second AM treatment and are therefore unlikely to provide an explanation of the results.

Similar effects have since been observed using other stressors and AM-induced behaviors. A single 20 min. period of footshock significantly sensitized both the intensity and duration of AM-induced sniffing 15 days later. Likewise, one four-hour period of immobilization by wrapping sensitized AM-induced polydipsia when this was examined two weeks later (Antelman and Chiodo, 1983). Robinson, et al. (1985) also found enduring, sensitizing effects of acute footshock or immobilization on AM-induced rotation in rats with a unilateral lesion of the substantia nigra.

Apomorphine

The influence of stressors on later administration of apomorphine is demonstrated in experiments in which low – presumably DA autoreceptor-specific – doses have been used. In the first such study, Chiodo and Antelman (1980) found that a single exposure to 0.7 sec. of electroconvulsive shock (ECS) seven days earlier attenuated the ability of a 4 μg/kg dose of apomorphine to inhibit the firing of single DA-containing neurons in the pars compacta of the rat substantia nigra. ECS administered only one hour before apomorphine was ineffective, suggesting that the effect of ECS grew (i.e., sensitized) with the passage of time. The long-lasting nature of one ECS exposure on apomorphine inhibition of DA-containing neurons was later confirmed by Tepper, et al. (1982). This consequence of ECS is almost certainly due to its action as a stressor since the same persistent effect was observed when the DA-inhibiting action of apomorphine was examined one or ten days following exposure to the stress of a four-hour period of immobilization (Antelman and Lucik, in Antelman, 1988).

Haloperidol and Fluphenazine

The first experiment which showed the persistent influence of a stressor on the impact of later haloperidol administration actually dealt with the question of whether a non-pharmacological stressor could sensitize rats to subsequent presentation of other stressors. The target stressor was tail-pressure-induced eating. Since this behavior typically occurs at or near asymptotic levels, and is therefore difficult to potentiate, it was first attenuated with haloperidol and we then observed whether a 72-hr period of food deprivation 15 days earlier could modify that antagonism. It did. Haloperidol was significantly less effective in those animals that had earlier been food-deprived (Antelman and Eichler, 1979). In other words, it would seem that food deprivation sensitized the organism to tail pressure and in so doing diminished the effectiveness of the unvarying dose of haloperidol. These early data turned out to have been a harbinger of something we've since observed with a number of other drugs. That is, the influence of pre-exposure to a stressor on later drug administration depends on whether the drug is itself followed closely by another stressful stimulus as was the case with haloperidol and tail pressure. In that event, the influence of the drug is diminished. Conversely, when the drug is the last stimulus presented, the impact of earlier stress is usually one of enhancement. For instance, a single injection of saline enhanced catalepsy up to at least four and eight weeks after haloperidol and fluphenazine hydrochloride, respectively (Antelman, et al. 1986). However, even when there is no additional stimulus, under some conditions prestress can inhibit the actions of neuroleptics. Degrees of stress may be an important variable in determining the directionality of effect. We recently found that while a single jab with an empty syringe needle potentiated haloperidol catalepsy two weeks later, an hour of immobilization inhibited the same behavior. Interestingly, even this inhibition reflected stress sensitization which developed with the passage of time, since it was observed two weeks but not one hour after immobilization (Antelman, et al. in preparation).

Diazepam

The point made above about the influence of stressors on later drug action when the drug is followed closely by another stressor, is particularly well-illustrated by our experiments with diazepam. In experiments showing that the convulsant and highly stressful stimulus, pentylenetetrazol (PTZ), significantly elevated rat nucleus accumbens and frontal cortical DA levels, it was found that whereas diazepam administered one hour earlier prevented this increase, a single injection of saline 28 days earlier completely obviated diazepam's action. That is, it restored DA levels to where they had been after PTZ alone (Antelman, et al. 1987). This finding has since been confirmed in the frontal cortex, substituting two hours of immobilization stress for the saline injection (Antelman et al. 1988). It has also been extended to hormonal and behavioral measurements. A single jab with a 26 gauge syringe needle 1-28 days earlier or immobilization 28 days earlier, prevented completely diazepam alteration of the plasma corticosterone levels achieved after PTZ. Moreover, up to a point the antidiazepam action of needle jab grew with the passage of time. Behaviorally, we found that PTZ administered two weeks before diazepam – which was itself given an hour before a second PTZ injection – significantly reduced its anticonvulsant activity (Antelman, et al. 1988).

Clonidine

There are two experiments which suggest that acute exposure to stressors might have a long-lasting influence on at least some of clonidine's actions. First, Tepper, et al. (1982) found that a single, brief ECS significantly reduced the ability of a low, i.v. dose of clonidine to inhibit the firing of norepinephrine-containing neurons in the rat locus coeruleus seven days later. More recently, Antelman, et al. (in press) demonstrated that one injection of PTZ significantly enhanced the hypokinetic effects of a presynaptic dose of clonidine (25 ug/kg, i.p.) administered up to at least two weeks later. The reduction in clonidine's effectiveness in the first instance (Tepper, et al. 1982) and its enhancement in the second (Antelman, et al. in press) may reflect the distinction discussed above between the influence of earlier exposure to a stressor on a drugs actions when that agent is or is not followed or accompanied by yet another stressful event. In the case of the Tepper study, the trauma associated with preparing an animal for electrophysiological recording may have served as the added stressor.

Tetracaine

Tetracaine is a highly toxic local anesthetic. It was used as part of a series of experiments investigating the effect of pre-exposure to pharmacological and non-pharmacological stressors on sudden cardiac death from cocaine and other local anesthetics. We found that 4 hours of immobilization 8 or 28 days earlier, virtually eliminated "instantaneous" death (within 5 sec.), which otherwise occurred in 50% of controls. That this protection is likely to reflect a time-dependent sensitizing influence of stress is suggested by the earlier finding that one exposure to a sympathomimetic, induced precisely such a growing time-dependent protection against sudden cardiac death from another local anesthetic, quinidine (Antelman and Chiodo, 1984).

The Effects of Acute Drug Treatment on Later Drug Administration

Imipramine

In single-unit electrophysiological experiments involving the effects of pretreatment with an antidepressant on substantia nigra DA-containing neurons, it was found that two days of imipramine administration (2 x 10 mg/kg, i.p./day) followed by ten drug-free days induced significantly less inhibition of DA firing after a low, i.v. dose of apomorphine than the same two-day treatment followed only by two drug-free days (Chiodo and Antelman, 1980b). In other words, the effect of imipramine pretreatment sensitized with the passage of time just as occurs after nonpharmacological stressors.

Phenelzine

Phenelzine is a monoamine oxidase inhibitor used clinically in the treatment of depression. We examined whether it could induce effects which grow with the passage of time by using both electrophysiological and behavioral procedures.

Electrophysiologically, it was found that pretreatment with this agent for two days induced greater subsensitivity of DA substantia nigra autoreceptors when this was measured (by iontophoretic application of DA) ten days later than when such testing was done only two days following drug administration (Antelman, et al. 1983). Moreover, the effect of two days of treatment followed by ten no-treatment days was indistinguishable from the reciprocal regime of ten-treatment and two no-treatment days. Behaviorally, two days of treatment followed by ten drug-free days also decreased the hypokinetic effect of apomorphine to the same extent as ten treatment and two no-treatment days (unpublished observations). In contrast to these data, actual measurement of monoamine oxidase inhibition correlated directly with duration of treatment (unpublished observations). The results described represent yet another instance of drug effects growing with the passage of time following minimal treatment and suggest that this growth is independent of the major pharmacological actions of the drug.

Bupropion

Bupropion is an atypical antidepressant which is thought to act by blocking DA reuptake. We tried to determine its likelihood of inducing time-dependent sensitization by noting whether two injections of the drug spaced at a three-week interval could induce sensitization of water drinking (in undeprived animals) to the same extent as daily drug treatment over the same period of time. The results indicated that it could (Antelman, 1988a). In addition, we inquired whether haloperidol administered prior to the first bupropion treatment could prevent the induction of sensitization and/or if it could block the manifestation of this phenomenon when injected before the last treatment. Haloperidol blocked the manifestation of sensitization but failed completely to interfere with its induction. This is an important finding since it points away from the pharmacology of a drug as a key factor in inducing sensitization. In so doing, it raises the possibility that the foreign or stressful aspect of drugs may be more important; a possibility which is, of course, more consistent with the ability of non-pharmacological stressors to trigger this phenomenon.

Desmethylimipramine (DMI)

Thusfar, in this section we have concentrated on measures of DA activity. With DMI we focused on central beta noradrenergic receptors. Administration on day 1 and day 17 – but not just on day 17 – significantly decreased Bmax to the same extent as that seen in animals receiving DMI daily for 17 days (Lace and Antelman, 1984). Although the effect was small in each case, the data are nevertheless consistent with drug-induced time-dependent sensitization of NE-containing neurons.

Amitriptyline

The tricyclic, amitriptyline, injected twice on one day enhanced the incidence of head twitch behavior following treatment with the serotonin (5HT) precursor, 5-hydroxytryptophan (5-HTP), 11 days but not 1 day later. Again, the 1 + 11 group did not differ significantly from animals receiving 10 days of treatment followed by two no-drug days prior to 5-HTP (Antelman, et al. 1983). Since head twitch is widely believed to be a serotonin-dependent behavior, these data suggest that like DA and NE, 5-HT systems can also sensitize in a time-dependent fashion after acute drug administration (for further discussion of this issue, see Antelman, 1988).

Chlorimipramine

The time-dependent effects of chlorimipramine (CIMI) were demonstrated in a small, open, clinical study of patients suffering from major depression. Patients received i.v. infusions of CIMI on two consecutive days following which they were left unmedicated for 10 days. During this interval plasma levels of drug were monitored and clinical ratings were done. Although drug levels were unmeasurable after three days, the clinical effects continued to grow so that scores on the Hamilton Depression Scale were significantly lower (i.e., better) than on the first no-drug day (Pollock, et al. 1986). This finding has since been replicated in a double-blind, placebo-controlled study (Pollock, et al. 1989).

Haloperidol and Fluphenazine Hydrochloride

The ability of haloperidol to show time-dependent sensitization was tested by injecting the drug twice at 1-4 week intervals and conducting catalepsy tests following each injection. Second injections, whether 1-4 weeks apart, invariably induced a significant elevation in the incidence of catalepsy relative both to first injections and to haloperidol in animals receiving saline as a first injection 1-4 weeks earlier. Moreover, the influence of second haloperidol injections grew as more time elapsed since the initial treatment. Thus, groups receiving their second haloperidol treatment 2-4 weeks after the first, showed a significantly increased incidence of catalepsy relative to those in which treatments were separated by only one week.

The same results seen following haloperidol were also observed in animals receiving two injections of fluphenazine hydrochloride spaced 2-8 weeks apart. In the latter case, we also found sensitization to the second fluphenazine treatment when the initial exposure was through the drinking water rather than injection and animals were tested for catalepsy only after the second treatment two weeks later (Antelman, et al. 1986).

Diazepam

The question of whether diazepam could induce time-dependent sensitization was examined in two ways. First, rats received two i.p. injections of a low dose of diazepam (0.5 mg/kg) separated by two or four weeks. Following the second injection they were administered a 40 mg/kg dose of the convulsant, pentylenetetrazol (PTZ). Pretreatment with diazepam two (but not four) weeks earlier significantly potentiated the anticonvulsive action of the same dose of this agent also administered one hour prior to PTZ.

Second, rats were administered a single injection of diazepam (2.5 mg/kg) or saline and were tested for catalepsy and ptosis after haloperidol administered two hours or two weeks later. When the interval between diazepam and haloperidol was two hours the response to haloperidol was the same as if there had been no pretreatment. On the other hand, when the interval was two weeks there was a marked potentiation of the actions of haloperidol (Antelman, et al. 1987; Antelman, et al. 1989b). In other words, the effect of the diazepam grew with the passage of time.

RO15-1788

RO15-1788 is a benzodiazepine antagonist which structurally also belongs to the benzodiazepine family. Its ability to induce time-dependent sensitization was shown by the finding that a dose too low to reverse diazepam antagonism of PTZ-induced convulsions was able to do so when preceded by a single injection of the same dose of RO15-1788 two weeks earlier (Antelman et al. 1989b).

Implications

The results reviewed suggest that drugs and nonpharmacological, stressful stimuli appear to be interchangeable in their ability to produce long-term effects that grow with time after acute exposure. In some instances, the end result is an increased responsiveness upon re-exposure to the original inducing stimulus or to another agent (sensitization), whereas in other cases, a reduced responsiveness to subsequent stimulation is observed. These latter examples may reflect sensitization to a third stimulus occuring in close proximity to, and masking the second stimulus. This may have occurred when a saline injection or immobilization completely obviated the anti-PTZ effects of diazepam, or when pre-exposure to electric shock attenuated shock-induced analgesia 10 days later. Alternately, decreased responsiveness to the second stimulus may, in some instances, suggest the growth of an indirect, inhibitory process. This latter possibility is suggested by the finding that, whereas a mild stressor, needle jab, increased haloperidol-induced catalepsy 2 weeks later, a more intense stimulus, immobilization, failed to affect catalepsy at 1 hour but suppressed it after 2 weeks. Whatever the explanation for why the effects of a prestressor are sometimes manifest as an enhancement in responsiveness to a later stimulus, and at other times as a decrement, the same general principle applies in each case. The effects of the initial stressor, be it a drug or nonpharmacological, environmental stimulus, grow with the passage of time.

These long-term, time-dependent effects of pharmacological and non-pharmacological stressors have now been shown using indices of dopamine, norepinephrine, serotonin and GABA function (see Antelman, 1988 for review). This suggests that the phenomenon is not unique to any particular transmitter system, and indeed may be characteristic of all transmitter systems. While the mechanisms that underlie this process are currently unknown, its long-lasting time course, coupled with the fact that time-dependent sensitization also has been reported for endocrine and immunological responses (Hennessey and Levine, 1977; Cooper, 1980; Burch and Evans, 1986), suggests that a genomic alteration may be involved. For example, the possible involvement of heat shock or stress proteins is currently being investigated.

The clinical implications of this research for the therapeutic use of drugs have been discussed elsewhere (Antelman, 1988). Briefly, these data suggest that the effects of a drug can be greatly influenced by a single, prior exposure to that drug, or by a stressful event, even when such experiences occur at a much earlier time in life. The possibility that a memory of an earlier stressful experience can persist and even grow over time, to influence a current event is strongly suggested in a number of clinical syndromes (e.g., post-traumatic stress disorder, panic and bulimia, see Antelman, 1988). With regard to pharmacotherapy, these results suggest the following: 1) it may not be necessary to give drugs on a daily or several-times-daily basis in order to obtain maximum therapeutic efficacy, 2) a widely intermittent regime of drug administration would exploit the capacity of drugs to trigger effects which grow with the passage of time, while minimizing the development of tolerance and its associated sequelae.

Finally, the observation that a vehicle injection or even a jab with an empty syringe needle can have powerful, and extremely long-lasting effects has important implications for how clinical and basic experiments should be conducted. For example, it can no longer be assumed that a vehicle control represents baseline, thus necessitating the need for additional controls. Moreover, drug actions can no longer be assumed to terminate when the drug leaves the body, indicating that extreme caution must be taken when employing "washout periods" in experimental designs assessing the therapeutic effects of drugs.

References

Antelman, S.M., Knopf, S., Kocan, D. and Edwards, D.J.: Persistent sensitization of clonidine-induced hypokinesia following one exposure to a stressor: Possible relevance to panic disorder and its treatment. *Psychopharmacology*. In press.

Antelman, S.M., DeGiovonni, L.A. and Kocan, D. (1989a): A single exposure to cocaine or immobilization stress provides extremely long-lasting, selective protection against sudden cardiac death from tetracaine. *Life Sciences 44*:201-207.

Antelman, S.M., Kocan, D., Edwards, D.J. and Knopf, S. (1989b): Anticonvulsant and other effects of diazepam grow with time after a single treatment. *Pharmacol. Biochem Beh. 33*.

Antelman, S.M. (1988): Time-dependent sensitization as the cornerstone for a new approach to pharmacotherapy: drugs as foreign/stressful stimuli, *Drug Development Research 14*:1-30.

Antelman, S.M. (1988): Is it possible that drugs have been given the wrong way for centuries? In Sen, A.K. and Lee, T. (eds): *Receptors and Ligands in Psychiatry and Neurology*. Cambridge:Cambridge University Press, pp. 484-503.

Antelman, S.M. and Eichler, A.J. (1979): Persistent effects of stress on dopamine-related behaviors; clinical implications. In Usdin, E., Kopin, I.J. and Barchas, J. (eds): *Catecholamines: Basic and Clinical Frontiers*, New York:Pergamon Press, pp. 1759-1761.

Antelman, S.M. and Chiodo, L.A. (1983): Amphetamine as a stressor. In Creese, I. (ed): *Stimulants: Neurochemical, Behavioral and Clinical Perspectives*. New York:Raven Press, pp. 269-299.

Antelman, S.M. and Chiodo, L.A. (1984): Stress: Its effects on interactions among biogenic amines and role in the induction and treatment of disease. In: Iversen, L.L., Iversen, S.D. and Snyder, S.H. (eds): *Handbook of Psychopharmacology*, Vol. 18. New York:Plenum Press, pp. 279-341.

Antelman, S.M., Eichler, A.J., Black, C.A. and Kocan, D. (1980): Interchangeability of stress and amphetamine in sensitization, *Science 207*:329-331.

Antelman, S.M., and Chiodo, L.A. and DeGiovanni, L.A. (1982): Antidepressants and dopamine autoreceptors: Implications for both a novel means of treating depression and understanding bipolar illness. In Costa, E. and Racagni, G. (eds): *Typical and Atypical Antidepressants, Molecular Mechanisms*. New York: Raven Press, pp. 121-132.

Antelman, S.M., DeGiovanni, L.A., Kocan, D., Perel, J.M. and Chiodo, L.A. (1983): Amitriptyline sensitization of a serotonin-mediated behavior depends on the passage of time and not repeated treatment. *Life Sci, 33*:1727-1730.

Antelman, S.M., Kocan, D., Edwards, D.J., Knopf, S., Perel, J.M. and Stiller, R. (1986): Behavioral effects of a single neuroleptic treatment grow with the passage of time. *Brain Res. 385*:58-67.

Antelman, S.M., Kocan, D., Edwards, D.J. and Knopf, S. (1987): A single injection of diazepam induces long-lasting sensitization. *Psychopharm. Bull. 22*:430-434.

Antelman, S.M., Knopf, S., Kocan, D. Edwards, D.J., Ritchie, J.C., and Nemeroff, C.B. (1988): One stressful event blocks multiple actions of diazepam for up to at least a month. *Brain Res. 445*:380-385.

Burch, J.B.E. and Evans, M.I. (1986): Chromatic structural transitions and the phenomenon of vitellogenin gene memory in chickens. *Mol. Cell. Biol. 6*: 1886-1893.

Caggiula, A.R., Antelman, S.M., Aul, E., Knopf, S. and Edwards, D.J (1988). One stressful event sensitizes the increased frontal cortical dopamine (DA) and plasma corticosterone activity, but attenuates the analgesic response to stress 10 days later. *Soc. for Neurosci*. Toronto, Canada Abst # 179.5.

Chiodo, L.A. and Antelman, S.M. (1980a): Electroconvulsive shock: Progressive dopamine autoreceptor subsensitivity independent of repeated treatment. *Science 210*:799-801.

Chiodo, L.A. and Antelman, S.M. (1980b): Repeated tricyclics induce a progressive dopamine autoreceptor subsensitivity independent of daily drug treatment. *Nature 287*:451-454.

Cooper, E.L.: Cell-mediated memory in invertebrates. In Manning, J.J. (ed) (1980): *Phylogeny of Immunological Memory*. Amsterdam:Elsevier/North Holland, pp. 35-46.

Davis, M. (1974). Sensitization of the rat startle response by noise. *J. Comp. Physiol. Psychol. 87*:571-581.

Domjan, M. and Burkhard, B. (1986). *The Principles of Learning and Behavior*. 2nd Ed. Brooks/Cole:Monterey, CA.

Eichler, A.J. and Antelman, S.M. (1979): Sensitization to amphetamine and stress may involve nucleus accumbens and medial frontal cortex. *Brain Res. 176*:412-416.

Hennessey, M.B. and Levine, S. (1977): Effects of various habituation procedures on pituitary-adrenal responsiveness in the mouse. *Physiol. Behav. 18*:799-802, 1977.

Kandel, E.R. (1976): *Cellular Basis of Behavior: An Introduction to Behavioral Neurobiology*. San Francisco:WH Freeman and Co.

Lace, J.W. and Antelman, S.M. (1984): Cortical B-adrenergic subsensitivity after desmethylimipramine may depend on the passage of time rather than daily treatment. *Brain Res. 278*:359-361.

McHugh, P.R., Black, W.C. and Mason, J.W. (1966): Some hormonal responses to electrical self-stimulation in the Macaca mulatta. *Am. J. Physiol. 210*:109-113.

Pollock, B.G., Perel, J.M., Schostak, M., Antelman, S.M., Brandom, B. and Kupfer, D.J. (1986): Understanding the response lag to tricyclics 1. Application of pulse-loading regimens with intravenous clomipramine. *Psychopharmacol. Bull. 22*:214-219.

Pollock, B.G., Perel, J.M., Nathan, R.S., and Kupfer, D.J. (1989): Acute antidepressant effect following pulse-loading with intravenous and oral clomipramine. *Arch. Gen. Psychiat. 46*:29-35.

Robinson, T.E. and Becker, J.B. (1986): Enduring changes in brain and behavior produced by chronic amphetamine administration: A review and evaluation of animal models of amphetamine psychosis. *Brain Res. Rev. 11*:157-198.

Robinson, T.E., Angus, A.L. and Becker, J.B. (1985): Sensitization to stress: The enduring effects of prior stress on amphetamine-induced rotational behivior, *Life Sci.* *37*:1039-1042.

Tepper, J.M., Nakamura, S., Spanis, C.W., Squire, L.R., Young, S.J., and Groves, P.M. (1982): Subsensitivity of catecholaminergic neurons to direct acting agonists after single or repeated electroconvulsive shock. *Biol. Psychiatry 16*:1059.

Terry, L.C. and Martin, J.B. (1978): Hypothalamic-pituitary responses in intracranial self-stimulation in the rat. *Brain Res. 157*:89-104.

Uretsky, E., Kling, A. and Orbach, J. (1966): Plasma 17-hydroxycorticosteroid levels following intracranial self-stimulation in rats. *Psychol. Rep. 19*:891-901.

GENOTYPE-DEPENDENT ADAPTATION OF BRAIN DOPAMINE SYSTEM TO STRESS

Stefano Puglisi-Allegra, Simona Cabib, Eliane Kempf* and Alberto Oliverio
*Istituto di Psicobiologia e Psicofarmacologia (C.N.R.) via Reno 1, I-00198, Roma, Italy and * Centre de Neurochimie (C.N.R.S./INSERM), 5 rue B. Pascal, 67084 Strasbourg, France*

Stress is considered to play a basic role in the etiology of some psychotic syndromes either as primary cause or as a precipitating factor. In both cases the individual genetic make-up is believed to be crucial in mediating the pathological effects of stress (Snyder et al., 1974; Crow, 1979; Stevens, 1979; Ciaranaello and Boheme, 1982). Twin studies on the genetic basis of schizophrenia give concordance rate for this disorder of less than 100 % in monozygotic twins, thus indicating a role of environmental factors (Kety, 1982; Diedern, 1983). Disturbed behavior is therefore considered to result from the interaction between environmental pressure (stress) and genetically susceptible individuals (Hirsh, 1979; Gottesman and Shield, 1982). According to this view a psychotic outcome will result from genotype-dependent responses of the organism to stressful experiences, thus not only from environmental pressure or genotype, but, also, from the interaction between genotype and stress.

Over the last two decades, the concept of stress as a non-specific adaptive response of the organism to physical factors has been modified as both experimental and clinical research have pointed to specific kinds of stress responses and to a major role of cognitive factors in determining stress. In the light of these new ideas stress can be viewed as one possible result of changes in the organism-environment interaction. An organism is an open system constantly interacting with its environment so that changes in the environment necessarily produce changes in the organism (Bertalanffy, 1968). When the environment is modified , the organism detects discrepancies between observed and expected events and attempts to restore the expected situation. This phase is characterized by a particular picture of physiological changes commonly termed *arousal* (Lindzay et al., 1975). If attempts to restore the expected situation fail, given the characteristics of the environment, the organism undergoes a crisis that we call *stress*. This crisis produces a number of significant alterations in the organism's functioning which lead either to a new equilibrium between the organism and the environment or to pathology. The stress response may involve different physiological systems depending on the characteristics of the stressor, on the hystory of the individual (i.e. previous stressful experiences) and on genetic makeup.

In the last decade a large body of evidence has accumulated which points to a major role of the brain dopamine (DA) system in the etiology of schizophrenia, and psychotic outcome has been considered to result from DA hyperactivity (Kety, 1982; Diedern, 1983). In view of these hypothesis and of that concerning the role of genotype in the etiology of these pathologies, a number of investigations have been carried out in laboratory animals on

S. Puglisi-Allegra and A. Oliverio (eds.), Psychobiology of Stress, 171–182.

genetic differences related to brain DA functioning. Studies conducted on inbred strains of mice have established that the concentration and turnover of DA, as well as the specific activities of its synthesizing degradative enzymes and receptor density, frequently show genotypic variations and regional selectivity in the brain (Fink and Reis, 1981; Severson et al., 1981; Ciaranello and Boheme, 1982; Kempf et al., 1982).

These studies based on neurogenetic or pharmacogenetic approaches tend to look mainly for single-gene effects and generally try either to control environmental influences or to ignore them (Plomin and Deitrich, 1982). However, in the light of the results of genetic studies of schizophrenia, investigations in laboratory animals should not only take into account the role of environmental factors as sources of variance, but should focus their attention on the influence of environmental factors on neural and behavioral plasticity. Consequently, the study of genotype-dependent differences (not only in terms of intensity but above all in terms of neural and behavioral 'strategies') in adaptive responses to environmental pressure may be of paramount interest in elucidating the role of genetic makeup in neural and behavioral adaptive responses of individuals subjected to stressful experiences.

In recent years a role of endogenous opioids in the development of schizophrenia was also envisaged (Oliverio et al., 1984, for a review) and a linkage of these peptides with DA systems has been hypothesized to be a neurochemical substrate of stress-induced psychotic outcome. In fact, opiate receptors located on DA terminals in the nigro-striatal and mesolimbic DA systems have been shown to decrease DA release (Oliverio et al., 1984, for a review). Therefore, it has been suggested that overproduction of endogenous opioids in response to recurrent stressors may lead to the development of altered sensitivity of DA receptors resulting in dopaminergic mulfunctioning (Amir et al., 1981).

Our work stems from this hypothesis and our purpose was to search for an animal model of alterations in brain DA system functioning resulting from recurrent endogenous opioids modulated changes in DA metabolism produced by stress. In particular we turned our attention to the role of genotype as a regulating factor in the adaptation of brain DA system to stressful experiences.

Genotype-dependent effects of Chronic stress on DA mediated behavior in the mouse

Mice of inbred C57BL/6 and DBA/2 strains were used throughout our studies. In fact, these two strains are characterized by different responsiveness to opiate-induced dopamine metabolism alterations (Oliverio et al., 1983). They were thus considerd a useful tool for investigating genotype-dependent differences in responsiveness of the DA system to opioid mediated environmental pressure (stress).

Mice of the DBA/2 strain subjected to 10 days of 2 h restraint and tested 24 h after the last stressful experience, presented a clear-cut increase of climbing behavior (commonly considered a reliable index of central DA system functioning (Cabib et al., 1984; 1985, for reviews) following the administration of low, presynaptic doses of the DA agonist apomorphine (APO). These results indicate that chronic stress affects DA system responsiveness, as shown by sensitization to APO, and are consistent with the hypothesis of DA 'hyperactivity' (Snyder et al., 1974).

By contrast, mice of the C57BL/6 strain subjected to the same experimental procedure exhibited a dramatic decrease in climbing behavior in comparison with non-stressed animals.

These results suggest that chronic stress induces opposite modifications of dopaminergic functioning in DBA and C57 strains. Further experiments were carried out in order to investigate the role of genotype in these effects of stress on sensitivity to APO. A quantitative genetic analysis involving F1 (B6D2 and D2B6) and F2 hybrids and the backcross populations (F1 x C57; F1 x DBA) showed that the alteration of behavioral effects of APO characteristic of chronically stressed C57 mice is inherited through a dominant mode of inheritance; in actual fact, a complete dominance was observed. A role of major genetic factors in the above-mentioned effects of stress was also envisaged. Moreover, a relatively high value of environmental variance was observed which indicates that the different genotypes respond to the same environmental constraints in different ways, that is, the environmental effects on phenotypic expression are not independent of genotype, but a genotype x environment interaction is present (Cabib et al., 1985). These results showed that ours could be considered as an animal model characterized by a potential heuristic value for the study of the interaction between environmental (stress) and genetic factors in alterations of brain DA systems.

Moreover they suggested the hypothesis that the opposite effects of repeated stressful experiences on responsiveness to APO are to be ascribed to the opposite effects of the endogenous opioids released during stressful experiences (restraint) on brain DA metabolism in the two strains of mice (Oliverio et al., 1983).

Effects of acute stress on brain DA metabolism in the mouse

To assess this hypothesis we carried out a study on the effects of restraint on brain DA metabolism measured in different brain areas, collected by micropunching, using High Performance Liquid Chromatography (HPLC) coupled with electrochemical detection (EC) as previously described (Cabib et al., 1988 a). The results showed that restraint stress induced an increase of dihydroxyphenylacetic acid (DOPAC)/ dopamine (DA) and homovanillic acid (HVA)/DA ratios and a reduction of 3-methoxytyramine (3-MT)/ DA ratio in the caudatus putamen (CP) and nucleus accumbens septi (NAS) in mice of both C57BL/6 (C57) and DBA/2 (DBA) strains. These effects were already evident after 30 min stress in the NAS, while in the CP it took 120 min the effects of stress to emerge. Restraint did not produce any effects on dopaminergic metabolism in the frontal cortex (FC) of the C57 strain either after 30 or after 120 min, stress while in mice of the DBA strain a time-dependent effect of stress on the HVA/DA ratio was evident (Cabib et al., 1988 a).

When B6D2F1 hybrids were considered, the effects produced by 120 min restraint in the CP and the NAS paralleled those observed in parental strains while in the FC 120 min stress induced the same increase in HVA as that observed in DBA mice, thus suggesting that the pattern of response in the FC characterizing the DBA strain may be inherited through a dominant pattern of inheritance.

The lack of relationship between the pattern of inheritance of FC response to acute restraint and that of the behavioral alterations induced by chronic exposure to stress in F1 hybrids rules out this possibility (Cabib et al., 1985; 1988 a).

Our results, indicating an increase of the DOPAC/DA and the HVA/DA ratios and a decrease of the 3-MT/DA ratio in the CP and in the NAS after 120 min restraint, suggest that this stress produces an inhibition of release in the two structures. In fact, the significant reduction of this metabolite in stressed animals could represent an indication of a functional reduction of released DA (for a review Westerink and Spaan, 1982).

These results do not account for the genotype-dependent differences in responsiveness to the behavioral effects of APO produced by repeated stressful experiences. In this study we were interested in the possible differences in strain sensitivity ti 120 min restraint which could be related to the opposite alterations found after chronic exposure to this type of stressor. The absence of relationship between the pattern of inheritance of FC response to acute restraint and that of the behavioral alterations induced by chronic exposure to stress in F1 hybrids rules out this possibility. In the light of the results presented so far the hypothesis arises that the different kind of sensitization effects to APO-induced stereotypic behavior described above have to be ascribed to different genotype-depenent adaptation of DA system following repeated restraint stress.

Involvement of the endogenous opioid system in the modulation of behavioral and biochemical effects of stress in the mouse

The data presented above are not consistent with the hypothesis of a different strain-dependent modulation of DA release by endogenous opioids during stress. In fact, reatraint produces similar effects on DA metabolites, in particular on 3-MT, in both C57 and DBA strain.

However, we subsequently observed that endogenous opioids are somehow involved in the sensitization effects produced by repeated restraint in C57 and DBA mice. In fact, the opioid receptor antagonist naltrexone (NTRX), injected before each stressful experience, completely prevented either the decrease or the increase of APO-induced climbing produced by chronic stress (FIG. 1), thus suggesting that endogenous opioids are involved in the development of behavioral alterations induced by chronic stress in the mouse, possibly through a modulatory effect on DA transmission. To test this hypothesis we investigated the effects of NTRX pretreatment on DA and its metabolites in mice (DBA) subjected to 2 h restraint. NTRX (2.5 and 5 mg/kg) was shown to prevent the decrease of 3-MT/DA ratio induced by restraint in the NAS while it did not affect the stress-induced decrease of the DA metabolite in the CP (Cabib et al., 1989).

These results point to a major role of the endogenous opioid system in inducing the effects of restraint on DA metabolism in mesolimbic system. Moreover, they strongly suggest that the antagonizing effects of NTRX on sensitization to APO produced by repeated stressful experiences reported previously are to be ascribed to the action of the opiate receptor antagonist on DA metabolism in the NAS.

Genotype-dependent adaptation of DA system to chronic stress

Different adaptation of the DA system, may possibly lead to different responsiveness to restraint throughout the chronic stressful experience. Moreover, it might be that at the moment of test for APO challenge, that is 24 h after the last stressful experiences, changes in DA activity were still present which would modify the response to APO. We observed, however, that in both C57 and DBA mice the effects produced by restraint on DA and its metabolites in CP and NAS of naive mice were still present in mice tested on the tenth day of repeated stress, a result indicating that neither habituation nor potentiation of the effects

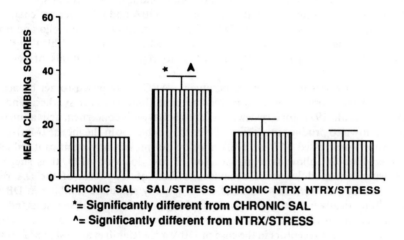

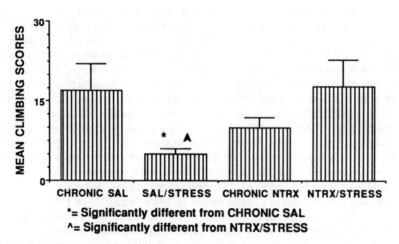

Figure 1. Effects of NTRX pretreatment on chronic stress-induced alterations of climbing behavior in two strains of mice.

of restraint were developed in these mice. Moreover, the effects of stress on DA metabolism disappeared 24h after the last stressful experience, thus ruling out any influence of metabolic alterations still present at the moment of the APO challenge in the climbing test (Cabib et al., 1988 b).

Therefore, the effects on APO-induced climbing behavior observed in C57 and DBA mice following chronic stress have to be ascribed to different genotype-dependent adaptations of DA receptors to the condition of altered transmission induced by restraint stress.

Both behavioral and biochemical studies indicate that DBA and C57 mice are characterized by different density and distribution of DA receptors in the brain (Ciaranello and Boehme, 1982; Cabib and Puglisi-Allegra, 1985; Seale et al., 1984). Thus, it is possible that these two strains develop different patterns of adaptation to repeated alterations of central DA transmission.

At the range of doses used in our experiments, APO is supposed to act by activating presynaptic DA receptors (autoreceptors) mediating inhibition of both synthesis and release of DA (Martres et al., 1977, for a review). As far as behavior is concerned, activation of these receptors is known to produce inhibition of climbing behavior in the mouse (Martres et al., 1977). However, unstressed C57 and DBA mice do not present any significant reduction in spontaneous climbing following administration of low doses of APO, thus suggesting a genotype-dependent modulation of behavioral effects of low doses of the DA receptor agonist. After chronic stress low doses of APO actually stimulate climbing in DBA mice while they dramatically inhibit spontaneous climbing in C57 mice. These results indicate an hypersensitivity of presynaptic DA receptors in the case of C57 mice and, viceversa, an hyposensitivity of these receptors in the case of DBA mice (Cabib et al., 1985; Martres et al., 1977). The analysis of the behavioral effects of a postsynaptic dose (1 mg/kg) of APO seems to confirm this hypothesis. In fact, at this dose the increase in APO-induced climbing in stressed DBA mice is no longer evident (Fig. 2), a result in agreement with others (Martres et al., 1977) concerning the behavioral effects of presynaptic DA receptor hyposensitivity. The C57 pattern is opposite and at the same time puzzling. In fact, the results shown by this strain, namely a reduction of climbing at both pre- and postsynaptic doses of APO following chronic stress seem to point to a presynaptic receptor hypersensitivity accompanied by a postsynaptic receptor hyposensitivity. It cannot be ruled out, however, that at higher doses of APO no differences between stressed and non-stressed mice would be evident, because stronger postsynaptic receptor stimulation would mask the effects on presynaptic receptors.

In order to further evaluate the involvement of DA presynaptic receptors in the sensitization to APO following chronic stress we investigated the effects of both a pre- and a postsynaptic dose of APO on DA metabolism in CP and NAS of stressed C57 and DBA mice. APO is known to decrease DA metabolite concentration in the abovementioned brain areas. This effect is the consequence of decreased release and synthesis induced primarily through presynaptic receptor stimulalation and secondly through a feedback loop when postsynaptic receptors are involved (Skirboll et al., 1979). Mice subjected to chronic intermittent stress were injected (20 min before sacrifice) with 0.25 or 1 mg/kg of APO 24 h after the last stressful experience and the effects of the DA agonist on DA metabolites were observed and compared to those produced by the drug in control animals.

APO (0.25 and 1 mg/kg) decreased all DA metabolites in the CP and NAS of both C57 and DBA mice (Fig. 3). After chronic stress, the effects of APO on striatal DA metabolism were unchanged as far as C57 mice were concerned, while a significant reduction of the effects of 0.25 mg/kg was evident in DBA mice.

By contrast, in the NAS C57 mice exhibited an increase of the effects of APO on DA metabolites at both doses of the DA receptor agonist while DBA mice showed a reduction of these effects only at the dose of 0.25 mg/kg.

It is worth noting that the results of the effects of APO on DA metabolites in the NAS of the two strains of mice parallel those referring to the effects of the DA agonist on climbing

CLIMBING SCORES

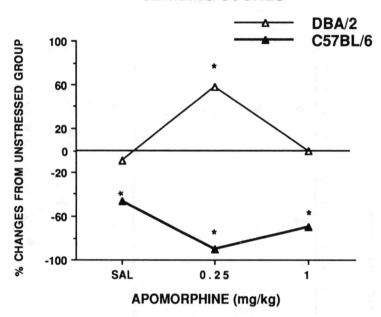

Figure 2. Effects of chronis stress on climbing behavior induced by different doses of apomorphine in two strains of mice. Results are expressed as percentage of change in comparison with unstressd group. * = significantly different in comparison with unstressed mice injected with the same dose of apomorphine (P<.01 by the Student's t test, two tailed).

behavior. In fact, as summarized in fig. 2, which shows the effects of 0.25 and 1 mg/kg of APO on climbing and DA metabolites in chronically stressed and non-stressed C57 and DBA mice, hyposensitivity to the effects of the DA agonist on 3-MT accompanies behavioral sensitization in stressed DBA mice following the administration of 0.25 mg/kg, while no stress effect is evident at the postsynaptic dose. Moreover, hypersensitivity to the effects of both doses of APO on DA metabolites accompanies behavioral hyposensitivity in stressed C57 mice.

The reduction of the ability of 0.25, but not 1 mg/kg, of APO to decrease DA metabolites concentration in the CP and NAS of DBA mice is in good agreement with the hypothesis of a hyposensitivity of DA presynaptic receptors. In fact, it has been shown that hyposensitivity of these receptors following APO administration results in a decreased ability of low doses of the agonist to reduce striatal HVA levels (Martres et al., 1977), in the absence of changes on basal levels of HVA or in the effects of high doses.

The results concerning C57 mice are more puzzling. Increased ability of APO to decrease DA metabolites may depend on hypersensitivity of both presynaptic and postsynaptic receptors (Martres et al., 1977). In the first case this effect is to be ascribed to an increase of presynaptic action (inhibition) on DA release, in the second, to increased activation of inhibitory neuronal feedback pathways (Skirboll et al., 1979).

Therefore, the increased ability to reduce DA metabolites in the NAS by either a low or a

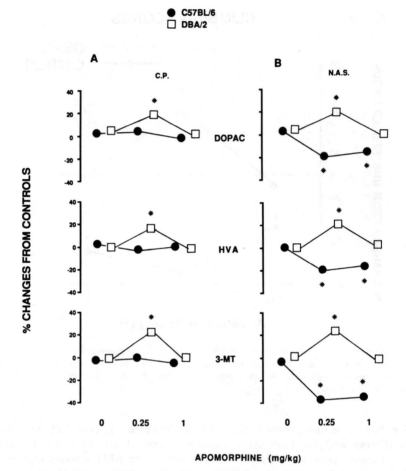

Figure 3. Effects of chronic stress on inhibition of DA metabolism induced by different doses of apomorphine in two strains of mice. Results are expressed as percentage of change in comparison with unstressd group. * = significantly different in comparison with unstressed mice injected with the same dose of apomorphine (P<.01 by the Duncan test).

high dose of APO may suggest a sensitization of postsynaptic DA receptor in the NAS of C57. It is worth noting, however, that the relative increase of the APO effects observed at the dose of 1 mg/kg in stressed mice compared with controls is no stronger than that observed at the dose of 0.25 mg/kg, thus possibly indicating that the postsynaptic dose (1 mg/kg) is not strong enough to mask the effects of APO on hypersensitized presynaptic receptors. In addition to this, we observed a decrease of the behavioral effects of 1 mg/kg of APO in C57 mice, which thus rules out a hypersensitivity of postsynaptic receptors. Moreover, post-synaptic D2 receptor sensitization is accompanied by an increase of HVA steady levels (Martres et al., 1977), an effect that we did not observe 24 h after the end of chronic intermittent stress treatment (Cabib et al., 1988 b).

It must be pointed out that 24 h after the last of 10 daily stressful experiences, drug-free (saline-injected) mice present a decrease in climbing behavior (Fig.2). Since no alteration in DA metabolism is evident 24 h after the last restraint session in NAS and CP (Cabib et al., 1988 b), one may speculate that hyposensitivity of postsynaptic receptors in the absence of significant changes in DA functioning, leads to decrease of spontaneous (drug-free animals) climbing behavior.

Further investigations with selective D1 and D2 receptor agonists or antagonists also using other experimental approaches such as binding studies could possibly point to an involvement of postsynaptic receptors in the genotype-dependent effects of chronic stress on APO responsiveness.

In fact, very recent studies have pointed to a complementary role of D1 and D2 receptors in the expression of DA agonist-induced behaviors. Since it has been shown that stimulation of both types of receptors is necessary in order to obtain the full expression of DA agonist-induced behaviors (Clark and White, 1987, for a review), it may be that imbalance between D1 and D2 receptors, due to different adaptation to chronic modification of DA release induced by stress, accounts for the altered behavioral effects of the mixed D1 and D2 receptors agonist APO. This hypothesis is not inconsistent with an imbalance between pre- and postsynaptic receptors, since several authors have suggested a different pre- and postsynaptic distribution of D1 and D2 receptors (Clark and White, 1987).

In order to evaluate this point, we studied the effects of chronic intermittent stress on D1 and D2 DA receptors in CP and NAS of C57 mice.

In vivo binding of 3H-spiperone (D2 antagonist) and 3H-SCH 23390 (D1 antagonist) was assessed according to Van Der Werf et al. (1984) and Frost et al. (1987). Chronic stress produced slight changes of Bmax values for both D1 ($+9\%$) and D2 (-11%) binding sites in the CP and a clearcut decrease of D2 (-64%) receptors in the NAS, while no effects on D1 receptors was evident in this brain area. These results show that chronic stress produces an increase in D1/D2 receptor ratio in the NAS due to the dramatic decrease of the D2 type, while it produces a very slight increase of the D1/D2 ratio in the CP.

The data presented above point to a major role of the NAS in the behavioral effects produced by chronic stress. In fact we observed that: 1) Acute stress produces stronger effects on DA metabolism in NAS than in CP; 2) NTRX, which antagonizes the behavioral effects produced by chronic stress, prevents the stress-induced decrease of 3-MT in NAS but not in CP; 3) Differences in the effects of APO on DA metabolism of stressed C57 mice compared to control mice are only evident in the NAS. These results strongly suggest that changes in D2 receptors in the NAS observed in the binding study are related to changes in climbing behavior, both in free-drug and in APO injected animals. In fact, an autoreceptor (presynaptic) 'reserve' has recently been shown which is activated by low doses of DA agonists or by irreversible antagonist binding sites inactivation (Meller et al., 1986). Therefore, it may be that the decrease in D2 receptors observed in the NAS depends on decreased postsynaptic receptors, the presynaptic receptors being unaffected because of the 'reserve'.

If so, and since there is no reason to hypothesize a reduction of autoreceptors located in the cell body in the Ventral Tegmental Area, these results are consistent both with the reduced effects of postsynaptic effects of APO (1 mg/kg) observed in the behavioral test and with the increased presynaptic effects (.25 mg/kg) observed in neurochemical assays.

Discussion

The results presented so far show that the genotype plays a crucial role in the determination of the pattern of adaptation an organism will adopt under environmental pressure. Restraint (up to 2 h) produces a decrease of DA release (measured in terms of 3-MT concentrations) in both C57 and DBA mice mostly in NAS. In the DBA strain, daily repeated (10 days) stressful experiences produce hyposensitivity of DA receptors (possibly presynaptic) modulating inhibition of DA release and metabolism in the NAS as shown by behavioral (climbing test) as well as biochemical (DA and its metabolites) results. Such a hyposensitvity of receptors regulating the decrese of DA release seems to be a sort of counter-regulation process preparing the mesolimbic system to compensate for the effects of restraint on DA release. Counter-regulation processes characterize homeostatic adaptation.

By contrast, the C57 strain exhibits hypersensitivity of those DA receptors which modulate the inhibition of DA release in the NAS. Such a sensitization, evident also on behavioral grounds, may be considered an over-regulation process which would allow the organism to favour the stress response: i.e. decrease of released DA. Moreover, the reduction of D2 receptors in the NAS of chronically stressed animals is a complementary aspect of decreased DAergic tone in the mesolimbic systems of this strain of mice.

According to this view, also the effects of chronic stress on the C57 genotype may be considered part of an adaptation pattern to stress and a process leading to a new equilibrium between the organism and the environment, namely an heterostatic adaptation (Silye, 1983) (Fig. 4).

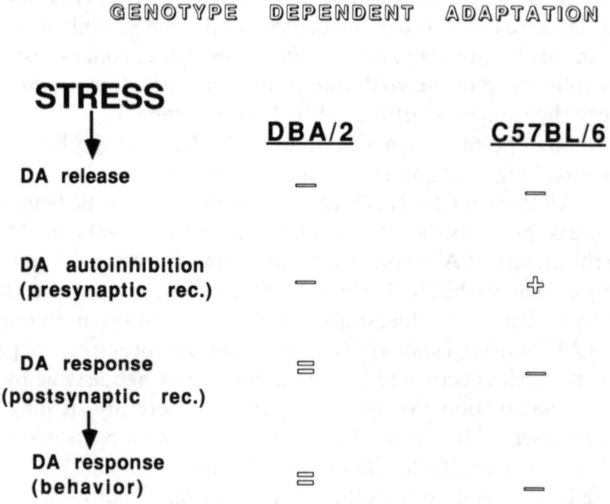

GENOTYPE DEPENDENT ADAPTATION

STRESS ↓	DBA/2	C57BL/6
DA release	−	−
DA autoinhibition (presynaptic rec.)	−	+
DA response (postsynaptic rec.)	=	−
DA response (behavior)	=	−

Figure 4. Scheme representing two different genotype-dependent adaptation patterns involving DA receptors in the NAS (see text for details).
Stress = chronic intermittent restraint
(−) : decrease
(+) : increase
(=) : no change

The animal model presented here shows that genotype may affect the interaction between the organism and the environment not only by modulating individual differences in responsiveness to stressors or in resistance to stress, but also by determining the kind of adaptation pattern to recurrent stressful experiences or, in other words, affecting the neurobiological 'strategies' of the organism facing environmental pressure.

Acknowledgements

Part of this work was supported by a grant from the Italian CNR and French INSERM (CNR/INSERM twining grant n.8; 1987-88)

References

Amir, S.Z., Brown, Z.W., Amit, Z. (1981) 'The role of endorphins in stress: Evidence and speculation', *Neurosciences Biobehavioral Review* 4, 77-86

Bertalanffy, L. von, (1968) 'General System Theory, Essays on its Foundation and Development', Braziller, New York (1968)

Cabib, S. and Puglisi-Allegra, S. (1985) 'Different effects of apomorphine on climbing behavior and locomotor activity in three strains of mice' *Pharmacology Biochemistry and Behavior* 23, 555-557

Cabib, S., Puglisi-Allegra, S. and Oliverio A., (1984) 'Chronic stress enhances apomorphine-induced steretyped behavior in mice: involvement of endogenous opioids', *Brain Research*, 298, 138-140

Cabib, S., Puglisi-Allegra, S. and Oliverio, A. (1985) 'A genetic analysis of Stereotypy in the mouse', *Behavioral and Neural Biology*, 44, 239-248

Cabib, S., Kempf, E., Schleef, C., Oliverio, A. and Puglisi-Allegra, S. (1988 a) 'Effects of immobilization stress on dopamine and its metabolites on different brain areas of the mouse: role of genotype and stress duration', *Brain Research*, 441, 153-160

Cabib, S., Kempf, E., Schleef, C., Mele A. and Puglisi-Allegra, S. (1988 b) 'Differents effects of acute and chronic stress on two dopamine-mediated behaviors in the mouse', *Physiology and Behavior*, 43, 223-227

Cabib, S., Oliverio, A. and Puglisi-Allegra, S. (1989) 'Stress-induced decrease of 3-methoxytyramine in the nucleus accumbens of the mouse is prevented by naltrexone pretreatment', *Life Sciences*, 45 (12), 1031-1037

Ciaranello, R.D. and Boheme, R. (1982) 'Genetic regulation of neurotransmitter enzyme and receptors: Relationship to the inheritance of psychiatric disorders' *Behavior Genetics* 12, 11-35

Clark, D. and White, F.J. (1987) 'Review: D1 dopamine receptors- the search for a function: A critical evaluation of the D1/D2 dopamine receptors classification and its functional implications', *Synapse* 1, 347-388

Crow, T.J. (1982) 'Biological basis of mental disorders: The case for viral aetiology'. In: Psychobiology of Schizophrenia , Namba, M. and Kaya, H. (eds.) Oxford, Pergamon Press pp. 249-264

Diedern, I. (1983) 'Genetics of schizophrenia'. In: Behavior Genetics: Principles and Applications, par Fuller, J.L. and Simmel E.C. (eds.), Hillside NJ, Erlboum Assoc. pp. 189-215

Fink, J.S. and Reis, D.J. (1981) 'Genetic variation in dopamine cell number: Parallel with differences in response to dopaminergic agonists and in naturalistic behaviors mediated by central dopaminergic system', *Brain Research* 222, 335-349

Frost, J.J., Smith, A.C., Kuhar, M.J., Dannalas, F.R. and Wagner, H.N. (1987) 'In vivo binding of 3H-Methylspiperone to dopamine and serotonine receptors', *Life Sciences* 40, 987-995

Gottesman I.I. and Shields, J. (1982) 'Schizophrenia: The Epigenetic Puzzle'. New York, Cambridge Univ. Press

Hirsh, S.R.(1979) 'Do parents cause schizophrenia?' *Trends in Neuroscience* 2, 1659-1664

Kempf, E., Mandel, P., Oliverio, A. and Puglisi-Allegra, S. (1982) 'Cyrcadian variation of noradrenaline, 5-hydroxitryptamine and dopamine in various brain areas of C57BL/6 and BALB/c mice', *Brain Research* 232, 472-478

Kety, S.S.(1982) 'Neurochemical and genetic basis of psychopathology: Current status', *Behavior Genetics* 12, 95-100

Lindzay, G., Hall, C. and Thompson, R.F. (1975) 'Psychology' Worth Publishers, N.Y.

Martres, M.P., Costentin, J., Baudry, M., Marçais, H., Protais P. and Schwartz, J.C. (1977) 'Long terme changes of pre- and post-synaptic dopamine receptors in mouse striatum evidenced by behavioral and biochemical studies', *Brain Research* 136, 319-337

Meller, E., Helmer-Matyjek, E., Bohmaker, K., Adler, C.H., Friedhoff, A.J. and Goldstein, M.(1986) 'Receptor reserve at striatal dopamine autoreceptors: Implications for selectivity of dopamine agonists', *European Journal of Pharmacology* 123, 311-314

Oliverio, A., Castellano, C. and Puglisi-Allegra, S. (1983) 'Psychopharmacogenetics of opioids', *Trends in Pharmacological Sciences*, 4, 350-352

Oliverio, A., Castellano, C. and Puglisi-Allegra, S. (1984) 'Psychobiology of opioids', *International Review of Neurobiology* 25, 277-337

Plomin, R. and Dietrich, R.A. (1982) 'Neuropharmacogenetics and behavioral genetics'. *Behavior Genetics*, 12, 111-121

Seale, T.W.K., McLanahan, K., Carney, J.M. and Rennert, M. (1984) 'Systematic comparison of apomorphine-induced behavioral changes in two mouse strains with inherited differences in brain dopamine receptors'. *Pharmacology Biochemistry and Behavior*, 21, 237-244

Selye, H. (1983) 'The stress concept: Past, present and future', In: Cooper, G.L. (Ed.) Stress Research, John Wiley and Sons, pp 1-20 (1983)

Severson, J.A., Randal, P.K. and Finch, C.E. (1981) 'Genotypic influences on striatal dopaminergic regulation in mice', *Brain Research* 210, 201-215

Skirboll T.R., Grace, A.A., and Bunney, B.S. (1979) Dopamine auto- and post-synaptic receptors: electrophysiological evidence for differential sensitivity to dopamine agonists, *Science* 206, 80-82

Snyder, S.H., Barenjee, S.P., Yamamura, H.I. and Greenberg, D. (1974) Drugs, neurotransmitter and schizophrenia, *Science*, 184, 1243-1253

Stevens, J.R. (1979) Schizophrenia and dopamine regulation in the mesolimbic system. *Trends in Neurosciences* 2, 102-105

Van Der Werf J.F., Van Het Schip, F, Sebens J.B. and Korf J. (1984) Quantification of in vivo spiperone binding in the rat striatum after lesions produced by kainate or decortication. *European Journal of Pharmacology*, 102, 387-399.

Westerink, B.H.C. and Spaan S.J. (1982) On the significance of endogenous 3-methoxythyramine for the effects of centrally acting drugs on dopamine release in the rat brain. *Journal of Neurochemestry* 38, 680-686

ADRENOCORTICAL AND CENTRAL MONOAMINERGIC SYSTEM RESPONSES TO DIFFERENT STRESSFUL SITUATIONS IN YOUNG AND SENESCENT RATS

S. Algeri, L. Biagini, P.Garofalo, M. Marconi, N. Pitsikas and G. Sacchetti.
Istituto di Ricerche Farmacologiche 'Mario Negri', via Eritrea 62, 20157 Milan, Italy

Activation of the hypothalamus-hypophisys-adrenal axis by the CNS, is central to the stress response. The experiments presented here, were aimed to study some of the responses to different kinds of stress in rats of different ages. Changes in corticosterone (Gc) secretion were studied as an index of adreno-cortical stimulation and changes in catecholamine metabolism were taken as an index of activation of some cerebral pathways. Aged rats presented higher basal Gc secretion and the increase wich followed application of a stressor was similar to or higher than that observed in young rats. No difference was observed in the kinetic of Gc secretion.

These results indicate that aged rats have a basal cortico-adrenal hypertonus. However the secretory response to stress stimulation does not appear modified in these animals.

After stress, dopamine and noradrenaline metabolism was enhanced in the limbic area and hippocampus and striatum of young rats. In old rats, the dopamine metabolism response to stress was reduced while that of noradrenaline was enhanced.

In another experiment we investigated the ability of adaptation to a repeated stress situation in rats of different ages. A mild stressor such as a novelty was used. Exposure of rats to a novel environment resulted in increased dopamine metabolism in the limbic area, which was significant in all the age group considered. When the stressor was repeated several times, a decrease was seen in the metabolic effect in young but not in old rats. This can be interpreted as an index of habituation.

Stress is a set of physiological and behavioral responses by which an animal reacts against the disruptive action of the environment. The pivotal role of the central nervous system (CNS) in the activation and control of these mechanisms is well documented, although not all the specific nuclei and pathways involved and their interrelations have been clarified. Apart from the hypothalamus, it is probable that different neuronal pathways are involved in the sequence of responses to stress.

A stressor has to be received as a single or multiple input to the CNS which are elaborated at different encephalic levels and then transduced into outputs activating the appropriate physiological and behavioral responses. For instance, activation by stress of the hypothalamic CRF releasing neurons which stimulate ACTH secretion by the adeno-hypophysis is well known but the intervention of other neuronal pathways is less documented. Activation of mesolimbic and mesocortical and, in some circumstances, striatal dopaminergic and hippocampal noradrenergic neurons has been reported in several laboratories (Curzon G. et al., 1979; Fadda F. et al., 1978; Glavin G. B., 1985, Thierry A.

S. Puglisi-Allegra and A. Oliverio (eds.), Psychobiology of Stress, 183–190.
© 1990 Kluwer Academic Publishers. Printed in the Netherlands.

M., et al., 1976) Since during aging there is a deterioration in several homeostatic and adaptive responses, we thought it interesting to investigate whether and how stress responses of senescent rats differ from those in young controls.

We assessed the responses to an acute psychological stress and the pattern of these responses after repetitive presentation of a mild psychological stressor, such as a noise or an unusual new cage, in other words, the animal's ability to adapt to new environmental situations. As biochemical parameters of the response to stress we measured plasma corticosterone (Cx), as an index of the activation of the hypothalamus-hypophysis- adrenal system and the changes in synthesis or metabolism of dopamine (DA) and norepinephrine (NE) as an index of activation of some central neuronal pathways.

Methods

Groups Sprague-Dawley rats (CD-COBS, Charles River Italy) of different age were studied. Rats were received when they were one month old and kept in standard animal housing conditions untill the age required for the study; the animals were divided, from their arrival into two sub-populations, both with free access to food but one fed a standard rodent diet while the other was fed a diet in which 50% of the carbohydrates and lipids was replaced with vegetable fibers.

Stress was induced either by the production of a brief and sudden unusual noise or by placing the rat for 30 minutes in a larger cage without the usual sawdust bedding.

Concentrations of plasma corticosterone and brain catecholamines and their metabolites were measured respectively by UV absorbtion and electrochemical detection after their separation on HLPC (Shimizu et al., 1983; Achilli et al., 1985).

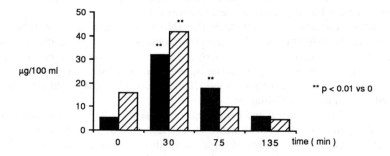

black column : 5 months
stripped column : 28 months

Fig. 1. Effect of sudden noise on the elevation pf plasma corticosterone in young (5 months) and old (28 months) rat.

Results and Discussion

Effect of a psychological stressor on corticosterone secretion

The time course of the increase in plasma Cx after a stress was not modified in old rats (fig. 1); the effect of a sudden noise on this parameter was compared in 5 and 28 month-old rats.

As reported by others authors (Sapolski et al., 1983), basal plasma Cx is often higher in old rats and it sometimes rises further in response to stress (fig 2). Sapolski et al. (1983) suggest this may be due to a deterioration of central feed-back controls as a result of degeneration of a Cx receptor-bearing neuronal population (Sapolski et al., 1983; Meaney et al., 1988).

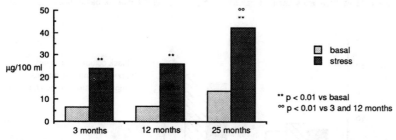

Fig. 2. Effect of age on the elevation of plasma corticosterone by exposure to a novel environment

This was not observed in a rat population fed a hypocaloric diet and placed in an unfamiliar environment; in the same situation normally fed rats presented higher basal Cx levels already at 14 months (fig 3).

In this experiment no age-related differences in stress-induced Cx increase were seen. In the light of Sapolski's hypothesis,it may be interesting to note that these rats performed significantly better than the age-matched normally fed rats, in some cognitive tasks (paper in preparation) where the hippocampus seems to play an important role (Gallagher and Pelleymounter, 1988).

Different effects of stress on the activation of some central dopaminergic systems in rats of different ages

Stress is known to activate mesocortical and mesolimbic dopaminergic sytems (Thierry et al., 1976; Fadda et al., 1978). In agreement with these findings we observed an increase in the concentration of dopamine acid metabolites (HVA and DOPAC) in limbic areas of rats exposed for 30 minutes to a novel environment, indicating an enhancement of DA turnover and perhaps also of the activity of mesolimbic neurons. When the same kind of stress was applied to rats 3, 12 or 25 months old, the effect was present in all three age groups but was less significant in the older rats (fig 4), suggesting that these animals are more resistent to the central stimulation by stress or that their mesolimbic neurons are less responsive.

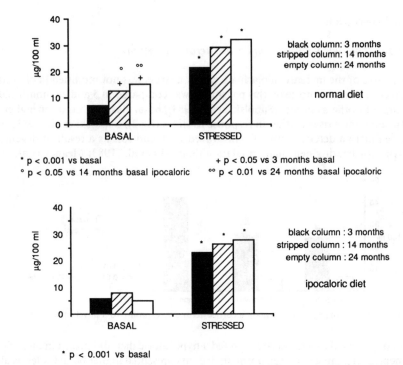

* p < 0.001 vs basal   + p < 0.05 vs 3 months basal
° p < 0.05 vs 14 months basal ipocaloric   °° p < 0.01 vs 24 months basal ipocaloric

* p < 0.001 vs basal

Fig. 3. Effect of open field stress on corticosterone secretion in rat of different ages fed normal or ipocaloric diet

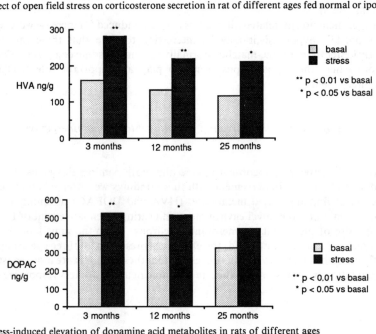

Fig. 4. Stress-induced elevation of dopamine acid metabolites in rats of different ages

Different effects of stress on the activation of hippocampal noradrenergic neurons in rats of different ages

Central noradrenergic systems have been reported to be stimulated by stress (Glavin, 1985). We observed stimulation of NE synthesis in hippocampal nerve terminals in 3- or 27- month-old rats under psychological stress due to restraint. NE synthesis was determined by inhibiting the enzyme DOPA decarboxylase and measuring accumulation of the precursor DOPA (Carlsson and Lindqvist, 1973) at different intervals during the stress procedure. Dopa basal levels were not affected by age (mean $\pm$ SEM: young 37.6 <7.9 ng/g, old 43.3 <1.5 ng/g). As discussed extensively elsewhere (Algeri et al., 1988) DOPA accumulation was biphasic in stressed rats but here we present only data obtained in the first 30 minutes of stress (fig 5) when accumulation of the metabolite was linear in all groups (non-linearity F test values: 0.48, 0.8, 0.9, 0.55 respectively for young controls, young stressed, old controls and old stressed). Regression analysis indicated that the accumulation rate in old rats was significantly higher than in the young group (angular coefficient 3 vs. 1.2, $p <0.05$). Analysis of the accumulation curves also showed that restraint stress significantly increased the rate of accumulation in both groups ($p<0.05$) but the increase was significantly more pronounced in old rats (angular coeficient 3.8 in old and 2.1 in young $p<0.05$).

A similar indication came from two-way analysis of variance of DOPA levels at the different intervals after stress and enzyme inhibition. At 3O min, young stressed rats had DOPA levels 25% higher ($p<0.05$) than the corresponding unstressed group; in old rats the difference appeared already after 15 min and was more pronounced (30 and 60% at 15 and 30 min respectively).

This greater enhancement by stress of noradrenergic tonus in old rats has also been observed in Wistar rats (Ida et al., 1982) and confirms our impression - based on our previous findings in the serotonergic system of aged rats (Algeri et al., 1982; 1987) – that aging does not always result in reduction of neuronal mechanisms. Of course, in view of the complex interneuronal balance this may reflect a deterioration of some of the normal CNS functions.

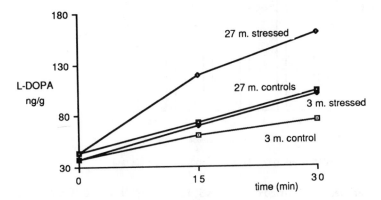

Fig. 5. Effect of stress on hippocampus of young and old rat

Adaptation of cortico-adrenal secretion and limbic area dopamine metabolism in rats of different ages after repeated stress

As presented above, exposure to a mild psychological stressor causes Cx hypersecretion and an increase in DA metabolism in limbic neurons. Many of the biochemical parameters that are increased by stress return to normal when the stress is repeated frequently and the animals become acquainted with the stressor. This adaptive mechanism is also evident for the increase in Cx secretion and DA metabolism caused by exposure to a novel environment: in fact Cx hypersecretion was progressively less intense as the number of exposures increased (fig. 6)

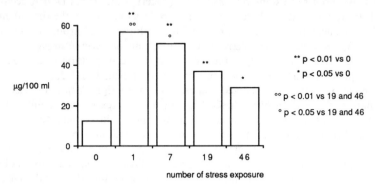

Fig. 6. Adaptation of corticosterone to chronic stress exposure in young rats

and the limbic HVA levels, which were significantly elevated after the first exposure, returned close to the levels in unstressed rats after the 7th exposure (fig. 7).

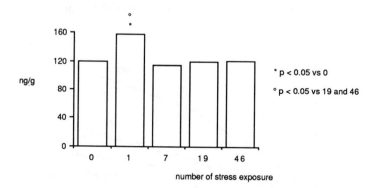

Fig. 7. Adaptation of HVA levels to chronic stress exposure in young rats

A similar experiment was carried out on rats 3, 12 and 25 months old. The effect of repetition of the stress on Cx secretion in the three age groups was to a certain degree unexpected (fig. 8): the rise after the first exposure was evident in all groups but the level remained constantly elevated in adults, while it returned to normal after the 25th exposure in young rats and even more rapidly in old rats (after the 7th).

The reasons for these fluctuations remain unclear. The data on limbic DA metabolism (fig. 9) which presumably indicate the status of central activation, show that in young and, although less significantly, in adult rats the levels of HVA are similar to the levels of unstressed rats but they continue to be elevated in old rats, suggesting that these animals continue to suffer the effect of stress. There seems to be a dissociation between the effects of stress on the hypothalamus-hypophysis-adrenal system and CNS activation.

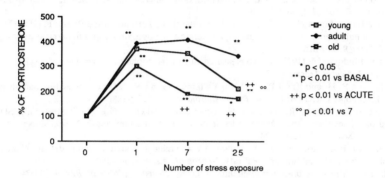

Fig. 8. Adaptation to chronic stress exposure of young (3 months), adult (12 months) and old (25 months) rats

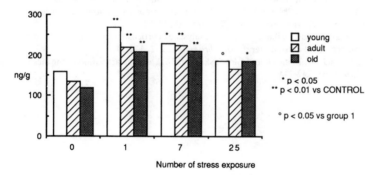

Fig. 9. Limbic HVA after chronic stress exposure in young (3 months), adult (12 months) and old (25 months) rats

Acknowledgements

This work was partially supported by the CNR (National Research Council, Rome, Italy), Contract No. 85.00421.56, S.P.M.P.R. 'Mechanisms of aging'.

References

Achilli G., Perego C. and Ponzio F. (1985) Application of the Dual-Cell Coulometric detector: a method for assaying monoamines and their metabolites. *Anal. Biochem.* 148, 1-9

Algeri S. , Calderini G., Lomuscio G., Vantini G., Toffano G. and Ponzio F. (1982) Changes with age in rat central monoaminergic system responses to cold stress. *Neurobiol Aging* 3, 237-242

Algeri S., Achilli G., Calderini G., Perego C., Ponzio F. and Toffano G. (1987) Age-related changes in metabolic responses to chronic monoamine depletion in central dopaminergic and serotonergic system of rats treared with reserpine. *Neurobiol Aging* 8, 61-66

Algeri S., Calderini G., Lomuscio G., Rocchetti M., Sacchetti G., Toffano G. and Ponzio F. (1988) Differential response to immobilization stress of striatal dopaminergic and hippocampal noradrenergic system in aged rats. *Neurobiol Aging* 9, 213-216

Carlsson A. and Lindqvist M. (1973) In vivo measurements of tryptophan and tyrosine hydroxylase activities in mouse brain. *J. Neural Trasm.* 34, 79-91

Curzon G., Hutson P. H. and Knott P. J., (1979) Voltammetry in vivo: Effect of stressfull manipulations and drugs on the caudate nucleus of the rat. *Br. J. Pharmacol.* 66, 127P-128P

Fadda F., Argiolas A., Melis M. R., Tissari A. H., Onali P. L., and Gessa G.L. (1978) Stress-induced increse in 3, 4-dihydroxyphenylacetic acid (DOPAC) levels in the cerebral cortex and in N. Accumbens: reversal by Diazepam. *Life Science* 23, 2219-2224

Gallagher M. and Pelleymounter M. A. (1988) An age-related spatial learning deficit: choline uptake distinguishes 'impaired' and 'unimpaired' rats. *Neurobiol. Aging* 9, 363-369

Glavin G. B. (1985) Stress and brain noradrenaline: a review. *Neurosci.Biobehav. Rev.* 9, 233-243

Ida Y.,Tanaka M., Kohono Y., Nakagawa R., Iimori K., Tsuda A., Hoaky Y. and Nagasaki N. (1982) Effect of age and stress on regional noradrenaline metabolism in the rat brain. *Neurobiol Aging* 3, 233-236

Meaney M. J., Aitken D. H., van Berkel C., Bhatnagar S., Sapolsky R. M. (1988) Effect of neonatal handling on age-related impairments associated with the hippocampus. *Science* 239, 766-768

Sapolsky R. M., Krey L. C., McEwen B. S. (1983) The adrenocortical stress-response in the aged male rat impairment of recovery from stress. *Exp. Gerontol.* 18, 55-64

Shimizu K., Amagaya S. and Ogihara Y. (1983) Analysis of corticosterone in the serum of mice and rats using high-performance liquid chromatografy. *J. Chromatog.* 272, 170-175

Thierry A. M., TassinJ. P., Blanc G. and Glowinski J. (1976) Selective activation of the mesocortical DA system by stress. *Nature* 263, 242-244.

APPROACHES TO STRESS IN MAN - PRESENT KNOWLEDGE AND FUTURE RESEARCH

Lennart Levi, M.D., Ph. D., Stockolm, Sweden

The evidence that environmental *physical* stressors can cause disease - in the sence that exposure, avoidance or manipulation of them increases, decreases, or removes the risk of becoming ill, or reverses ill health when it occurs - is established for a large number of stressors and diseases.

The role of extrinsic *psychosocial* stressors is not so clear, cf. Levi (1971, 1972, 1975 a & b, 1978, 1981 and 1987).

The Club of Rome correctly conceives the contemporary environmental and human problems of our planet as an 'untidy tangle' of interrelated issues. No doubt, the interactions between the different physical, psychosocial and health threads of this tangle are many and varied but only dimly understood.

A theoretical model

Our approach - in terms of research as well as environmental and health action - to this 'untidy tangle' human eco-system is facilitated, and its components can be more easily described, analyzed, interrelated and chosen as targets for coordinated environmental and health actions, if based on, and introduced into, the following reasonably heuristic human ecological model (Kagan and Levi, 1975; Figure 1).

As indicated in Figure 1 we are surrounded by nature. Humans modify nature through a variety of social structures and processes (box 1). Examples of social structures are a family, a neighbourhood, a school, a place of work, a hospital, a city, or a country. A social process is what takes place in such a structure - for example, care of children, education, work, medical care, etc.

These structures and processes are perceived and appraised. If these appraisals indicate significant and/or pronounced and/or persistent discrepancy between the subject's abilities, needs and expectations on the one hand, and environmental demands, opportunities and outcomes, as perceived, on the other, psychosocial stimuli arise (box 2).

The propensity to react - or not to react - to these stimuli depends on the organism's psychobiological 'programming' (box 3), conditioned by genetic factors and previous environmental influences.

Pathogenetic reactions (box 4) to these stimuli are of four types: emotional reactions (e.g. anxiety, depression, hypochondriasis), cognitive reactions (e.g. restrictions of scope of perception), behaviours (e.g. increased consumption of tobacco, alcohol, and certain foods), and physiological (e.g. neuroendocrine 'stress') reactions. These four types of closely

S. Puglisi-Allegra and A. Oliverio (eds.), Psychobiology of Stress, 191–201.

192

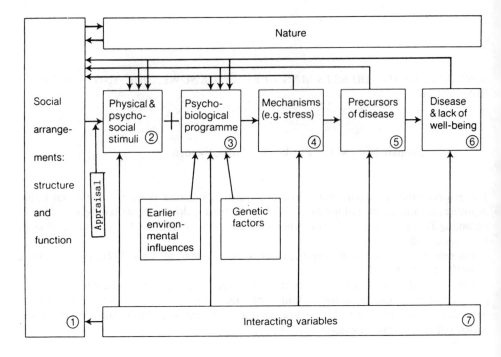

Fig. 1. Human ecological system. Human element detailed (Kagan and Levi, 1975).

interrelated and potentially pathogenic reactions may in turn lead to transitory disturbances of a number of mental and physical functions (box 5), and to functional and eventually also organic damage – for example, to the cardiovascular system (box 6).

The pathogenic process is modified by interacting variables (box 7), for example, the ability for a needed combination of emotion- and problem-oriented coping, and the availability and utilization of social support. The process is cybernetic. A person who sustains, say, a reactive depression, or a stress-induced myocardial infarction, influences people around him, who in turn influence him in a different way than previously. His propensity to react and the quality of his reactions, i.e. his programming, also change. In the worst case, a vicious circle develops, with successive reinforcement of the pathogenic process.

Whereas essential parts of the mechanisms (box 4) are genetically determined and have remained largely unchanged over the last few dozen of millennia, environmental demands and opportunities (box 1) have changed dramatically, as reflected in man's transition from hunter to farmer, craftsman, industrial worker, process supervisor, information processor (and unemployed). This may have created a potentially pathogenic person-environment mismatch.

Equally pathogenic is the *lack* of environmental change in a number of respects. At least 430 million human beings are malnourished or undernourished, due to mass poverty. About 60 per cent of the population in the developing countries has no access to safe potable water. 75 per cent has no access to basic hygiene – simple latrines, ways to dispose of household rubbish, and personal and food hygiene. Only 19 per cent has access to housing of acceptable quality. Four out of every five households in developing countries live in miserable huts in the countryside, or in shacks made of burlap, cardboard and sheet metal in the slums of the large cities.

These and related conditions act as extremely potent physical and psychosocial stressors (box 2), in the latter respect by posing threats, e.g., to survival, health, and well-being, to self-esteem, to close attachment with significant others, and to the sense of belonging to a valued group. These threats, in turn, can provoke potentially pathogenic reactions (mechanisms, box 4) which subsequently lead to superimposed increases in existing morbidity and mortality (boxes 5 and 6).

In practice, many stressful and potentially pathogenic psychosocial situations (box 1) fall into four general categories: uprooting; dehumanization of societal institutions; side-effects of the spread of innovations; and constraints on environmental and health programs and activities.

What we need to know is the 'content' of each 'box': which influences are pathogenic, in which individuals, by which mechanisms; which diseases does this lead to; and which interacting variables modify the pathogenic process. Also needed is the understanding of how the various components of the system interact, and how we best can describe, analyze and intervene in this entire, complex human eco-system.

There is little direct evidence of a causal relationship between such social structures and processes and their change (or lack of change), and the incidence and prevalence of stress-related morbidity and mortality. But, a substantial body of indirect evidence strongly suggests that such associations exist and emphasizes the need to better understand their role. Much of this association may be mediated by health-related *behaviors*. An example would be tobacco dependence, quoted by WHO as the primary cause of one third of all cases of cancer, 75 per cent of chronic bronchitis and 25 per cent of myocardial infarction in the U.S. Other example are excessive risk behaviors among youth, e.g. experimenting with drugs, sexual activity without precautions against sexually transmitted diseases or adolescent pregnancies, driving at excessive speed, alcohol abuse, and suicidal or homicidal behaviors.

More closely related to *neuroendocrine* 'stress' mechanisms (in Selye's sense) is the high prevalence of somatic symptoms probably resulting from exposure to psychosocial stressors, i.e. cases with no ascertainable organ pathology, or with complaints of discomfort disproportionate to the physical problem. This is said to account for 30-50 per cent of all consultations in developed countries – the largest single complaint category in primary care.

These new notions about social determinants of health are reflected and summarized in some recent Swedish and international initiatives. Here follow some examples.

A Swedish initiative

The new Swedish Public Health Service Bill (1984/85: 181) declares that 'our health is determined in large measure by our living conditioning and lifestyle'.

The Bill further indicates that 'the health risks in contemporary society take the form of, for istance, work, traffic and living environments that are physically and socially deficient, unemployment and the threat of unemployment; abuse of alcohol and drugs, consumption of tobacco, unsuitable dietary habits, as well as psychological and social strains associated with our relationships – and lack of relationships – with our fellow beings.'

'These health risks... are now a major determinant of our possibilities of living a healthy life.

This is true of practically all the health risks which give rise to today's most common diseases, e.g. cardiovascular disorders, mental ill health, tumours, and allergies, as well as accidents'. Accordingly, 'care must start from a comprehensive approach, (by which is meant) that people's symptoms and illnesses, their causes and consequences, are appraised in both a medical and a psychological and social perspective.'

This would be particularly important if and when – as is often the case – an increased vulnerability in the individual coincides with an increased exposure to pathogenic environmental influences.

The relationship between environmental and life style factors and manifestations of ill health are summarized in Figure 2. This Environment/Life-style/Health matrix shows the magnitude and strength of the relationships, without attributing the causal nature of the relationship.

Environment/life-style/health Matrix * indicates some, and ** strong relationships		Cardiovascular diseases	Mental illness	Skeletomuscular disease	Tumours	Injuries	Respiratory diseases
	Social upbringing environment	*	**				
	Social work environment and unemployment	**	**			*	
	Physical work environment		*	**	**	**	**
Environments	Social living environment		*				
	Physical living environment				*	**	*
	Air/water pollutants				*		*
	Traffic				*	**	*
Behaviors	Diet	**			**		*
	Alcohol and drugs		**		*	**	*
	Tobacco	**			**		*

Fig. 2. Correlation between health hazards and illness groups. (Source: SOU 1984, 40.)

Health for All Strategies; intersectoral action

Utilizing a related approach in their Health for All Strategy, the Member States of the World Health Organization (WHO) committed themselves to creating the conditions which will enable all people to enjoy a reasonably healthy life by the year 2000. Primary health care became the principal means of achieving this objective. This strategy reordered the priorities in the health sector and moved from a perspective which was predominantly disease-oriented and curative, to one in which the main focus was on the *prevention* of ill health, the maintenance and promotion of good health and the capacity to resist disease. It follows that an essential component in this strategy is promotion of health by intervening in social systems. The latter also means broadening the perspective to include intersectoral action.

According to Report of the Technical Discussion of the 39th World Health Asembly (A 39/Technical Discussions/4; 15 May 1986) 'it has become evident that health goals defined in these terms, cannot be realised through the services delivered by the health sector alone. They require the combined effort of several other sectors whose activities have a major impact on health. From the inception, the primary health care strategy therefore recognized that *intersectoral* action is a vitally important condition for achieving its goals. It identified the main sectoral elements, including food and nutrition, water and sanitation, housing and education. In this approach, improvements in health are perceived as a multisectoral responsibility, in which the main development sectors would need to collaborate with the health sector.'

A number of *recommendations* were then agreed upon, emphasizing the need to
- include health objectives as an integral part of other sectoral policies;
- promote action-oriented multidisciplinary research on socio- economic and environmental determinants of health to identify cost-effective intersectoral health actions;
- train health professionals to make them aware of relationships between environments, life-styles, and health;
- increase awareness in, and train, other professional groups for intersectoral health-related action;
- develop mechanisms for intersectoral actions;
- include equity-oriented targets in Health for All strategies;
- identify vulnerable groups and health hazards; and monitor health conditions;
- utilize health status of disadvantaged groups as an indicator of development; and
- protect the most disadvantaged groups in implementing economic adjustment policies.

Such formulations are based mostly on *epidemiological* studies, who – whether retrospective, cross-sectional or prospective; aggregate or individual – provide associations between exposures to various environmental stressors and subsequent ill health i.e., circumstantial evidence.

Although animal experiments do provide evidence of *causal* relationships, they must be interpreted with caution, because of the artificiality of many experimental settings and the difference between mice and men. Most experiments with human subjects – in laboratory settings and in real life – provide data relating to the influence of psychosocial stressors on potentially pathogenic mechanisms, but (with few exceptions) not to morbididy and mortality.

This bird's-eye view of social environment and human health illustrates that there are probably many interacting and strongly pathogenic influences (stressors) in the human ecosystem. If one variable is influenced, effects will occur in many others, with complex interactions and many feed-back loops.

Consequently, it is usually impossible to cope successfully with current and future environmental, behavioral, and health problems by consideration – in research, therapy and/or prevention – of just one or two of the components of the total human ecosystem. Success is more probable if as many as possible of the critical components can be taken into account.

Critical etiological factors in disease onset and course

But which components are the critical ones? Do we really know which environmental exposures are (a) necessary, (b) sufficient, or (c) contributory, in
- causing a certain disease,
- accelerating its course, and/or
- triggering its symptoms?

As already indicated, the thruth is that there are many assumption and suspicions, but a general lack of scientific proof. As decision makers are under pressure from the public to act even without evidence that their actions will achieve desired goals, they act anyhow and – more often than not – without ever evaluating systematically, interdisciplinary, and intersectorally, the effects and side-effects of their actions (cf. DHHS, 1987).

This is why evaluation of controlled interventions in social system becomes so crucially important. Of course, we do not anticipate that all questions policy makers want to ask can be answered in a few studies. Yet such studies are an important complement to descriptive and analytic approaches.

Application of existing knowledge

The three major problem areas in the present context – enhancement of health care, prevention of mental and physical ill health, enhancement of well-being – are all complex and affected by a large variety of factors. For each of the three areas, there is a high level of probability that application of present knowledge would be beneficial, and for each there is certainty that greater undertanding is required. Much of the interaction is mediated through a relatively limited number of social arrangements. The first approach would be to introduce existing knowledge of psychosocial factors, or new ideas where appropriate, into the assessment of health problems and social actions, and evaluation of their impact on health and well-being. The relation of acceptance, availability, and use of health and social actions to the mode of administration on the one hand and to health and well-being on the other would be assessed, as well (Kagan and Levi, 1975).

Future research

In many instances, however, our present state of knowledge simply does not allow rational health action, even if combined with evaluation. To close such critical gaps, new knowledge must be acquired regarding: high-risk situations; high-risk groups; and high-risk reactions.

To provide the necessary data, 'stress' research projects can often be carried out in *three complementary steps*:
- problem identification with survey techniques and morbidity data, to describe the size of the problems and to find environmental and other *correlates* of health problems;
- longitudinal, multidisciplinary intensive studies of the intersection of high-risk situations and high-risk groups as compared with controls, to identify *temporal* relationships between environmental exposures and pathogenic mechanisms (and between such mechanisms), as modified by interacting factors;
- controlled intervention, including laboratory experiments as well as therapeutic and/or preventive interventions in real- life settings (for example, natural experiments; inter-

disciplinary evaluation of health action).

These approaches could, and should, include testing hypotheses that will lead to increased understanding of the system of psychosocial factors that affect health in general. Our strategy should be to identify situations and carry out projects, where applied and basic research can be combined efficiently. It is often possible, and desirable, to approach simultaneously the problems of enhancement of health care, prevention of ill health, increase of well-being, and testing of key hypotheses, using not much more resources for all four than each would require separately (Kagan and Levi, 1975).

Hypothesis testing and evaluation of health action

In response to a greater-than-ever demand, considerable numbers of studies on health and its social and other environmental determinants have been carried out all over the world. Advances have been made at the molecular, cellular, and organ level but very little at the family, neighbourhood, or community level. Referring to this gap, Kagan (1981) has pointed out that once attention moves from the laboratory to the community, reports on hypothesis testing studies are hardly ever found. Some notable early exceptions are reviewed by Eisdorfer and Stotsky (1977), Estes and Freeman (1976), Gunn et al. (1983), and Murphy and Schoenborn (1987).

As soon as one looks beyond clinical trials, it is rare to find an evaluation of community-oriented health action.

Innumerable studies have shown that there are *associations* between psychosocial and other environmental factors, and health, have speculated on ideas for health or social action, or have put forward hypotheses in relation to the community spread and control of disease. In rare, important exceptions a hypothesis has been tested or action has been evaluated, showing that it is possible to carry out controlled, community-oriented social intervention studies (cf. Kagan, 1981; Levi et al., 1982). Reasons given for not doing so are that it is: unethical, technically impossible, too expensive, or takes too long. Whilst there is an element of truth in all four objections, the first three can often be addressed adequately, and something can be done to reduce the effects of the fourth (Kagan, 1981; Levi, 1987b). Once such research is considered, however, it becomes unethical, and probably more costly, to impose an environmental or health action of still-to-be-proved value, possible danger, and high cost, without evaluation, or to accept a hypothesis without prior test (Kagan, 1981). It is a philosophical and political question whether controlled intervention studies should await additional fundamental research, or be a way to close these gaps (Levi, 1989).

Based on descriptive studies, *measures* that may be presumed to prevent disease and promote health can then be proposed, that are (a) likely to be of greatest causal importance, (b) accessible to change, (c) feasible and (d) acceptable to all concerned.

At a third stage, these measures should be evaluated in an interdisciplinary, experimental model study.

Fourth, depending on the outcome in terms of benefits, side effects, costs etc., a wider application may be implemented, continuously monitored, evaluated, and modified as necessary.

Desirable characteristic of research approaches

Briefly, then, a future overall research program on stress in man should aim at being:

- *system-oriented*, analyzing health-related interactions in the human ecosystem (e.g., family, school, work, hospital, old people's home);
- *interdisciplinary*, covering medical, physiological, emotional, behavioral, social, and economic aspects of these interactions, both basic and applied;
- *problem-solving oriented*, and comprehensive, including epidemiological identification of health problems and their environmental and other correlates, followed by longitudinal, interdisciplinary field studies of exposures, reactions, etc., and by subsequent experimental evaluation under real-life conditions of health-promoting and disease-preventing interventions;
- *health-oriented* (and not only disease-oriented), trying to identify what constitutes and promotes positive health and counteracts ill health (cf. Antonovsky, 1987);
- *intersectoral*, evaluating environmental and health actions administered in other sectors (e.g., work, housing, nutrition, traffic, education);
- *participative*, in close interaction with potential care givers, care receivers, planners and policy makers.
- *international*, including transcultural, collaborative, complementary projects with centers in other countries.

Such a research program would be of great benefit in making distinctions among stressful social structures and processes (Figure 1, box 1), reactions to such stressors (box 4), the health consequences of such reactions (boxes 5 and 6), and the interacting factors (box 7) that modify the flow of events. Information is also needed on the critical determinants that make some events and conditions stressful (appraisal, and psychobiological program), on the effects of the resulting stimuli on a broad range of possible pathogenic mechanisms, and on the resulting health consequences. Information is similarly needed on the components of health-promoting (salutogenic) processes and how they interact (cf. Elliott and Eisdorfer, 1982; Antonovsky, 1987).

Examples of experimental interventions in social systems

As already indicated in this chapter, most existing health interventions favour attempts to change individual rather than social variables. However, in the following, a few studies of the latter type will be cited. These studies, drawn primarily from the Nordic experience, illustrate if and how persons at different points in the life course can benefit from such interventions (Levi, 1989).

Unemployment and ill health

Work is often important for structuring and giving meaning of life. Thus, jobs give a person identity and self-esteem, friends and acquaintances, and, of course, material assets. Unemployment deprives people of these opportunities (Jahoda, 1979). In a controlled intervention (Levi et al., 1986) with 200 unemployed female factory workers in Olofstrom in the south of Sweden, a multidisciplinary longitudinal evaluation was made of efforts to make good this loss and prevent its possible effects, in the first place by finding or creating new jobs cooperatively and idividually (which usually turned out to be impossible), in the second with meaningful, unpaid, collective, self-administered activities.

Results from the study demonstrate: (a) pronounced mental and physiological stress reactions (high blood levels of cortisol) one month preceding job loss; (b) a diminution of these reactions after the first month of actual unemployment; (c) a subsequent, successive rise in self-rated depression and in cortisol levels over the first year of unemployment, following the initial period of relaxation and relative optimism; and (d) high mean levels of depression and cortisol and decreased immune function one year after the start of unemployement. The attempt to 'replace' real employment by offering various unpaid collective self-administered activities turned out *not* to influence significantly the potentially noxious psychosocial and physiological effects of becoming and remaining unemployed.

The study illustrates what to expect before and after the onset of unemployment and provides health reasons for promoting gainful employment for all.

Eliminating night shift for industrial workers

Another study (Akerstedt and Torsvall, 1978) concerned 400 industrial workers on a three-shift rota. For some half of them, the night shift was eliminated – some switched to a two-shift system, others to continuous day-shift – and the other half went on as before.

In a follow-up one year later, those workers who changed from 3- or 4- shift working to 2-shift working (i.e., without night work) showed a significant increase in well-being with respect to sleep, mood, gastrointestinal functioning, and social factors, along with an improvement in attitude to their work schedules. Those who were switched from shift to day work reported greatly increased social well-being, a shortening of sleep-length during free days, and considerably improved attitudes towards their work schedule; their sickness absence rates were also reduced. Questionnaire scores remained at their original levels in those workers, whose schedules were left unchanged. It is concluded that the abolition of night work results in a substantial improvement in mental, physical and social well-being.

Increased autonomy and social interaction for institutionalized pensioners

All 30 pensioners in one section of a senior citizen apartament building – but not the 30 in another, comparable section – were encouraged to take control of their own lives to a greater extent, with the hope that secondarily, this would lead to an increase of their social activities (Arnetz, 1983).

As a result of the intervention, social activities (planned and spontaneous) increased three times over, restlessness decreased, carbohydrate turnover (hemoglobin Alc) improved, anabolic hormones (plasma testosterone, estradiol and dehydroepiandrosterone) increased, and there was no decrease in body weight and stature. The study demonstrates the importance of active participation in care programs and shows that simple measures to this end, besides having social and psychological benefits, favoured health and well-being.

Summary

Psychosocial stressors originate in social structures and processes. They affect the human organism through psychological processes, including experience and appraisal of person-environment interactions in terms of environmental demands and opportunities on the one hand, and man's abilities, needs, expectations, customs and culture on the other. They influence health and well-being through four types of closely interrelated mechanism –

emotional, cognitive, behavioral and physiologic. The outcome is modified by situational factors (e.g., social support) and individual ones (e.g., personality characteristics and coping repertoire). The system is a dynamic one with many feedback loops.

Accordingly, approaches in future stress research should aim at being: system-oriented; interdisciplinary; problem-solving oriented; health- (and not only disease-) oriented; and participative.

The objectives of such approaches include establishing: which psychosocial stressors are of etiological significance, under what circumstances (high-risk situations), and for whom (high-risk groups); which emotional, cognitive, behavioral and/or physiologic mechanism ensuing from psychosocial stimuli (high-risk reactions) are implicated in the pathogenesis of various diseases; and which modifications of the social environment and/or human behavior will promote health and well-being, and which should be avoided, and by whom (evaluation of health action).

With such knowledge as a basic, environmental and/or individual measures can be proposed that may be presumed to prevent disease and promote health, which in a third step are evaluated interdisciplinarily, on a model scale.

Whenever possible the approaches should combine testing of key hypotheses (basic research) with evaluation of health action (applied research), with particular emphasis being placed on studies of fundamental areas of social life and institutions undergoing rapid changes. Tactics include several complementary steps with both traditional epidemiological studies; interdisciplinary, longitudinal, intensive studies; and controlled interventions, in the laboratory as well as in real-life. A basis for this is further elaboration on definitions and standardization of nomenclature, and preparation of cross-culturally applicable methodology 'kits' for research and environmental and health monitoring.

References

Akerstedt, T. and Torvall, L. (1978), Experimental changes in shift schedules: their effects on well-being. *Ergonomics,* 21: 849-856.

Antonovsky, S. (1987), Unraveling the mystery of health. How people manage stress and stay well. Jossey-Bass, San Francisco.

Arnetz, B. (1983), Psychophysiological effects of social understimulation in old age. (Doctoral Thesis). Karolinska Insituted, Stockolm.

DHHS: Prevention '86/'87 (1987): Federal Programs and Progress. U.S. Government Printing Office, Washington, D. C.

Eisdorfer, C. and Stotsky, B.A. (1977), Intervention, treatment, and rehabilitation of psychiatric disorders. In: Birren, J.E. and Schaie, K.W. (eds), Handbook of the Psychology of Aging. Van Nostrand Reinhold, New York.

Elliot, G.R. and Eisdorfer, C. (eds) (1982), Stress and human health. Analysis and implication of research. Springer, New York, pp. 724-748.

Estes, C.L. and Freeman, H.E. (1976), Strategies of design and research for intervention. In: Binstock, R.H. and Shanas, E. (eds), Handbook of Aging and the Social Sciences, Van Nostrand Reinhold, New York, pp. 536-560.

Gunn, W.J., Orenstein, D.R., Iverson, D.C. and Mullen, P.D. (eds) (1983), An Evaluation Handbook for Health Education Programs in Alcohol and Substance Abuse. IOX Assessment Associates, Culver City, California.

Jahoda, M. (1979), The impact of enemployment in the 1930s and the 1970s., *Bull. Br. Psychol. Soc.,* 32: 309-314.

Kagan, A.R. (1981), A community research strategy applicable to psychosocial factors and health. In: Levi, L. (ed), Society, Stress and Disease, vol. 4, Working Life, Oxford University Press, Oxford, pp. 339-342.

Kagan, A.R. and Levi, L. (1975), Health and environment: psychosocial stimuli. A review. In: Levi, L. (ed), Society, Stress and Disease, vol. 2, Childhood and Adolescence, Oxford University Press, London, pp. 241-260.

Levi, L. (ed) (1971), Society, Stress and Disease: The Psychosocial Environment and Psychosomatic Diseases, vol. 1, Oxford University Press, London, New York, Toronto.

Levi, L. (1972), Stress and distress in response to psychosocial stimuli, *Acta Medica Scand.*, Suppl. 528.

Levi, L., (ed) (1975a), Society, Stress and Disease: Childhood and Adolescence, vol.2, Oxford University Press, London, New York, Toronto.

Levi, L., (ed) (1975b), Emotion: Their Parameters and Measurement, Raven Press, New York.

Levi, L., (ed) (1978), Society, Stress and Disease: The Productive and Reproductive Age. Male/Female Roles and Relationships, vol. 3, Oxford University Press, Oxford, New York, Toronto.

Levi, L., (ed) (1981), Society, Stress and Disease: Working Life, vol. 4, Oxford University Press, Oxford, New York, Toronto.

Levi, L, (ed) (1987), Society, Stress and Disease: Old Age, vol. 5, Oxford University Press, Oxford, New York, Tokyo.

Levi, L. (1989, in press), Intervening in social systems to promote health. In: Abeles, R. and Ory, M. (eds), *Aging, Health and Behavior.*

Murphy, L.R. and Schoenborn, T.F. (eds) (1987), Stress Management in Work Settings, U.S. Departement of Health and Human Services, National Institute of Occupational Safety and Health, Cincinnati, Ohio.

Swedish Public Health Service Bill n— 1984/85 (1985): 181. Ministry of Social Affairs, Stockholm.

WHO (1986): Report of the Technical Discussions of the 39th World Health Assembly (A 39/Technical Discussion/4; 15 May 1986).

BEHAVIOUR IN SITUATIONS OF CONFLICT
A hypothesis on pleasure and stress

Michel Cabanac
Département de physiologie
Faculté de médecine
Université Laval
Québec, GlK 7P4
Canada

In Hans Selye's (1956) words, stress is "the sum of all nonspecific changes caused by function or damage". This definition includes recovery and adaptation to various "factors" such as: "infections, intoxications, trauma, nervous strain, heat, cold, muscular fatigue or roentgen-ray irradiation. However, it has been found that stimuli lead to a stress reaction only if the subject shows an emotional response (Mason et al., 1976) or if muscular exercise reaches exhaustion (Suzuki, 1983) which is also the source of strong emotion. Thus stress now defines less organically and more behaviourally as "the prolonged inability to remove a source of potential danger, leading to activation of systems for coping with danger beyond their range of maximal efficiency" (Archer 1979). Indeed a list of stress stimuli, both acute and chronic, used in recent experimental studies would include mostly confinement, transportation, forced muscular exercise, thermal discomfort, discomfort of litter, chronic tethering, social aggression or subordination, pain and pain avoidance, etc. When techniques separate the psychological and physical factors in stressful conditions, the psychological factors are the main cause of stomach ulcers and other disorders (Weiss 1972). This can be summarized with Mason's (1971) statement that: "psychological stimuli are among the most potent of all stimuli to affect the pituitary adrenal cortical system". As a result of this evolution in thinking, stress is now viewed mainly as a general biological response to environmental demands. Since behaviour is, par excellence, the response to the environment, it is appropriate to study stress from a behavioural point of view (Bohus et al. 1987, Wiepkema 1987). In turn it is also appropriate to analyze the determinism of behaviour.

Experiments with human subjects provide a unique tool to study behavior; they allow not rely the measuring of the behavioural response, as with animal behaviour, but also of the verbal response describing the subject's introspect. It then becomes possible to compare motivation with actual behaviour.

Sensory pleasure has been repeatedly demonstrated as the appropriate signal of a useful stimulus and displeasure of a useless or noxious stimulus (see review in Cabanac 1979) in the cases of temperature and of alimentary stimuli. Pleasure occurs when a stimulus which facilitates the recovery from a deviation of the milieu intérieur is presented to the subject. In turn subjects have been repeatedly recorded as seeking pleasure. Figure 1 shows such an example of a subject switching his preference from hot to cold water when his deep body

S. Puglisi-Allegra and A. Oliverio (eds.), Psychobiology of Stress, 203–209.

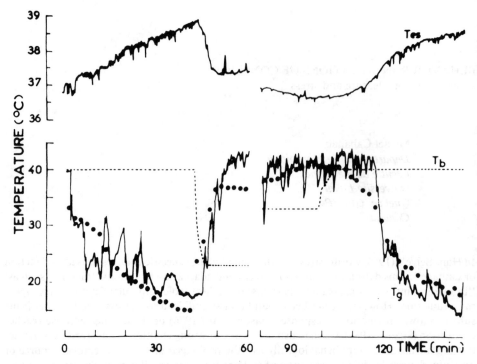

Figure 1. Time course of body-core temperature (Tes), bath temperature (Tb) and preferred temperature (Tg) of a subject immersed up to his chin in water. Preferred temperature was recorded in a glove perfused with thermostated water on the subject's left hand. The subject himself commanded with his right hard the thermostat regulating the temperature of the water flowing into the glove. The dot indicate the theoretical temperature preferred by the subject as computed periodically knowing bath temperature and core temperature from the following model: Tpref = − 0.3 Tbath (Tes − 36.3) + 44, all temperatures in °C.
This shows that it is possible to predict from physiological variables the pleasure aroused by a temperature applied to a subject's skin.

temperature evolved from hypo- to hyperthermia and reciprocally. In this case, the pleasure depends so closely on body temperatures that it is possible to produce simple mathematical models predicting the preferred skin temperature of the subject as shown in the example of Figure 1 (Cabanac et al. 1972).

This is the situation found in the cases of thermoregulatory behaviour and food intake associated with sensory pleasure aroused by thermal, gustatory and olfactory sensations. Other sense vision and audition, are able to arouse a different pleasure that is more cultural than sensory. Finally tactile and mechanical stimuli may arouse pleasure, especially in relation to reproductive behaviour. Modern science has not yet started to study the pleasures of these sensations, but common sense tells us that the laws of pleasure in temperature and alimentary stimuli are very likely to be applicable also to the pleasures of mechanical stimuli.

Thus pleasantness is both the sign of a useful stimulus and a motivation to approach this stimulus when it is needed. The reverse is also true with unpleasant stimuli. The great advantage of pleasure is that it short-cuts any other cognitive analysis by the subject, any rationalization as to causes and effect. Thus further cognition may improve behaviour, but is not necessary, and the subject can adjust perfectly to its environment without any information other than sensory pleasure.

A priori there is no reason why the seeking of pleasure should induce stress or be related to stress in the situations involving only one motivation such as in the experiments briefly summarized above. However, pleasure or the lack of pleasure might be a causal factor in stress, as will be developped below.

First evidence of the pathogenic influence of conflict situations was brought by Pavlov and his pupils. In an experiment which became the archetype of experimental neurosis, Chenger-Krestovnikova created a conditionned salivary reflex in dogs and used a circle as conditionned stimulus. In a second step the circle was differenciated from all other geometric frames including elipses. In the last part of the experiment, the dogs were exposed to conditionned stimuli that tended to merge i.e. cercles tended to become elipses and elipses were increasingly circular. Then some of the dogs presented a syndrom described by Pavlov (1927) as experimental neurosis. Other historical studies confirmed that situations of conflicting motivations, generated maladaptive behaviors, stereotyped responses, neurosis. A classical case was that of the cat that, once conditionned to bar-press for a food reward was faced with this same behaviour providing not only the reward but simultaneously an unpleasant stimulus such as an air puff or electric shock Masserman (1956). Similarly rats were exposed to problems without solutions (Maier 1956). It is therefore of interest to extend the study of pleasure to situations of clashing motivations in humans. We have seen above that subjects tended to maximize pleasure, when only one motivation was presented at a time to the subjects. In situation of conflict the following motivations were studied: thermal comfort, gustatory pleasure, fatigue, pain. One example will be analyzed here where fatigue clashed with thermal discomfort (Cabanac & Leblanc 1983). The general method as well as the results are typical of this series of experiments on various conflicts of motivations.

Each subject served in three series of sessions where he was his own control. All three sessions took place in a climatic chamber and the half naked subject walked on a treadmill at 3 km h^{-1} for one hour . On the first series of sessions, the ambient temperature was fixed by the experimenter at 5, 10, 15, 20 or 25° C and the subject could periodically adjust the slope of the treadmill. Thus, at low ambient temperature, the subjects could raise the treadmill slope, and did so to warm themselves by increasing their metabolic heat production, a by-product of their muscular work. On the other hand when ambient temperature was lukewarm, the subjects could choose to walk without ascending and did so, thus limiting metabolic heat production in their muscles.

For the second series of sessions the subjects found the treadmill fixed at a constant slope of 0, 6, 12, 18 or 24%. During the hour of walk they could periodically adjust the ambient temperature. Thus, when walking without ascending they could raise ambient temperature, and in so doing, they limited their heat loss. On the other hand when the slope of the treadmill was steep the subjects could choose to lower their ambient temperature. They did so thus facilitating the loss of heat produced intensely in their muscles.

As a result of the first two series of sessions the subjects produced a behaviour that was clearly thermoregulatory. Figure 2 shows as dots the ambient temperatures and treadmill slopes chosen by a subject at the end of one hour sessions when slope or ambient temperature was imposed by the experimenter. The negative linear correlation of slope and ambient temperature is the outcome of an obvious thermoregulatory behaviour. When the subjects were placed in a situation where ambient temperature and muscular exercise presented

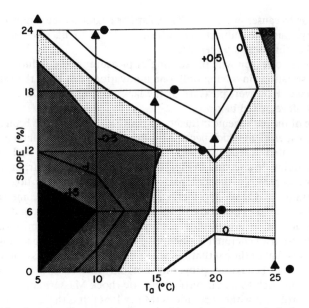

Figure 2. Results of 3 series of one hour sessions in one subject platted on one single figure. First series of 5 sessions (Δ): the ambient temperature (Ta) was imposed by the experimenter and the subject could adjust the slope of the treadmill on which he was walking 3km/h. Second series of 5 sessions (o): the slope of the treadmill was imposed by the experimenter and the subject could adjust the ambient temperature of the climatic chamber. Third series of 25 sessions (iso hedonic lines): both treadmill slope and ambient temperature were imposed and the subject gave two ratings to describe pleasure or displeasure, one rating for thermal comfort and overeating for fatigue. Positive isolines and white areas: pleasure. Negative isolines and grey areas: displeasure. It can be seen that the subject moved to pleasure areas as defined bidimensionally.

simultaneously produced a conflict of motivations fatigue *vs* thermal discomfort, they made their decisions by chosing a useful behaviour as judged from the point of view of their temperature regulation.

The question raised by this striking coincidence of behavioural choice and usefulness concerns the mechanism by which, spontaneously, subjects are able to optimize their behaviour. The results obtained in the last series of sessions answer that question. In that series, each session consisted of a one hour walk on the treadmill at imposed conditions of slope and ambient temperature. The five slopes combined with the five ambient temperatures provided 25 conditions. At the end of the session, the subject gave a quantitative rating on his perceived thermal comfort and perceived rate of exercise. The rating did not describe the intensity (or magnitude of the perception) but rather whether the subject liked or disliked it. Pleasure was rated positive and displeasure as negative. Thus the subject gave two ratings, one related to thermal comfort, and the other related to fatigue. Now, if pleasure and displeasure motivate the subjects in their behavioural choice, ratings and behaviour should coincide.

Figure two presents the ratings obtained in this series of experiments, as isohedonic lines of equal pleasure or displeasure. The lines were obtained, by interpolation between the nodes of the matrix, from the algebraic sum of the two ratings obtained to describe pleasure/ displeasure aroused by muscular work and thermal environment.

Figure 2 also presents as dots the behavioral choice made at the end of the one hour walk. It can be seen that all dots but one fall in the area of bi-dimensional pleasure as obtained in other sessions. The only dot not in the pleasure area was nevertheless at the location of minimal possible bi-dimensional displeasure, as judged from the ratings.

This experiment shows that the conflict fatigue vs thermal discomfort was solved by maximizing the pleasure obtained not in one dimension only of the motivational space, i.e. fatigue *or* discomfort, but simultaneously in both dimensions, i.e. fatigue *and* discomfort. Such a result has been described in some detail because it exemplifies the clash of two motivations, both physiological in nature but completely different in their mechanism. The general conclusion has been confirmed in other situations where sweet taste was pitted against sour or against temperature of the tasted stimuli (Cabanac & Ferber 1987), and where slope was pitted against speed of a treadmill (Cabanac 1986). On both occasions, subjects manipulated the variable, that they were permitted to adjust in such a way as to maximize, not necessarily the pleasure accused by that variable, but rather the algebraic sum of pleasures on both conflicting variables. It is therefore safe to hypothesize that physiological conflicts reach a solution from the trend to maximize sensory pleasure.

The experiments reported above were limited to physiological motivations and conflicts. One may question whether it is possible to extend the conclusions to domains other than biology. The notion of "behavioral common pathway" is especially enlightening from this perspective. Paraphrasing the image of the motoneuron final common pathway of all motor responses, McFarland and Sibly (1975) pointed out that behavior is also a final common pathway on which all motivations converge. This image incorporates all motivations into a unique category since behavior must satisfy not only physiological motivation but also social, moral, esthetical, ludic, etc. Indeed, it is often the case that behaviors are mutually exclusive; one cannot work and sleep at the same time. Therefore, the brain, responsible for the behavioral response, must rank priorities and determine tradeoffs in the decisions concerned with allocating time among competing behaviors. It can be accepted that, at each instant, a subject responds to the motivation that will provide the greatest additional pleasure for any given cost. It can be expected that the brain operates this ranking by using a common currency (McFarland 1985).

Moral or esthetical motivations must share a common currency with physiological motivations. It may be hypothesized that pleasure is also the common currency of non-physiological motivation. Such a hypothesis was verified experimentally.

Money was used as a non-physiological motivation and pitted against two physiological motivations: cold discomfort in a climatic chamber (Johnson & Cabanac 1983) and pain from isometric contraction in the thighs (Cabanac 1986). In these experiments, human volunteers could earn money against duration of exposure to these unpleasant sensations. Thus, the longer they stood cold discomfort, or pain, the more money they earned. In several sessions, the rate of monetary reward was varied. In both cases it was found that the sensation, as described by the ratings given by the subjects, increased linearly with time. In both cases it was found that the subjects tolerated higher discomfort, or pain, for a longer time when the monetary reward was higher. This finding is in conformity with common sense expectation. In addition, however, it experimentally measures and demonstrates that an unpleasant sensation was reliably and quantitatively matched against money, a non-physiological motivation. The relation between reward and duration of tolerated displeasure was logarithmic. It can be assumed that the subjects decided to end a session just at the instant when the displeasure of the sensation became greater than the pleasure of the monetary reward. This result opens up the field of all other motivations to the hypothesis that decisions involving non-physiological motivations are made by following the tendency to maximize pleasure.

Thus pleasure seems to be at the center of behaviour. Both a result of useful behaviour and a motivation, pleasure is the key to optimal behaviour. This paramount importance of pleasure in life leads to the hypothesis that pleasure could not only be useful but also necessary and that the lack of pleasure could be the source of stress.

As underlined earlier, recent experimental studies present problems without solutions to the animals where the animal can do nothing to improve its situation. Such a situation is pathogenic and the animal develops apathy, depression, disease, impairment of the immune system (Seligman 1975), and impairment of future learning in that situation (Maier & Seligman 1976). Many examples will be found elsewhere in this book. This experimental evidence is confirmed by the epidemiological information available on humans: pathology is closely correlated to helplessness (Levi 1990). It may be hypothesized that these situations where subjects are unable to solve their problems can be equated to their incapacity to obtain pleasure. If this were the case, pleasure would not only be the key to optimal behaviour and the answer to conflicts of motivations but also a necessary experience the absence of which is pathogenic.

References

Archer (1979). The organization of agression and fear in vertebrates. In: Perspectives in Ethology. Vol. 2. P.P.G. Bateson & P.H. Klopfer. eds. Plenum Press: New York.

Bohus B., Koolhaas J.M., Nyakas C., Steffens A.B., Fokkema D.S., Schevrink A.J.W. (1987). Physiology of stress: a behavioral view. In: Biology of stress in farm animals: an integrative approach. P.R. Wiepkema & P.W.M. Van Adrichem. eds. Martinus Nijhoff Publ.: Dordrecht.

Cabanac M. (1979). Sensory pleasure. *Q. Rev. Biol.* 54: 1-29.

Cabanac M. (1986). Performance and perception at various combinations of treadmill speed and slope. *Physiol. Behav.* 38: 839-843.

Cabanac M. (1986). Money *vs* pain: experimental study of a conflict in humans. *J. Exp. Anal. Behav.* 46: 37-44.

Cabanac M., Ferber C. (1987). Sensory pleasure and preference in a two-dimensional sensory space. *Appetite* 8: 15-28.

Cabanac M., Leblanc J. (1983). Physiological conflict in humans: fatigue *vs* cold discomfort. *Am. J. Physiol.* 244: R621-R628.

Cabanac M., Massonnet M., Belaiche R. (1972). Preferred skin temperature as a function of internal and mean skin temperature. *J. Appl. Physiol.* 33: 699-703.

Levi L. (1990). Approaches to stress in man, this volume.

Maier N.R.F. (1956). Frustration theory: estatement and extension. *Psychol., Rev.* 63: 370-388.

Maier S.F., Seligman M.E.P. (1976). Learned helplessness. Theory and evidence. *J. exp. Psychol. general.* 105: 3-46.

Mason J.W. (1971). A re-evaluation of the concept of "non-specificity" in stress theory. *J. psychiat. Res.* 8: 323-333.

Mason J.W., Maher J.T., Hartley L.H., Mougey E.H., Perlow M.J., Jones G.J. (1976). Selectivity of corticosteroid and catecholamine responses to various natural stimuli. In: Psychopathology of human adaptation. G. Serban ed. Plenum Press: New York.

Masserman J. (1956). Principes de psychiatrie dynamique. Presses universitaires de France. Paris.

McFarland D. (1985). Animal behaviour. Pitman publ. ltd.: London.

McFarland M., Sibly R.M. (1975). The behavioural final common path. *Phil. Trans. R. Soc. (Series B)* 270: 265-293.

Pavlov I (1927). Conditionned reflexes: an investigation in the physiological activity of the cerebral cortex

(Transl. G.V. Anrep) Oxford Univ. Press: London.
Seligman M.E.P. (1975). Helplessness: On depression, development and death. Freeman: San Francisco.
Selye H. (1956), p. 36. General physiology and pathology of stress. In: Fifth annual report on Stress. 1955-56. H. Selye & G. Henser. eds. M.D. Publications: New York.
Suzuki T. (1983). Physiology of adrenacortical secretion. Frontiers. *Horm. Res.* 11. Karger: Basel.
Weiss J.M. (1972). Psychological factors in stress and disease. *Scientific Amer.* 226: 104-113.
Wiepkema P.R. (1987). Behavioural aspect of stress. In: Biology of stress in farm animals: an integrative approach. P.R. Wiepkema & P.W.M. Van Adrichem. eds. Martinus Nijhoff Publ.: Dordrecht.

STRESS AND DISTRESS FROM FLUORESCENT LIGHTING

Arnold J. Wilkins
MRC Applied Psychology Unit,
15 Chaucer Road, Cambridge CB2 2EF, England

Conventional fluorescent lighting pulsates rapidly in brightness. Although it is not usually visible as flicker, the high-frequency pulsation is resolved by the visual system and can give rise to an unnatural pattern of nervous excitation. The pulsation appears to have little consequence for vision or eye movements, but is responsible for a large proportion of the eye-strain and headaches suffered by office workers.

Fluorescent lighting has long been associated with complaints. People dislike it for reasons they find hard to express. Some complain that it gives them a headache, others complain of eye-strain. Similar complaints have been directed towards computer displays. This chapter reviews evidence suggesting that the complaints have a common basis, and that they are attributable to rapid imperceptible pulsations in light intensity.

Before reviewing the evidence concerning the adverse effects of fluorescent lighting, it will be helpful to describe its physical characteristics in some detail.

Physical characteristics of fluorescent lighting

A fluorescent lamp is lit by a current passed between two electrodes, one at each end of a long glass tube containing a gas, usually mercury vapour. The current is the result of a discharge created through the gas. The visible radiation from the discharge is mainly at the blue end of the spectrum but the ultra-violet radiation is converted to visible light by a fluorescent coating of phosphor on the inner surface of the tube.

Because the electrodes are connected directly to the alternating current supply, the voltage across the ends of the tube varies sinusoidally, one electrode carrying alternately a positive and then a negative potential compared to the other, with the result that two discharges occur with each cycle of the supply. Over most of the length of the tube the light therefore pulsates with a frequency twice that of the supply voltage: in Europe the light pulsates 100 times per second and in America 120 per second.

The discharge does not occur evenly along the length of the tube: there are dark spaces in front of the negative electrode. At the ends of the tube the light is therefore flickering with only 50 or 60 flashes per second, and the flicker is sometimes visible. For this reason fluorescent tubes are usually housed in a box or "luminaire" with reflective inner surfaces so that the light emitted by the tube is diffused within the luminaire, "diluting" the low-frequency 50- or 60-per-second component.

S. Puglisi-Allegra and A. Oliverio (eds.), Psychobiology of Stress, 211–221.

Luminaires of this kind are effective at diluting pulsation at the frequency of the electricity supply, provided the tube is new. Unfortunately, as the tube ages, one electrode often burns more rapidly than the other so that the discharge from one end of the tube becomes brighter than that from the other end. The luminaire is unable to reduce the effects of this asymmetric discharge, and as a result the light may fluctuate slightly at a frequency of 50 or 60 per second. This frequency component typically accounts for less than 2% of the light output, unless the lamp is faulty.

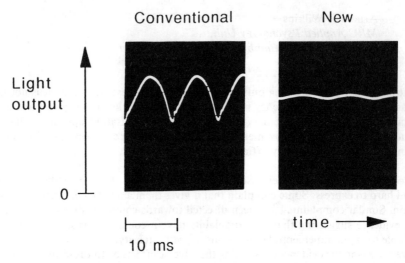

Figure 1. Variation in light output by a typical fluorescent lamp shown as a function of time. The measurements were made using a photodiode with a spectral sensitivity approximating that of the human eye (V γ). At the blue end of the spectrum the variation is greater than that shown.

The light varies between a maximum and about 60% of that maximum with each half-cycle of the supply. Typical variation in light output is shown as a function of time in Figure 1. The measurements were obtained using a photodiode with spectral sensitivity that approximated that of the eye (i.e. a sensitivity that followed the V γ curve). The pulsation is greater at the blue end of the spectrum than at the red because most fluctuation comes not from the phosphor but from the visible part of the gas discharge itself. These differences are not shown in Figure 1. which shows only an average over the range of wavelengths in the visible spectrum weighted on the basis of psychophysical data for the sensitivity of the eye (i.e. the V γ curve). Note that the light is never completely extinguished, even though the voltage across the ends of the tube repeatedly drops to zero. This is because the phosphor coating on the inner surface of the tube retains some of the light from one discharge to the next. Different mixtures of chemicals are used to produce different spectral composition, and the pulsation therefore differs from one type of lamp to another (Eastman and Campbell, 1952).

The above characteristics of fluorescent lighting have a bearing on its biological effects, some of which will now be discussed.

Neurophysiological effects of flicker

Most people are not aware of the pulsations in the light from a fluorescent lamp because the pulsations are too small and too rapid to be seen. The pulsations may nevertheless affect the brain. Eysel and Burandt (1984) recorded from visual neurons in the cat when the animal was looking at a surface subtending 50 degrees at the eye, a far larger stimulus than is usually used in physiological studies of single neurons. The surface was illuminated by fluorescent light, by incandescent light with the same time-averaged luminance, or by daylight. Neurons from the optic tract and lateral geniculate nucleus of the thalamus fired twice as strongly under fluorescent illumination than under the incandescent illumination or daylight. Some cells responded to each flash from the fluorescent light, some responded less frequently, but the firing of all cells was phase-locked to the light pulsation: they all tended to fire just before the light reached its peak. This pattern of activity occurred whether the fluorescent lighting pulsated 100 or 120 times per second. It was only when the supply was altered so that the pulsation in the light reached a frequency of 160Hz that the neurons failed to show phase-locked responding. The luminance of the light did not seem to make much difference over a range of one log unit (a factor of ten). Comparatively little phase-locking was seen in response to incandescent illumination with the same luminance. Incandescent lighting fluctuates less than fluorescent because the hot filament from which the light is emitted does not have long to cool during the cycle time of the ac supply.

Eysel & Burandt did not record from cells in the visual cortex. Cortical cells are thought not to respond to variation in contrast at frequencies in excess of about 20-per-second (Movshon et al., 1978). Further, Eysel & Burandt's experiments were performed on cats. They have not been repeated with primates and it is impossible to know whether cells in the human visual system show a similar response to fluorescent light. Nevertheless electrical potentials recorded from electrodes attached to the human eye (electroretinogram) show responses to intermittent light when the frequency is more than 100-per-second (Greenhouse et al., 1988). There is also psychophysical evidence that the human visual system can resolve very high temporal frequencies of intermittent light, frequencies that are far higher than those at which the light appears to flicker.

Brindley (1964) stimulated the human retina electrically so as to produce the appearance of flashes of light known as phosphenes. At high frequencies of electrical stimulation the phosphenes ceased to flicker and became continuous. Brindley then stimulated the eye with intermittent light, again at frequencies *above* those at which flicker was perceived. When the frequencies of the electrical stimulation and the light stimulation were similar but not identical, the beats between the two forms of stimulation could be seen as flicker. The beats could only have been seen if the human visual system resolved the electrical and visual stimulation. Brindley generated flickering light using a sectored disc. The fluctuation from fluorescent light is less than that from such a disc, but it remains possible that the high-frequency pulsations from a fluorescent tube are resolved at some level of the human visual system, although they are not perceived as flicker.

Consequences for eye movements and visual perception.

Eysel and Burandt (1984), who observed phase-locked responding in cat's lateral geniculate nucleus, pointed out that any sub-cortical structures connected by short neural chains to that nucleus should also show phase-locked responding. Such structures include the superior colliculus, a body associated with the control of eye movements. For this reason Wilkins (1985) studied the eye movements of human observers.

He recorded movements of the eyes across text illuminated by two types of fluorescent light, one conventional, exhibiting the usual 100Hz pulsation, and the other using different circuitry that produced relatively steady light. The observer was asked to direct his gaze alternately between two specified letters on the same line of text and the size of the rapid eye movements (saccades) was measured. The saccades were slightly (3-5%) larger under the conventional fluorescent light. The increase in saccade size, though statistically significant, was weak: it accounted for only 4% of the experimental variance.

The effect of pulsating light on eye movements may help to explain the small impairments in visual search observed in a study by Rey and Rey (1963). Over the course of four weeks five subjects with good vision performed a complex visual search task in which certain letters from a list had to be cancelled if they occurred in a context defined by other neigbouring letters. Before and after periods of 45 minutes at this task a variety of other measures were taken. These included the simple reaction time to a visual stimulus, and the frequency at which a series of brief flashes appeared continuous. Under conventional fluorescent lighting (50Hz circuitry), performance of the visual search task was slightly poorer than under similar lighting with high frequency circuitry. The measures taken before and after the task showed a greater decrement under the conventional lighting.

Rey and Rey review a number of early studies of the effects of fluorescent lighting on vision and visual perception. The effects of the pulsations are sometimes inconsistent but are generally detrimental. The effects are uniformly small.

If fluorescent lighting had major effects on visual perception or the control of eye-movements these would have been noticed and would have given rise to complaints. Complaints there certainly have been, but concerning headaches and eye-strain rather any visual or perceptual effects.

Links with headaches and eye-strain

At the anecdotal level there is much to suggest that a few people, particularly those with migraine, find that fluorescent lighting can induce headaches. Some are severely affected and have to avoid this form of lighting as much as possible.

There are several pointers towards pulsating light as being one possible cause for these complaints. When an observer looks at a flickering light, the electroencephalograph (EEG) measured with electrodes over the back of the head shows a response at the frequency of the light. Golla & Winter (1959) showed that in persons suffering episodic headaches, but not in controls, the amplitude of this response to intermittent light is greater at flash frequencies of around 20 per second than at lower frequencies. Their findings have been replicated by Jonkman & Lelieveld (1981) using more up-to-date recording techniques. Brundrett (1974) extended these findings by using higher flash rates. Above 20Hz, the amplitude of the averaged EEG response at the frequency of the light (the steady-state evoked potential) tends to decrease as the frequency increases. Brundrett measured the amplitude of the evoked potential in a small sample of headache sufferers. He showed that the decrease with frequency was less than in controls, suggesting that the headache sufferers were in some way unusually sensitive to high-frequency flicker.

Evidence that fluorescent lighting causes headaches

A large scale field study by Wilkins and colleagues (1988) has recently shown that the pulsating light from conventional fluorescent lighting is responsible for a substantial proportion of the eye-strain and headaches suffered by office workers. The study compared the incidence of complaints of headache and eye-strain when the offices were lit by conventional fluorescent lighting or by a new form of fluorescent lighting controlled by a high-frequency solid-state ballast. In this form of circuitry the mains voltage is transformed by a high-frequency transistor oscillator. The discharge across the ends of the tube is made 32,000 times a second and the light pulsates 64,000 times per second. The depth of the pulsation fluctuates slightly at the much lower frequency of 100Hz (ie there is low-frequency amplitude modulation of the high frequency waveform). This is because ripple from the ac supply affects the voltage on the transistor oscillator. The resulting 100-per-second variation is typically less than 7% of peak light output, however, and it does not give rise to phase-locked responses in cat neurons (Eysel, personal communication).

The two forms of lighting differ only with respect to their circuitry and their outward appearance is identical. Because of this it was possible for Wilkins and colleagues to undertake a double-blind study: a study in which all participants, including those collecting the data, were unaware of which type of fluorescent lighting they were experiencing.

Staff of a government legal department participated in the study. The department had a large number of small offices receiving little daylight and the staff undertook close visual work almost entirely without the use of computer displays. They were asked to keep weekly diaries of the incidence of headaches or eye strain.

The conventional and the new lighting differed in the speed with which they ignited: the new lighting ignited instantaneously whereas the conventional lighting lit only after a few preliminary flashes. A third type of lighting was therefore introduced at the same time as the new high-frequency lighting. This pulsated in the conventional manner but incorporated electronic ignition so that it ignited rapidly. The group exposed to this type of lighting can therefore be thought of as a control group: they experienced a visible change in the speed of ignition of the lighting but no change in the 100-per-second pulsations. By comparing this group with the other two it was therefore possible to know whether the speed of lamp ignition was important and whether subjects' expectations were affecting their reports of headache and eye-strain.

It cost too much to change the lamps, so these were left in place. Some were "white" and some "cool white". The white pulsated between maximum and 43-47% of maximum and the cool white 49-50%. As it turned out, there were no differences in the incidence of complaints under the two types of lamp.

The three types of fluorescent lighting circuitry were allocated at random and were changed over in mid winter, halfway through the period during which the data were collected.

There was no difference in the incidence of eye-strain and headaches under the two forms of conventional lighting circuit (slow ignition and rapid ignition), suggesting that the speed of ignition or subjects' observation of it had little, if any, effect.

Under the new steady type of lighting, however, the incidence of eye-strain and headaches was more than halved. Presumably participants used the term "eye-strain" to refer to pain in and around the eyes. Those who suffered frequent eye-strain tended to suffer frequent headaches as well. The correlation between the two types of pain is shown in Figure 2.

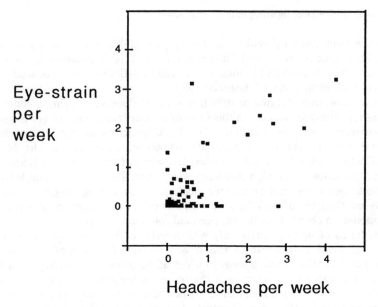

Figure 2. Scattergram showing the correlation between complaints of eye-strain and those of headache. Each point shows the data for an individual subject.

Although the correlation is significant, it is not large, and for this reason eye-strain was analysed separately from headaches.

Figure 3 shows the mean incidence of eye-strain in the group that changed from

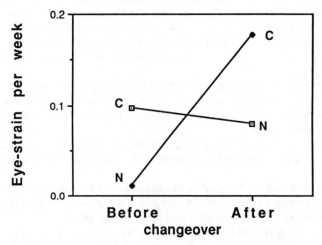

Figure 3. Mean incidence of eye-strain under the two lighting conditions shown separately for those subjects who changed from conventional to new lighting (C-N) and those who experienced the conditions in the reverse order (N-C).

conventional lighting (C) to new (N) and the group that changed from new to conventional. These means are based on very small samples, as can be seen in Figure 4a which shows the incidence of episodes of eye-strain under the two lighting conditions. Most participants suffered no eye-strain, and their data fills the lowest left-hand cell. The few participants who suffered relatively frequent eye-strain tended to do so under the conventional lighting (the data points tend to lie below the diagonal). Figure 4b shows the corresponding data for headaches. Although headaches are more common than eye-strain, again there is a tendency for the data points to lie below the diagonal.

Although the change in the incidence of eye-strain and headaches as a function of lighting was statistically significant, it was based on only a few participants. Nevertheless the sample size can be enlarged by considering the relatively large number of participants whose lighting remained unaltered. These people experienced conventional lighting throughout the period of the study and served as a baseline control. When these individuals are included the size of the sample of people experiencing conventional fluorescent lighting before the changeover is increased to 91. The distribution of the incidence of eye-strain and headache in this group is shown in Figure 5. The distribution is highly skewed: a few people suffer frequently, most do not suffer at all. This figure also shows for comparison the corresponding distributions for the group who experienced the new lighting. Notice that the "tail" of the distribution is shorter: there are now fewer people who suffer frequently. The distributions after the changeover in lighting were highly similar to those in Figure 5. Evidently the imperceptible pulsations in conventional fluorescent + lighting can provoke eye-strain and headaches.

The building was six stories high and although the offices all had windows they overlooked buildings of similar height. The amount of daylight entering the windows therefore increased with the height above the ground, and on a sunny day the increase averaged about 80 lux per storey, measured at the occupant's desk. The incidence of headaches amongst those exposed to conventional fluorescent lighting decreased with the height of the office above the ground. In this particular office building the height of the office was not correlated with the age or status of the occupant or with any other factor that might otherwise have affected the reporting of eye-strain or headache. Eye-strain and headaches may therefore have been affected by the availability of natural light. Even when the fluorescent lighting is turned on daylight has the effect of "diluting" the pulsations in illumination (i.e. of reducing the depth of modulation).

Another aspect of the study provided a completely independent measure of preference for steady lighting. Timers were inserted in the light fittings to measure the total time the lights were turned on. The office occupants had the opportunity of turning the lights on or off by means of wall switches and pull cords on the luminaires. They turned on the new steady fluorescent lighting for 30% longer than the conventional lighting. Evidently people preferred it.

In describing the above study, reference has been made to the 100-per-second pulsation for convenience. As already mentioned however, conventional fluorescent lighting fluctuates not only at 100Hz, but also to a very much lesser extent at 50 per second. The above study does not help to distinguish which of these frequency components is responsible for headaches. Although the 50Hz component is very much smaller than the 100Hz component the sensitivity of the human visual system to 50Hz flicker is very much greater. On the other hand, Eysel and Burandt's (1984) data strongly suggest that the 100Hz component may be worth as much attention as the 50Hz component. Certainly neither the 100-per-second nor the 50-per-second fluctuation is usually visible. It is therefore interesting that pulsating

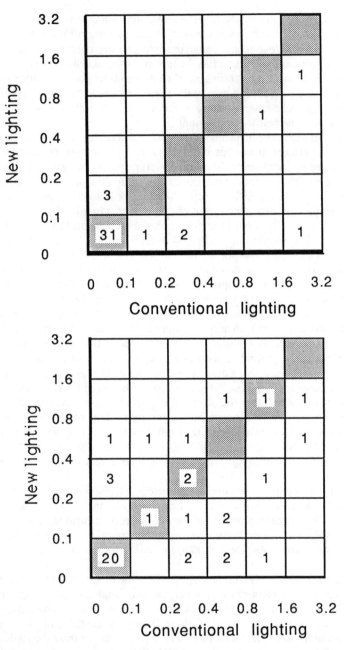

Figure 4. a. Scatterplot showing the weekly incidence of eye-strain under new and conventional lighting for 40 subjects who experienced both conditions. Data tend to be below the diagonal: the few subjects who experienced frequent eye-strain tended to do so more under the conventional lighting than under the new. *b.* Corresponding data for reports of headaches.

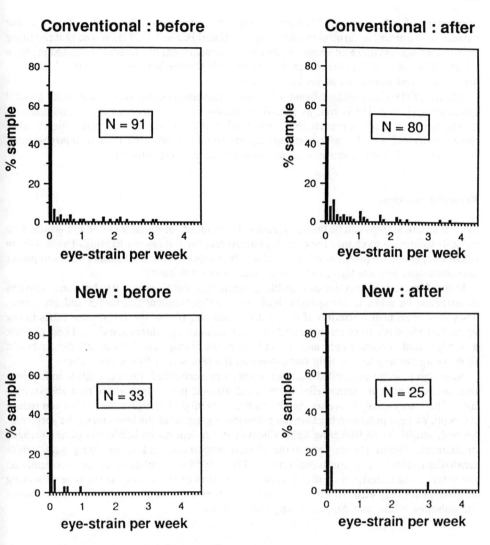

Figure 5. Histograms of the weekly incidence of eye-strain among subjects exposed to new and conventional fluorescent lighting.

illumination is capable of inducing headache even when people are unaware of the fluctuation.

Reasons for the pain

Why should pulsating light induce headaches and eye-strain? The reasons for the pain are far from clear, but perhaps a limited degree of speculation will be excused. Pain is usually a component of a homeostatic mechanism responsible for informing an organism that the current environment is harmful in some way. Eysel and Burandt (1984) showed that the firing rate of neurons was doubled under conditions of conventional fluorescent illumination. Such a high firing rate might make metabolic demands that in the long term are difficult to meet, and fatigue and headaches might be a consequence.

Attacks of classical migraine begin with visual phenomena suggestive of abnormal cortical functioning (Lance, 1986), but, as already mentioned, cortical neurons in cats and primates do not resolve frequencies much above 20Hz, when their frequency response is measured in the conventional way. No measurements appear to have been carried out with large whole-field flicker and free eye movements across a structured field, however.

Remedial measures

Several practical steps can be taken to reduce the incidence of headaches and eye-strain. One obvious solution is to change the circuitry controlling the fluorescent lighting. The new form of circuitry (high-frequency solid-state ballast) is cheaper to run: it consumes 40% less power and emits only 5% less light, so it will eventually pay for itself.

In the short term it may not be possible to change the lighting circuitry, but some help can nevertheless be given to those individuals who suffer frequent headaches and eye-strain. They should sit near a window if daylight is available. If not, the fluorescent tubes in the immediate vicinity can be removed and tungsten-halogen uplighters installed. These provide a steady broad-spectrum light and reflect it from the ceiling. The light is sufficiently bright to make up for any loss of light occasioned by the removal of fluorescent tubes.

Although placebo-controlled studies remain to be performed, there are initial indications that sufferers may get some relief from tinted glasses, particularly those that absorb blue light. There are two obvious reasons why such glasses might be helpful. First, they can reduce the depth of light pulsation because the pulsation is greatest at the blue end of the spectrum. Second, sensitivity to flickering light is heavily dependent on its brightness (time-averaged luminance). Tinted glasses reduce the overall brightness, and in so doing may reduce sensitivity to those fluctuations that remain. The reduction in pulsation depends critically on the detailed characteristics of the absorption spectrum of the lenses. At the time of writing an attempt is being made to generate a tint with an absorption characteristic that minimises the pulsation whilst avoiding a strongly coloured tint.

Conclusions

The complaints with which fluorescent lighting has been associated appear to have physiological foundation. The pulsating light that it produces can cause pain and discomfort. Perhaps the complaints with which visual display terminals have been associated may be traced to similar mechanisms. Harwood and Foley (1987) compared steady and intermittent displays and reported more discomfort with the latter, and a field study by Laübli, Hünting and Grandjean (1981) showed that complaints were related to the modulation depth of flicker from the display.

References

Brindley G.S. (1962) Beats produced by simultaneous stimulation of the human eye with intermittent light and intermittent or alternating electric current. *J Physiol (Lond)* 164, 157.

Brundrett G.W. (1974) Human sensitivity to flicker. *Lighting Res Tech* 6(3) 127.

Eastman A.A. and Campbell J.H. (1952) Stroboscopic and flicker effects from fluorescent lamps. *Journal of the Illuminating Engineering Society*, 47, 27.

Eysel UT and Burandt U (1984) Fluorescent tube light evokes flicker responses in visual neurons. *Vision Res* 24 943.

Golla F.L. and Winter A.L. (1959) Analysis of cerebral responses to flicker in patients complaining of episodic headache. *Electroenceph clin Neurophysiol* 11 539.

Greenhouse D.S., Berman S.M., Bailey I.L. and Raasch T.W. (1988) Paper presented at ARVO.

Harwood K. and Foley P. (1987) Temporal resolution: an insight into the video display terminal (VDT) "problem". *Human Factors*, 29(4), 447.

Jonkman E.J. and Lelieveld M.H. (1981) EEG computer analysis in patients with migraine. *Electrenceph clin Neurophysiol* 52, 652.

Lance J.W. (1986) Visual hallucinations and their possible pathophysiology. In Pettigrew JW, Sanderson KW and Levick W.R. (ed). Visual Neuroscience Cambridge University Press, Cambridge, 374.

Laübli T., Hünting W. and Grandjean E. (1981) Postural and visual loads at VDT workplaces II. Lighting conditions and visual impairments. *Ergonomics*, 24, 933.

Movshon J.A., Thompson I.D. and Tolhurst D.J. (1978) Spatial and temporal contrast sensitivity of neurones in areas 17 and 18 of the cat visual cortex. J Physiol (Lond) 283 101.

Rey P. and Rey J.P. (1963) *Ergonomics* 6, 393.

Wilkins A.J. (1986) *Human Factors* 28(1) 75.

Wilkins A.J., Nimmo-Smith I., Slater A.I., and Bedocs L. (in press) Lighting Res. Tech.

PSYCHOPATHOLOGICAL SYMPTOMS AFTER OPEN HEART SURGERY: EFFECTS OF A MAJOR STRESSOR

D. Naber[1], M. Bullinger[2], R. Holzbach[1], G. Oliveri[1], T. Schmitt[1],
E. Klein[1], U. Preuss[1] and P. Schmidt-Habelmann[3]
[1] Psychiatric Hospital of the University of Munich, FRG
[2] Institute for Medical Psychology of the University of Munich, FRG
[3] German Heart Centre, Munich, FRG

Psychiatric symptoms and in particular post-operative psychoses are frequent complications in patients undergoing open-heart surgery (Blacher, 1982; Heller and Kornfeld, 1986; Mayou, 1986). Depending on patient population, type of operation and diagnostic criteria, the incidence of psychopathological complications varies between 10 and 50% (Egerton and Kay, 1964; Heller et al., 1970; Speidel et al., 1971; Sveinsson, 1975; Kornfeld et al., 1978; Dubin et al. 1979). Postoperative psychosis, also referred to as post-cardiotomy delirium (Blachy and Starr, 1964), develops between the second to the fourth day after surgery, consists of different subtypes and is usually of short duration. Most frequently, a delirious state and paranoid-hallucinatory symptoms occur lasting for about three days. Symptoms of major depression with a duration of several weeks have also been observed.

A variety of factors has been discussed with regard to the etiology of such exogenous psychoses. These include sex, age, cardiac status, psychiatric and neurologic history as well as perioperative data such as time on bypass and operative complications (Blachy and Starr, 1964; Hazan, 1966; Layne and Yudofsky, 1971; Heller et al., 1978; Speidel et al., 1979). Physchological investigations have focussed on personality traits (e.g. depression and anxiety), situational factors (e.g. presurgical information, ward milieu) as well as subjective appraisal of and coping with the operation such as perception of threat or avoidance strategies (Kimball et al., 1973; Heller et al., 1974; Kornfeld et al., 1974; Davies-Osterkamp and Moehlen, 1978; Meffert et al., 1979; Kinchla et al., 1985).

Research on endocrine variables during open heart surgery documents a marked increse of stress related hormones such as growth hormone, cortisol, beta-endorphin and norepinephrine (Lee et al., 1969; Naber and Bullinger, 1985; Salter et al., 1972). Although somatic and psychological variables have been reported to be associated with the incidence of psychopathological symptoms, results regarding such predictors are not unequivolcal. They do, however, indicate the stressful nature of the operation and its potential relevance to transient psychiatric disturbance.

Within such a stress-theoretical frame-work, open research questions pertain to the type of psychopathological symptoms presenting, to factors associated with their incidence and to consequences of psychopathological symptoms for patients quality of life after cardiac surgery (Magni et al., 1987; Folks et al., 1986; Schonberg et al., 1985). Within an ongoing longitudinal study on psychosocial adjustment in open heart surgery, stress- related somatic

223

S. Puglisi-Allegra and A. Oliverio (eds.), Psychobiology of Stress, 223–229.

and psychological variables were assessed in order to determine their influence on psychopathological outcome of open heart surgery and its implications for patients postoperative status.

Methods

In cooperation with the German Heart Centre, Munich, a prospective observational cohort study was carried out in patients scheduled for open heart surgery. Inclusion criteria were age (18 to 65 years), indication for either aortic valve replacement (VR) or coronary artery bypass graft (CABG) surgery and informed consent. Exclusion criteria were emergency operations and lack of literacy. Variables assessed included personal characteristics auch as personality, coping, self- control, social support as assessed via psychometrically tested german questionnaires.

Neuropsychological function was assessed with eight standard mental tests related to attention, recall, learning and concentration. Clinical data included psychopathological symptoms, medical history and diagnosis as well as different indicators of health status before, at and after surgery including peri-operative information. Quality of life data pertained to mood state (State Anxiety STAI, Profile of Mood States POMS and visual analogue VAS scales related to appraisal of the operation), to physical state in terms of a disease – and treatment – specific complaint list as well as to the Beth Israel scale for physical ability, and to daily acrivities in terms of work, social relations, recreation and self-actualization. Additionally, overall quality of life was assessed with the Psychological Well-Being Index (PGWI), a life satisfaction list and a 5-point global indicator. Spouse ratings involved the POMS and different visual analogue scales related to patients as well as own feeling and behaviour. Data were collected prior to surgery (2 weeks and 1 day pre-op), during a post-operative in- patient period (days 4, 7, 10 post-op) as well as during a one- year follow up period with assessements at 3, 6 and 12 months after surgery (table 1).

All eligable patients scheduled for surgery were informed about the psychosocial study and asked for their cooperation. Two weeks prior to surgery, questionnaires relating to personal characteristics and quality of life as well as the pre-operative appraisal list were mailed to participating patients and their spouses. The day before operation the patients responded to mood scales, were neuropsychologically tested and psychiatrically rated. These expert ratings as well as mood self-assessments were repeated at days 4, 7 and 10 after the operation; at day 10, patients underwent the second neuropsychological testing. Statistical analyses included descriptive statistics, t-test, analyses of variance and correlation analyses using SPSS computer package.

Until now, 130 consequtive patients were approached; 15 denied participation, 1 patient died and 4 patients dropped out of the study before the 3 month follow up. The present sample consists of 71 male and 39 female patients (n = 110) with a mean age of 57 ± 11 years, 48 of whom received valve and 63 bypass surgery. Results cover the time-span from both presurgical assessements to the 3-month follow-up and include the inpatient period with measurepoints at the 4th, 7th and 10th postoeprative day.

Table 1: Variables and measurepoints

	Pre-op period		inpatient period			follow-up		
	−14d	−1d	+ 4d	+ 7d	+ 10d	+ 3m	+ 6m	+ 12m
CLINICAL								
Psychiatric AMDP-rating		x	x	x	x			
Medical history	x							
Diagnosis, pre-op Status		x						
Peri- and post-op Data			x	x	x	x		
NEUROPSYCHOLOGICAL								
Vocabulary		x			x			
Benton		x			x			
Trailmaking		x			x			
D2-Test		x			x			
Color Stroop		x			x			
Earl Ray		x			x			
Maze		x			x			
QUALITY OF LIFE								
Mood POMS	x	x	x	x	x	x	x	x
STAI		x						
Stress VAS	x					x	x	x
Function Everyday Life	x					x	x	x
Physical Complaint List	x					x	x	x
Beth Israel	x					x	x	x
Overal PGWI	x							
Life satisfaction	x							
PERSONAL								
Personality FPI	x						x	
Social Support SSQ	x							x
Coping FEKB	x							x
Self Control SKF	x						x	
Sociodemographic	x							
SPOUSE RATINGS	x					x	x	x

Results

The day prior to the operation (OP), psychiatric ratings as well as self-rating scales revealed minor symptoms of anxiety, agitation and depression in 34% of the patients. After the operationm psychotic symptoms weree apparent in 24 of 110 patients (21%), valve patients were over represented (see table 2). Most frequently, a delirious state occured (n = 11), followed by a paranoid-hallucinatory syndrome (n = 8) and symptoms of major depression such as suicidal ideas or delusions of guilt (n = 5). Psychotic symptoms usually developed within two to four days after OP and did not last longer than three days. Only in the group of depressed patients, symptoms had not ceased within 10 days after OP.

Table 2: Psychopathological symptoms

	Bypass	Valve	Total
No sypmtoms	44%	28%	37%
Subclinical	39%	45%	42%
Psychoses	17%	27%	21%

Preopertative psychopathology was significantly correlated with post-operative AMDP scores: total score r = .54, p<.001; delirious syndrome r = .48, p<.001, paranoid-hallucinatory syndrome r = .62, p<.005, depressive syndrome r = .49, p<.001. Prior to OP, neuropsychological functioning was normal in 83% of patients, slightly reduced in 12% and markedly impaired in 5%. Ten days after OP, the mean score in all tests did not significantly change.

However, as compared to the individual preoperative scores, six patients considerably improved (more than 1 standard deviation) in at least three tests and 14 patients worsened. Cognitive abilities were not related to psychopythological or other variables and did not differ between both types of operations.

Preoperative anxiety, as measured with the STAI, was above average reference scores in 67% of the patients. VAS ratings also indicated that the majority of patients perceived the surgery as a major stressor. In comparison to the preoperative scores, mood as measured with the POMS improved at days 4, 7, and 10 after OP with the exception of slightly increasing fatigue scores. Three months after the operation, slight deteriorations in mood in the totale sample were aparent but failed to reach significance. Physical and functional quality of life measures showed significant improvements (p<.05) from two weeks before surgery to the 3-month follow-up. Similarly, judgements of overall quality of life including the PGWI indicated a positive perception of surgical outcomne for the majority of patients. Postoperative quality of life ratings were correlated with preoperative scores as well as with personality and coping characteristics reflecting optimistic future outlook (r = .43 to r = .87) but neither with pre- operative neuropsychological function nor with changes in cognitive abilities after OP.

Analyzing groups of patients without symptoms (n = 40), minor symptoms (n = 46) and major symptoms (n = 24), significant differences were found in appraisal of the operation, in coping strategies and in mood one day prior surgery, but not in preoperative clinical data. None of the perioperative variables such as duration of the operation, intraoperative complications or time on extracorporal circulation was related to psychopathological outcome.

Specifically, preoperative anxiety as measured by the STAI differentiated significantly between the groups (p<.05) indicating that extremely anxious perso ns perceiving the operation as a dramatic strain (VAS) are at risk in developing a psychosis. When accounting for the coping style denial as a covariate, however, patients with low anxiety and high denial scores also appeared to be at risk, while average anxiety scores and low denial scores did not have an effect. In line with these results it should be mentioned, that of the 15 patients not participating in but seen during the study, had developed psychopathological symptoms as did one of the 5 drop out patients.

In addition, a high internal locus of control orientation as well as the preference of an problem-oriented coping style correlated with more unfavourable AMP ratings (r = .46), while external or chance control orientation as well as palliative coping orientations correlated with lower scores in emotional quality of life indicators 3 month after OP (r = .-41 to r = .-64).

Discussion

In agreement with the large literature on psychiatric disturbances after open heart surgery (Blachy and Starr, 1964; Hazan, 1966; Speidel et al., 1979; Dubin et al., 1979; Heller and Kornfeld, 1986; Mayou, 1986); 21% of the patients in this study suffered from post-operative psychosis and an additional 42% from minor psychopathological symptoms. As previously reported, psychopathological outcome was worse after valve replacement than after bypass surgery (Rabiner et al., 1975), but otherwise more related to pre-operative psychopathology than to somatic variables such as duration of operation, minimal blood pressure etc. (Kornfeld et al., 1974; Flemming et al., 1979; Dubin et al., 1979; Speidel et al., 1979; Kinchla et al., 1985; Naber and Bullinger, 1985). Neuropsychological tests did not show significantly reduced functioning in the total group, but deterioration in a subgroup of patients. These results are in accordance with the sparse literature (Savageau et al., 1982; Hammeke and Hastings, 1988). The suggested bad prognosis of patients with pre-operative cognitive deficits (Willner et al., 1976) is not supported by the present data but will be re-analyzed when the follow-up data of 12 months are complete and intellectual functioning has fully recovered (Aaberg et al., 1983),

In accordance with the growing recognition of cognitive-emotional processing in stressfull situations (Endler and Edwards, 1983; Holroyd and Lazarus, 1983), e.g. the influence of subjective appraisal and coping behaviour on neuroendocrine activation (Greene et al., 1970; Frankenhaeuser, 1980; Bullinger et al., 1984; Magni et al., 1986), appraisal and coping with open-heart surgery were found to be related to pre- and post-operative psychopathology. Of particular importance was the perception of the OP as stressfull and the – often not successful – attempt to maintain self-control in the hospital situation. Thus, a high level of arousal, anxiety and a strong effort to influence the course of illness might be associated with to psychiatric complications (Kimball et al., 1973; Flemming et al., 1978; Speidel et al., 1979). Coping styles such as denial or avoidance, not useful in some diseases (Fefel et al., 1987), might be adaptive strategies in the experience of open-heart surgery (Lair and King, 1976; Kimball et al., 1973; Naber and Bullinger, 1985), implying a reconsideration of the pre-operative information process (Kendall, 1982). It is doubtful and not well investigated yet whether pre-operative psychological or psychiatric intervention could decrease the incidence of post-operative psychopathology (Surman et al., 1974; Kornfeld et al., 1974).

The present study, indicating improved quality of life, is in agreement with most (Stanton et al., 1984; Schoenberg et al., 1985; Folks et al., 1986), but not all similar studies (Gundle et al., 1980). Again, complete follow-up data are needed to analyze predictors for psychosocial adjustment (Heller et al., 1974; Magni et al., 1987) and to evaluate whether psychiatric symptoms in the immediate post-operative period are unrelated to long-term psychopathological outcome (Rabiner et al., 1976).

The detailed analysis of the total population of 163 patients will hopefully contribute to unravel the complex network of somatic and psychological variables in the etiology of psychiatric symptoms after oper-heart surgery and their consequences for patients quality of life, a fascinating area of interdisciplinary research.

References

Aaberg T., Aalund P., Kihlgren M. (1983), 'Intellectual function late after open-heart surgery". *Ann Thor Surgery* 36, 680-683.

Blacher R.S. (1972), 'The hidden psychosis of open-heart surgery. With a note on the sense of awe". *JAMA* 222, 305-308.

Blachy P.H., Starr A. (1964), 'Post-cardiotomy delirium". *Am J Psychiatry* 121, 371-375.

Bullinger M., Naber D., Pickar D., Cohen R.M., Kal in N.H., Pert A., Bunney B.E. (1984) 'Endocrine effects of the cold pressor test: relationships to subjective pain appraisal and coping". *Psychiatry Res* 12, 227-233.

Davies-Osterkamp S., Moehlen K. (1978) 'Post-operative course after cardiac surgery as correlated with pre-operative fear and its control". *Med Psychol.* 4, 247-260.

Dubin W.R., Field H.L., Gastfriend D.R. (1979) 'Postecardiotomy delirium: a critical review". *J.Thor Cardiovasc. Surg.* 77, 586- 594.

Egerton N., Kay J.J. (1964) 'Psychological disturbance associated with open-heart surgery". *Br. J. Psychiary* 110, 433-439.

Endler N.S., Edwards J. (1983) 'Stress and personality". In: Goldberg L., Bresnits S. (Eds) Handbook of Stress: Theoretical and Clinical Aspects. MacMillan Press, New York, pp.36-48.

Feifel H., Strack S., Nagy V.T. (1987) 'Coping strategies and associated features of medically ill patients". *Psychosom. Med.* 49, 616-625.

Flemming B., Dahme B., Goetze P., Huse-Kleinstoll G., Kalmar P., Meffert J., Spreidel H. (1978) 'Some psychological predictors for psychosis after open heart surgery". *Thoraxchirurgie* 26, 458-462.

Folks D.G., Blake D.J., Fleece L., Sokol R.S., Freeman A.M. (1986) 'Quality of life six months after coronary artery bypass surgery: a preliminary report". *South Med.* J. 79, 397-399.

Frankenhauser M. (1980) 'Psychoendocrine approaches to the study of stressful person – environment transaction". In: Selye H. (ed) Selye's Guide to Stress Research. Van Nostrand, New York, pp.330- 354.

Greene W.A., Conron G., Schalch D.S., Schreiner B.F. (1970) 'Psychologic correlates of growth hormone and adrenal secretory responses of patients undergoing cardiac catheterization.' *Psychosom. Med.* 32, 599-614.

Gundle H.J., Reeves B.R., Tate S., Raft D., McLauren L.P. (1980) 'Psychosocial outcome after coronary artery surgery.' *Am J. Psychiat.* 137, 1591-1594.

Hammeke T.A., Hasting J.E. (1988) 'Neuropsychological alterations after cardiac operation' *J.Thorac. Cardiovasc: Surg.* 96, 326- 331.

Hazan S.J. (1966) 'Psychiatric complications following cardiac surgery' Part I: a review article. *J. Thorac. Cardiovasc. Surg.* 51,307-319.

Heller S.S., Kornfeld D. (1986) 'Psychiatric aspects of cardiac surgery.' *Adv. Pshychosom. Med.* 15, 124-139.

Heller S.S., Frank K.A., Kornfeld D.S., Malm J.R., Bowman F.O. (1974) 'Psychological outcome following oper-heart surgery.' *Arch. Intern. Med.* 134, 908-914.

Heller S.S., Kornfeld D.S., Frank K.A., Hoar P.F. (1978) 'Postocardiotomy delirium and cardiac output.' *Am. J. Psychiatry* 136, 337-339.

Holroyd K.A., Lazarus R.S. (1983) 'Stress, coping and somatic adaptation.' In: Goldberg L., Bresnitz S. (Eds) Handbook of Stress: Theoretical and Clinical Aspects. MacMillan, New York, pp.21-35.

Kendall P.C. (1982) 'Stressful medical procedures: cognitive behavioural strategies for stress management and prevention.' In: Meichenbaum D., Jahremko M.E. (Eds) Stress Reduction and Prevention, Plenum Press, New York, pp. 159-191.

Kimball C.P., Quinlan D., Osborne F., Woodward B. (1973) 'The experience of cardiac surgery.' V.Psychological patterns and prediction of outcome. *Psychother. Phychosom.* 22, 310-319.

Kinchla J., Weiss T. (1985) 'Psychological and social outcome following coronary artery bypass surgery.' *J. Cardiopulmonary Rehabil.* 5, 274-283.

Kornfeld D.S., Heller S.S., Frank K.A., Edie R.N., Barsa J. (1978) 'Delirium after coronary bypass surgery.' *J.Thorac. Cardiovasc: Surg*: 76, 93-96.

Kornfeld D.S., Heller S.S., Frank K.A., Moskowitz R. (1974) 'Personality and psychological factors in postcardiotomy delirium.' *Arch. Gen. Psychiatry* 31, 249-253.

Lair C.V., King G.D. (1976) 'MMPI Profile predictors for successful and expired open heart surgery patients.' *J. Clin. Psychol.* 32, 51-54.

Layne O.Z., Yudofsky S.C. (1971) 'Post-operative psychosis in cardiotomy: the role of organic and psychiatric factors.' *N. Engl. J. Med.* 284, 518-520.

Lee W.H., Miller W., Rowe J. (1969) 'Effects of extracorporeal circulation on personality and cerebration.' *Ann Thorac Surg.* 7, 562-570.

Mayou R. (1986) 'The psychiatric and social consequences of coronary artery surgery.' *J. Psychosom. Res.* 30, 255-271.

Magni G., Canton G., Valfre C., Polesel E., Cesari F. (1986) 'Anxiety, hostility and blood pressure variation during heart surgery.' *Psychosomatics* 27, 362-369.

Magni G., Unger H.P., Valfre C., Polesel E., Cesari F., Rizzardo R., Paruzzolo P., Gallucci V. (1987) 'Psychosocial outcome one year after heart surgery. "A prospective study. *Arch. Intern. Med.* 147, 473-477.

Meffert H.J., Dahme B., Flemming B., Goetze P., Huse-Kleinstoll G., Rodewald G., Speidel H. (1979) 'Open-heart surgery from the psychological point of view and resulting therapeutic considerations.' *Psychother. Psychosom.* 23, 148-156.

Naber D., Bullinger M. (1985) 'Neuroendocrine and psychological variables relating to post-operative psychosis after open-heart surgery.' *Psychoneuroendocrinology* 10, 315-324.

Rabiner C.J., Willner A.E. (1976) 'Psychopathology observed on follow-up after coronary bypass surgery.' *J. Nerv. Ment. Dis.* 163, 295-301.

Rabiner C.J., Willner A.E., Fishman J. (1975) 'Psychiatric complications following coronary bypass surgery.' *J. Nerv. Ment. Dis.* 160, 342-348.

Salter C.P., Fluck D.C., Stimmler L. (1972) 'Effect of open-heart surgery on growth-hormone levels in man.' *Lancet II*, 853-855.

Savageau J.S., Stanton B.A., Jenkins C.D., Klein M.D. (1982) 'Neuropsychological dysfunction following elective cardiac operation.' I.Early assessment. *J. Thor. Cardiovasc. Surg.* 84, 585-594.

Schoenberg B., Zuercher M., Baur H.R. (1985) 'Lebensqualitaet nach Herzklappenersatz. *Schweiz med Wschr* 115, 239-241.

Speidel H., Dahme B., Flemming B., Gotze P., Huse-Kleinstoll G., Meffert H.J., Rodewald G. (1979) 'Psychische Stoerungen nach offenen Herzoperationen.' *Nervenarzt* 50, 85-91.

Stanton B.A., Jenkins C.D., Savageau J.A., Thurer R.L. (1984) 'Functional benefits following coronary artery bypass graft surgery.' *Ann. Torac. Surgery* 37, 286-291.

Surman O.S., Hackett T.P., Silverberg E.L., Behrendt D.M. (1974) 'Usefulness of psychiatric intervention in patients undergoing cardiac surgery.' *Arch. Gen. Psychiatry* 30, 830-835.

Sveinsson I.S. (1975) 'Postoperative psychosis after heart surgery.' *J.Thor. Cardiovasc. Surg.* 70, 717-726.

Willner A.E., Rabiner C.J., Wisoff B.G., Hartstein M., Struve F.A., Klein D.F. (1976) 'Analogical reasoning and postoperative outcome. Predictions for patients scheduled for open heart surgery.' *Arch. Gen. Psychiatry* 33, 255-262.

ASSESSING PATTERNS OF ADJUSTMENT TO THE DEMANDS OF WORK

G. Robert J. Hockey and Marion Wiethoff
MRC/ESRC Social and Applied Psychology Unit,
University of Sheffield S10 2TN

Animal models of stress have been developed largely on the basis of data from studies which impose rather severe changes on the animal's environment. As many of the other papers to this volume clearly demonstrate, stressors such as the separation of young from their mothers, imobilization or enforced swimming are associated with profound changes in both peripheral and central components of stress mechanisms. In the study of human stress, such manipulations are, of course, impossible, although natural disasters and trauma provide important evidence on the effects of severe stressors (eg, Marshall, 1947). More often, however, when we talk of stress in relation to people, it is in the context of normal social activity: work, family life and leisure.

Stress may be considered to result from a failure to adjust satisfactorily to the constant changes in the person's environment. There is a problem of definition, however. Clearly, we cannot assume that the psychobiological consequences of resolving an argument with a spouse, or of meeting the demands of a busy work day, are comparable with those found in many experimental studies of animal stress. The difference may be one of degree, though it is likely that there are also qualitative differences in the patterning of the stress response.

In most cases of human stress there are a range of appropriate responses available for resolving the problem, and far greater significance is attatched to the use of coping responses by individuals. The human response to stress is mediated through a process of cognitive appraisal of the nature and extent of threat, and an evaluation of the efficacy of different available actions (Lazarus and Folkman, 1984). It is perhaps only at extreme levels of human stress, such as those involving disasters, that the human and animal models come together, since the range of appraisal and coping options is greatly reduced in such circumstances.

The Study of Occupational Stress

For human beings, one of the most significant and persistent sources of stress is their job. Although relatively few genuine insights can be claimed for the study of human adjustment to work demands, some broad generalisations have emerged from research efforts in recent years. For example, jobs which make heavy work demands (eg, those which are machine-paced or which require frequent action) may be particularly susceptible to impaired well-being and health (Caplan et al., 1975; Cooper and Marshall, 1976). This is a generalisation that accords well with a 'commonsense' view of stress (as resulting from environmental overload), but there are some important qualifications.

S. Puglisi-Allegra and A. Oliverio (eds.), Psychobiology of Stress, 231–239.

An influential approach to understanding the relationship between stress and strain has been that of the 'Michigan' group (eg, Caplan et al., 1975; Harrison, 1978). This is characterized in terms of the 'fit' between an individual and his environment.

In the 'person-environment (P-E) fit' model stress is defined as a function of the discrepancy between human abilities and needs, on the one hand, and environmental demands and opportunities for skill use, on the other. Thus, high job demands may be offset by high levels of skill, or even by a need to feel busy. The critical factor in the development of strain may not be the level of work demands *per se*, but the balance of work demands and human resources.

A more specific qualification concerns the extent to which the work environment is *controllable* (ie, it provides the individual with opportunities to manage his or her workflow, for example through regulation of the timing or choice of task actions). Karasek (1979), in a major study of the impact of job characteristics on health and well-being, demonstrated that the controllability of the workplace (Karasek calls it 'job decision latitude') is a major moderating factor, providing an effective buffer against the putative effects of high demands. Karasek found that only jobs which had both high workload and low controllability were associated with patterns of psychological and physical ill-health.

Approaches to the study of the relation between work stress and strain have implicitly assumed that individuals respond passively to work demands, in supplying what is needed to do the job rather than being an active component of the demand-supply process. By contrast, a transactional stress model (Cox, 1978; Lazarus and Launier, 1978) considers the individual to 'manage' workload through imposed personal goals, needs and strategies. Karasek's results demonstrate that individuals in controllable jobs are able to manage their workload more effectively in order to reduce any imbalance in the P-E fit between external demands and internal resources.

It is also clear that individuals are able to exercise choice in their strategy for resolving stress states. For example, Hockey (1986) has distinguished between two broad modes of control in the context of task performance under stress: (a) active: high-cost (effortful), high effectiveness, or (b) passive: low-cost, low effectiveness. The high priority of task performance goals, both in the laboratory and at work, means that performance is often 'protected' against disruption by external stressors or high workload, though the compensatory effort involved in this may be observed as costs in other systems, such as increases in catecholamine levels or subjective strain (Hockey, 1984; Mulder, 1986). Additional mental involvement (in the form of effort) may also be detectable in phasic changes in the sympathetic/ parasympathetic balance, for example in pupil dilation (Beatty, 1982) or heart rate variability (Mulder, 1986).

From the point of view of transactional models, whether general job characteristics are demanding or not will depend less on their objective description, than on how they are interpreted and acted upon. Similarly, requirements to exercise control or discretion in decision-making (job decision latitude) may provide opportunities for workload management only for those who interpret such job characteristics as additional resources for exercising active control. For others, such apparent control opportunities may in fact be additional (and unwanted) demands, or encourage strategies which are sub-optimal.

Other Approaches to Human Stress

The occupational stress area has focused necessarily on adjustment to the demands of work. In doing this, however, other sources of stress may be ignored. For an analysis of individual adjustment this may be misleading, since difficulties of adjustment to domestic or social stressors may markedly affect both appraisal and coping at work. Research on life events and daily hassles, notably the work of Lazarus and his colleagues, gives considerable prominence to personality factors in coping behaviour (eg, Lazarus and Launier, 1978). Lazarus has been primarily responsible for the acceptance of differential coping activity, in the form of cognitive transactions, as an essential moderating factor in the stressor-strain relationship (Lazarus and Folkman, 1984). Recent developments in this area (eg, DeLongis et al., 1988) have advocated the use of longitudinal, rather than cross-sectional designs, in order to investigate the covariation between fluctuations in demands and health measures, using within-subject correlations. This method allows the investigation of the efficacy of individual coping strategies by removing variability between individuals in the scaling of demands, and is a central component of the design strategy outlined here.

A rather separate tradition, having its origins in the psychobiology of motivation and emotion (Frankenhaueser, 1979; Ursin et al., 1978), emphasises the role of the endocrine and autonomic stress systems in the coping response, following the fundamental work of Cannon and Selye. Frankenhaeuser has argued that examining the impact of emotional demands on neuroendocrine activity provides direct insights into adaptive processes. She has also proposed that the controllability of work may also be reflected in the patterning of endocrine response (Frankenhaeuser and Johansson, 1986). Active (effortful) management of demands (in controllable work environments) is associated with increased levels of catecholamine levels (adrenaline and noradrenaline), and reduced adrenocortical activity (cortisol). Passive coping (in low controllability jobs) is, by contrast, associated with distress and increased levels of both catecholamines and cortisol.

These two patterns of hormonal activity have been interpreted in terms of normative responses to situational factors, in terms of the differential availability of effective coping responses. A broader question, which we are currently investigating, is whether such active and passive coping patterns are found in different individuals in the *same* situation. From our earlier discussion, we might expect the use of active coping to vary across individuals, even in controllable environments, as a function of a willingness to tolerate the required high effort problem-solving state, and the relative priority of task and other goals. Such variability may be found in all work settings, though it is likely to be most marked at relatively low levels of controllability, where the effort required to sustain work goals under high workload is greatest. The two studies summarized below offer some preliminary analyses of coping styles in junior doctors, and reinforces the value of pursuing this kind of approach.

An Alternative Methodology for Studies of Occupational Stress

A major purpose of this paper is to propose an alternative methodology for research on occupational stress which brings together these frameworks. The general rationale is outlined more specifically below, and illustrated with reference to two current studies of naturally-occuring demands in junior hospital doctors.

(1) Cross-sectional research designs provide insights into normative relationships between variables, though at the expense of analysis of variation between individuals. This requires repeated measures on the same group of individuals over a range of stressor conditions, in order to provide reliable response patterns on an individual level.

(2) Individuals employ different characteristic coping methods in managing threats to work effectiveness or well-being. It is likely that a significant proportion of the variability in stressor effects may be accounted for by systematic differences in coping activity. This requires the assessment of both dispositional (enduring) coping style, and current (situational) coping strategy. These may then be used as a basis for classifying individual patterns of responses to stressful environments, including heavy work demands. Such an approach needs, however, to be based on relevant theory, rather than rely on general psychometric methods or broad-band tests. From our above discussion this means basing our coping measures on behaviour such as work-goal orientation and effort regulation.

(3) In addition to subjective reports (of mood or symptoms) the assessment of strain needs to include measures of work performance, as well as independent biochemical or physiological strain indicators. We have argued previously (Hockey, 1986) that the advantage of multi-level measurement is not confined to the greater generalisability of the findings. As we have observed above, particularly in situations where task performance goals are afforded high priority, decrements in effectiveness of work may be prevented by increased recruitment of cognitive resources (compensatory effort). Because of this 'trade-off' between performance and effort, the costs of coping attempts may only be detected in other components of the adaptive response (subjective effort and fatigue, high levels of endocrine activity, etc, as well as in more subtle indices of performance). Assessment of several levels of the adjustment response is necessary in order to interpret the overall *pattern* of coping activity.

By way of illustration, we present a summary of two studies we have carried out with junior doctors into patterns of adaptation to naturally-occurring changes in workload. In addition to providing preliminary information for the use of longitudinal, multi-level methods in our larger research programme, these studies allowed us to explore the detailed relationships between demands, moods and endocrine activity on an individual basis. Although only two subjects took part in the first study some valuable implications may be found for the use of longitudinal sampling in studying the impact of workload in larger samples. The second, while retaining the longitudinal, repeated measures design, tests a larger sample (n = 32) with correspondingly less observations for each subject. This increased sample allows us, however, to begin to explore the basis for individual differences in coping.

Junior Doctor Study 1

The subjects in this study were two junior doctors, TD and CB (both single males, aged 23), working in a cardiothoracic unit of a large teaching hospital. They were tested on each working day over a period of 5-6 weeks at the end of their pre-registration year. This provided 27 complete days of data for one subject (CB) and 32 days for the other (TD). Subjective ratings of work demands and affective state (mood) were obtained twice a day: at the end of the morning period (around 1230), and again at the end of the afternoon (around 1630). Since continuous urine colllection proved impossible over the 6-week period, creatine-corrected estimates were made of adrenomedullary and adrenocortical concentrations by analysis of urine samples collected on three occasions, around 0800, 1230 and 1630.

As part of a larger study, subjects also carried out performance tests (visual search and working memory) at the end of each working day, accompanied by subjective ratings of mental effort. In addition, continuous ECG data were collected for later spectral analysis of heart rate variability (an independent measure of mental effort).

The two main subjective measures were of the pattern of work demands experienced, and current mood state. The work demands questionnaire asked subjects to indicate the extent to which their work over the previous few hours had involved time pressure, high levels of concentration and effort, emotional adjustment, and so on, as well as the degree of support they received from colleagues, and the degree of personal control experienced. Mood was assessed using a shortened (12-item) version of a checklist developed in Sheffield. This allows us to measure two (orthogonal) affective dimensions: alert – tired and relaxed – tense, and express these in terms of negative affect (or 'strain') scores, called 'fatigue' and 'tension'.

Results.

Prior to multivariate analysis, all data were subjected to a 'detrending' procedure to remove sequential dependencies and periodic effects associated with single-subject time series designs. For this purpose only data for complete 5-day weeks (with no weekend days) were used, resulting in 25 sets of data for each individual. The resulting normalised residuals, indicating relative changes in each of the variables, were used as the data set for the correlational and regression analyses. These were conducted for each subject separately. For the present purposes morning and afternoon data are combined to give whole-day measures.

A principle components analysis of work demands revealed slightly different solutions for the two subjects. Each included one major factor, corresponding to a composite of the majority of the demand variables, with two secondary factors for CB (physical demands, emotional demands), and one for TD (knowledge/skill demands). Resource variables (support and control) were negatively loaded on the primary demand factor for both subjects. In order to allow comparison between individuals a common demand classification

Table 1. Correlation coefficients (decimal points omitted) of work demand factors with mood and endocrine variables (based on $n = 25$ days for each subject). * $p < .05$, + $p < .001$

CB (top) TD (bottom)	CogWL	EmotWL	Knowl	PhysWL
Fatigue	−25	19	−24	45*
	41*	28	−37*	01
Tension	39*	38*	03	48*
	85+	88+	06	12
Adrenalin	35*	−11	28	−21
	36*	38*	−06	−19
Noradrenalin	04	−10	−10	13
	39*	27	22	19
Cortisol	29	36*	04	06
	−12	16	−34*	−12

was adopted, which combined these two sets of factors: cognitive workload (Cog WL), emotional workload (Emot WL), physical workload (Phys WL) and knowledge demands (Knowl).

A summary of the correlation analyses for each subject is given in Table 1. For simplicity this shows only correlations between work demand variables and strain indices. Both sets of data show an impact of workload on moods and biochemical state, though the patterning is somewhat idiosyncratic. Tension is affected by the level of both cognitive and emotional workload, for both CB and TD. Fatigue, however, correlates for CB only with physical workload, while TD shows a correlation with cognitive load. In addition, but only for TD , demands on the use of knowledge are associated with *reduced* levels of fatigue.

Correlations with endocrine activity support this general pattern of differences. CB shows positive correlations of both adrenaline and cortisol with cognitive load, and cortisol with emotional load. For TD, cortisol, also, is lower for greater demands on the use of knowledge: otherwise the pattern is primarily of an increase in the level of both catecholamines with workload.

Discussion.

This study was carried out primarily as a pilot study of adaptation patterns across an extended period of normal work, as a basis for the development of this kind of methodology in our larger research programme. The value of being able to carry out within-subject correlation and regression is evident from the data, since such a design allows for the assessment of reliable individual patterns of adjustment to demand variations within normal everyday work. If data such as these were available from a large number of individuals, it would raise serious doubts about the generality of conclusions based on cross-sectional studies.

Despite the natural limitations of a study involving only two individuals, the differences between them in the patterning of the work demands – strain relationship illustrate the strong influence of interindividual moderating factors in coping activity (Frese, 1986). The differential relationships of work demands with endocrine patterning are surprisingly consistent with the extension of Frankenhaeuser's active/passive coping theory, discussed earlier, though this is, of course, fortuitous. CB appears somewhat 'passive'. He shows increases in both catecholamines and cortisol with higher workload levels, and no effect of the psychological workload factors on fatigue (which we hypothesise as resulting from the mental effort involved in work management). TD presents altogether a more 'active' profile of catecholamine involvement and sensitivity of fatigue to external demands, with reduced cortisol levels. These observations are only of passing interest here, but such factors are currently being examined in a second study with a much larger sample. A preliminary summary of this study is reported here.

Junior Doctor Study 2.

Thirty-two junior doctors participated in a semi-longitudinal study. Data were collected for each doctor for a complete working day, once every two weeks or so, over a 5 – month period (between 8 and 13 occasions each). Endocrine activity (catecholamines and cortisol) for each half-shift was assessed from (approximately) 4-hour urine samples. Subjective measures, as in the study 1, comprised assessment of workload and affective states over the same half-shift periods, with additional reports of experienced effectiveness, sleep quality, life events and mental and physical health. Subjects also carried out performance tests (a computerised drug prescription task, involving heavy working memory and planning). These were completed at the end of each working day, accompanied by subjective ratings of mental effort and measurement of ECG.

In order to examine the role of individual differences in coping activity more directly a questionnaire was devised to assess coping styles. These are assumed to be relatively stable (dispositional) factors, which influence the choice of strategy in particular situations, dependent on the perception of work goals. The questionnaire was based on Lazarus' analysis of the two primary functions of coping as 'problem-focused' (concern with task goals) and 'emotion-focused' (concern with one's own state), combined with our own emphasis on the distinction between active (effortful, goal-oriented) and passive (low effort, goal-relinquishing) strategies.

Measurement of Coping strategies.

Four different 'pure' coping patterns, representing combinations of these two dimensions were defined, with examples, and presented to the subjects, accompanied by a list of 15 stress events. These were based on reports of 'stressful experiences' encountered in the course of their hospital work (e.g. "death of a patient on your ward", "making a mistake in diagnosis or treatment"). They were asked to rate each item for its frequency of occurrence and stressfulness, whether it provided a threat to work goals, emotional stability or self-esteem, and the perceived controllability of the event. They were also asked to indicate how each event was dealt with at the time, by endorsing one or more of the four coping patterns. A profile of coping styles was obtained for each person by summing their responses across all stress events. Separate scores were then obtained for active-passive and problem-emotion focused, and used as the basis for splitting the group of 32 into high and low sub-groups for the two dimensions. Since the two dimensions were correlated for this sample of subjects, a combined basis of classification was adopted for the analysis presented below, selecting extreme sub-groups (n = 6 for each) of individuals who scored high on both + dimensions (active/ problem-focused; A/Pr), and those who were low on both (passive/ emotion-focused; P/Em).

Results.

Figure 1 summarizes the effects of individually-defined periods of high and low workload on subjective strain (tension and fatigue) for the two sub-groups. For simplicity the data for cognitive and emotional workload are combined, and expressed as mental workload, though they show similar patterns. Clearly, both groups show increases in tension on days when demands are high, as we would expect. However, whereas the A/Pr group also shows an effect of workload on fatigue, no such effect is found for P/Em individuals. These different patterns are striking, especially since there are no mean differences between the sub-groups in either the level or variability of work demands. Furthermore, although they respond more to work demands, the active copers show evidence of somewhat better overall mood states.

238

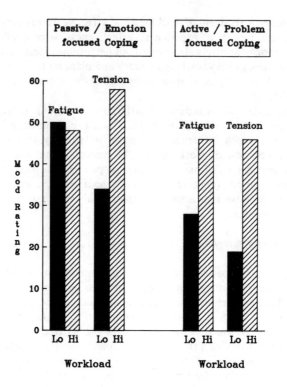

Fig. 1.

General Discussion

This is a necessarily crude preliminary analysis The use of more sophisticated multivariate methods is planned in order to adequately analyse the very large data set obtained from this study. In addition, we do not yet have access to endocrine or performance data, which will provide a more complete picture of these differences between coping types. Nevertheless, there appear to be clear moderating effects of coping styles on the work demands – strain relationship, and this is differentially related to the development of work-related fatigue rather than tension. Such a result fits well with our model of active coping in terms of the sustained use of effort mobilization in order to maintain work goals. Fatigue is the result of increased mental effort. Tension, by contrast, is associated with the appraisal of high levels of demand, rather than the response to demands.

The fatigue incurred by active/problem-focused copers in attempting to overcome the threat to their work goals is predicted to be a short-term cost, and may serve to prevent the build-up of minor problems into much larger ones. For more passive or emotion-focused individuals experienced fatigue is shown to be less directly influenced by the demands of work, though it is equally subject to variability. We may hypothesis that this is due to residual effects of previous work or unresolved difficulties arising outside the workplace. The short-term advantages for personal state of not responding to demands may, therefore, be offset by longer-term strain arising from the unresolved problem or conflict. To investigate such relationships requires daily measurements over a prolonged period, using a design such as that employed in our first (n = 2) study, combined with samples at least as large as that used in our second study, in order to allow for the clear separation of sub-groups. This would permit the analysis of lagged effects of both work and non-work demand variables on strain measures and health indices, for different sub-groups. Such a design is expensive in terms of both time and resources. It would, however, allow a more systematic and precise analysis of the relationship between work demands and strain, through a focus on the role of individual coping strategies.

References

Beatty, J. (1982). Task-evoked pupillary responses, processing load and the structure of processing resources. *Psychological Bulletin*, 91, 267-292.

Caplan, R.D., Cobb, S., French, J.R.P., Van Harrison, R., & Pinneau, S.R. (1975). *Job demands and worker health*. NIOSH Research Report. Washington, DC:USHEW, US Government Printing Office.

Cooper, C.L. and Marshall, J. (1976). Occupational sources of stress: a review of the literature relating to coronary heart disease and mental ill health, *Journal of Occupational Psychology*, 49, 11-28.

Cox, T. (1978). *Stress*. London: Macmillan.

Delongis, A., Folkman, S. and Lazarus, R.S. (1988). The impact of daily stress on health and mood: psychological and social resources as mediators. *Journal of Personality and Social Psychology*, 54, 486-495.

Frankenhaeuser, M. (1979). Psychoneuroendocrine approaches to the study of emotion as related to stress and coping. In: H. E. Howe and R. A. Dienstbier (Eds), *Nebraska Symposium on Motivation*. Lincoln: University of Nebraska Press.

Frankenhaeuser, M. and Johansson, G. (1986). Stress at work: psychobiological and psychosocial aspects. *International Review of Applied Psychology*, 35, 287-299.

Frese, M. (1986). Coping as Moderator and Mediator between Stress at Work and Psychosomatic Complaints. In: M. H. Appley and R. Trumbull (Eds), *Dynamics of Stress*. New York: Plenum Press.

Harrison, V. R. (1978). Person-environment fit and job stress. In: C. L. Cooper and R. L. Payne (Eds), *Stress at Work*. Chichester: Wiley.

Hockey, G. R. J. (1984). Varieties of attentional state: the effects of environment. In: R. Parasuraman and D. R. Davies (Eds), *Varieties of Attention*. New York: Academic Press.

Hockey, G. R. J. (1986). A state control theory of adaptation to stress and individual differences in stress management. In: G. R. J. Hockey, A. W. K. Gaillard and M. G. H. Coles (Eds), *Energetics and Human Information Processing*. Dordrecht: Martinus Nijhoff.

Karasek, R. A. (1979). Job demands, job decision latitude, and mental strain: implications for job redesign. *Administrative Science Quarterly*, 24, 285-308.

Lazarus, R. S. and Folkman, S. (1984). *Stress, Appraisal and Coping*. New York: Springer.

Lazarus, R. S. and Launier, (1978). Stress-related transactions between persons and environment. In: L. A. Pervin and M. Lewis (Eds), *Perspectives in Interactional Psychology*. New York: Plenum Press.

Marshall, S. L. A. (1947). *Men Against Fire*. New York: Morrow.

Mulder, G. (1986). Mental effort and its measurement. In: G. R. J. Hockey, A. W. K. Gaillard and M. G. H. Coles (Eds), *Energetics in Human Information Processing*. Dordrecht: Martinus Nijhoff.

Ursin, H., Baade, E., and Levine, S. (1978). *Psychobiology of Stress a Study of Coping Men*. New York: Academic Press.

ENVIRONMENTAL STRESS: EFFECTS OF AIR POLLUTION ON MOOD, NEUROPSYCHOLOGICAL FUNCTION AND PHYSICAL STATE.

Monika Bullinger, Ph.D.
Institute for Medical Psychology at the University of Munich,
Goethestr. 31, D 8000 Munich 2, FRG.

Air pollution represents a major evironmental concern in industrialized countries. While its health effects have extensively been studied in epidemiological and toxicological investigations (Goldsmith and Friberg, 1978, Lave and Seskin, 1977), research on its psychological consequences is generally underrepresented (Evans and Jacobs, 1981, Evans and Campbell, 1983; Robinson et al., 1983). Even though a recent review of the literature from 1961 to 1988 identified 170 psychologically oriented studies, most of them focus on perceptions of air pollution and annoyance reactions in laboratory and survey research (Bullinger and Schlehaider, 1988). The few studies on psychological well-being pertain mostly to psychiatric symptomatology (Strahilevitz et al., 1979, Rotton and Frey, 1984).

This lack of investigations does not only stem from the priority of mortality and morbidity endpoints in environmental research, but also from the failure to recognize the importance of psychological impairments for mental and physical health. An underlying problem is how to conceptualize relationships between air pollution and psychological variables as well as their basic mechanism (Baum et al., 1982, Stockols, 1983).

A theoretical concept allowing relation of environmental conditions to psychological state comes form stress research (Lazarus and Cohen, 1978, Folkman et al., 1979). Air pollution can constitute a major stressful stimulus to exposed persons and can lead to a variety of emotional, mental and physical changes not only by direct toxic effects e.g. on CNS functions, but also by an indirect cognitive mediation in terms of a negative appraisal of pollutants (Cohen, 1980). This implies a new look at air pollution from a psychobiological perspective in terms of a stressor – appraisal – effect sequence (Baum et al., 1981, Campbell, 1983, Cohen et al., 1986). A corresponding heuristic model specifies air pollution in terms of pollutant – weather constellations as environmental stressor, appraisals as evaluations of air quality (quality ratings and annoyance reactions) and effects as impairments of psychological well-being (eg. mood), neuropsychological function (eg. concentrational ability) and physical state (eg. bodily complaints).

In order to identify such potential effects of air pollution on the basis of a stress-theoretical approach, a naturalistic setting, a longitudinal design and a statistical methodology allowing causal inferences are necessary. The present paper describes two studies, whose general aim was the identification of relationships between ambient pollutants and biopsychosocial variables in healthy residents of a polluted as compared to a non-polluted area (study 1) and in chronic respiratorily ill as compared to chronic, but not respiratorily ill residents of a highly polluted area (study 2). Research questions in each study pertained to

S. Puglisi-Allegra and A. Oliverio (eds.), Psychobiology of Stress, 241–250.
© 1990 *Kluwer Academic Publishers. Printed in the Netherlands.*

1. differences between groups in terms of psychological well- being, neuropsychological function and physical state,
2. relationships between air pollution and the above mentioned variables with special emphasis on the quality, intensity and patterning of these relationships as well as on the role of cognitive processing of environmental stimuli and
3. the characterization of subgroups of pollution sensitive persons.

Specific questions with regard to study 1 included comparisons in terms of pollutant effects on well-being, function and physical state between healthy residents of a polluted as compared to a non polluted area. It was hypothesized that these measures would indicate increased impairment and more pronounced associations to observed pollutants in residents from a polluted area. Likewise it was assumed in study 2, that such impairments and their relationships to air pollution would be more pronounced in chronic respiratorily ill residents from a polluted area as compared to chronically ill residents of the same area, not suffering from respiratory disease.

Both studies represent a new approach to research into human responses to air pollution in terms of a continuous monitoring of observed pollutants and biopsychosocial variables in a field setting over time and in terms of the use of time-series analysis as a statistical tool for the identification of pollutant effects on well-being (temporal cause-effect estimations).

Methods

Design

Both studies followed a prospective quasiexperiemental design over a period of two month with continuous recordings of pollutants and weather variables and parallel assessments of psychological, neuropsychological and somatic variables in residents of differentially air polluted areas in the state of Bavaria, FRG. Using the Bavarian State Environmental Ministeries Pollution Monitoring Network (PMN), structurally and geographically similar areas differing in air pollution levels were identified. The city of Ingolstadt (polluted area) and the city of Schrobenhausen (non-polluted area) were selected for study 1; for study 2, the city of Nurnberg was selected as an area with high pollutant concentrations according to previous annual recordings. Both studies were conducted in autumn/winter periods with an expected high probability of increased air pollution levels.

Study samples

To obtain homogeneous and comparable study populations likely to be exposed to pollutants, nonworking, nonsmoking women between 50 and 70 yers, residing in the area for over 2 years of a distance less than 5 kilometers from a PMN- Station were selected. Additional inclusion criteria for study 1 included good health status, i.e. the absence of impairing acute or chronic diseases and for study 2 the diagnosis of either chronic bronchitis or diabetes mellitus according to WHO-criteria. Eligible subjects for study 1 were screened with the aid of the city administrations of Ingolstadt and Schrobenhausen, each providing a random sample of 50 addresses for subsequent telephone-based selection of volunteers (resident refusal rate was 30%). In each area, 11 residents fulfilled the inclusion criteria and were willing to participate. In study 2, a random sample of 60 Nurnberg general practitioners was approached for recruitment of age-matched pairs of bronchitis and diabetes patients (physician refusal rate was 62%). Of 25 physicians willing to participate, 17 recruited 20 bronchitis and 20 diabetes patients into the study.

Variables and procedures

In both studies, data regarding general environmental perceptions, health status, sociodemographic characteristics and personality traits were collected via interview and questionnaires prior to and after the 2 month period of investigation. During that period, daily ambient concentrations (means of half-hourly recordings over 24-hours) of sulfur dioxide (SO_2), particulate matter (dust), nitrous oxides (NO and NO_2), carbon monoxide (CO), hydrocarbons with and without methane (CnHm and $CnHm-CH_4$) and ozone (O_3) for all areas were recorded from the PMN-stations. In addion, climatic variables (temperature, humidity, air pressure) and indices such as the stagnation index as a measure for inversion situations were obtained daily.

Parallel to the environmental recordings, subjects daily rated their psychological well-being in terms of mood via a standardized adjective checklist (BF-S, von Zerssen, 1976) and in terms of perceived stress via a 100 mm visual analogue scale. Appraisal of air quality was assessed via a 14 adjective polarity profile, annoyance via Likert-type ratings regarding overall pollution, odours, dust, weather and automobile exhaust. Subjects rated their somatic well-being with a standardized complaint checklist (BL, von Zerssen, 1976) in study 1 at weekly intervals, in study 2 daily together with desease specific symptom list and a 100 mm visual analog scale reflecting physical well-being. As measures of stress-related physiological arousal, 24-h urinary concentrations of catecholamines and cortisol as well as blood pressure recordings were obtained at weekly intervals from study 1 subjects. In both studies assessments of neuropsychological function were carried out via weekly tests of concentrational ability (standardized D2 Test, Brickenkamp, 1967) and of reaction times (using a specifically designed optic/accoustic stimulus and response recording equipment).

The studues were conducted with the help of trained assistants who carried out the interviews, introduced the data collection procedures in a trial session and visited the respondents at identical time points during the course of the investigations in order to collect diaries, neuropsychological recordings and physioclogical data.

Statistical analyses

Daily assessments of environmental and psychological data were analysed using time series modelling of the ARIMA-type (Box and Jenkins, 1976). Basis of this analysis is the dependence between consecutive observations in a series and the goal is the estimation of mathematical models underlying this dependence. In univariate ARIMA models, the series actual and future behavior can be described in terms of its autoregressive component (AR: the series is influenced by its past), of moving average components (MA: the series is influenced by random external shocks) and of its non-stationarity (I: an integrated series has to be differentiated to fulfill the stationarity criterion). Additionally, this information can be used to test the difference in the course of two series by including the ARIMA parameters in the T-test equation (ARIMA-adjusted T-test: Keeser, 1978); Bullinger and Keeser, 1985). In multivariate ARIMA modelling, relationships between two and more series can be estimated and the temporal sequence of events can be used to infer statistical causality in one series preceeding and influencing another (Tiao and Box, 1979).

In the studies presented, ARIMA modelling was carried out with the standard statistical BMDP package (Dixon and Brown 1983) and the Wisconsin Multiple Time Series Program (WMTS, Tiao et al., 1981). For data collected at weekly intervals, analyses of variance with repeated mesures and covariates were performed (ÁNOVA program, BMDP Package).

Results

According to the interview results from study 1, residents of polluted areas felt more distress due to environmental conditions than the residents of the non polluted area in study 2; Bronchitis patiens appeared to be especially affected by air pollution. In polluted areas, the air was generally descibed as unpleasant and unhealthy, odours and dirt were noted more often and somatic complaints were attributed to the air quality. In addition, with regard to their respective controls, healthy residents from the polluted area Ingolstadt (study 1) appeared to be more annoyed by air pollution as were the chronically ill residents of Nurnberg (study 2). Thus the general environmental evaluations indicated that air pollution was perceived as a stressor in polluted areas.

During the period of investigation daily means of the ambient pollutant concentrations were low and daily maxima reached less than 50% of official air quality threshold standards in both studies. The pollutants followed simple ARIMA models primarily at early lags. ARIMA – adjusted t – testing for differences between areas in study 1 yielded for the polluted area Ingolstadt (ING) as compared to Schrobenhausen (SCHR) significantly higher daily mean concentrations of SO_2 (mg/m^3: 0.029 ING vs. 0.010 SCHR; p = 0.0006), CO (ppm: 1.028 ING vs. 0.521 SCHR; p = 0.0003) and CnHm -Ch4 (ppm: 0.59 ING vs 0.35 SCH; p = 0.008). Pollutant concentrations for the Nurnberg area in study 2 were higher (eg. SO_2 0.065 mg/m^3), but still below critical levels.

Univariate ARIMA analyses for daily self-ratings (means per day over 57 days for each groups) yielded more complex models and significant differences between groups in terms of mood, perceived stress, bodily symptoms and odour annoyance. Specifically, the Ingolstadt residents from study 1 as compared to the Schrobenhausen controls weree more impaired in terms of mood and perceived stress and they also reported more annoyance to odours. Similar impairments of mood and stress were also found in study 2 for bronchitis patients as compared to diabetes controls; additionally the bronchitis patients had a higher bodily complaint score.

Pollutants and psychological well-being.

Multivariate ARIMA modelling was used to assess the influence of pollutants on psychological variables. Preliminary identification procedures suggested that of all pollutants only SO_2 and dust were associated with emotional effects. Therefore, multivariate time series estimations including SO_2, dust, mood, stress and the stagnation index as a measure for inversion situation were carried out for healthy residents of Ingolstadt and Schrobenhausen in study 1 as well as for residents of Nurnberg suffering from bronchitis and diabetes patiens in study 2. In both studies, complex patterns of variable interrelationships were apparent that consisted of synchronous relationships between variables at lag 0, time-lagged autocorrelations of variables indicating that actual observations are dependent on previous ones and cross correlations reflecting an influence of one variable on another.

In Ingolstadt residents from study 1, a multivariate model was accepted with significant cross-correlation paramaters showing time delayed effects of SO2 on mood (BF-S) at lags 1 and 3. For Schrobenhausen residents, cross-correlations between SO2 on stress at lags 3 and 4 were apparent. In the bronchitis patiens of study 2, multivariate ARIMA- modelling yielded similar effects of SO2 on mood (BF-S) at lag 1 and on stress at lag 2. For the diabetes control no significant cross correlations between pollutants and psychological variables were found. However, SO2 was synchronously associated with stress.

To further explore individual differences in emotional reactivity to pollutants, single-case multivariate ARIMA-models including the variables SO2, dust, mood and stress were calculated for each of the volunteers from both studies. Emotional reactivity was defined as the presence of at least one cross-lagged relationship btween pollutants and emotional state. In 9 of the 11 Ingolstadt respondents from study 1 such cross-correlations were found as compared to 3 of 11 Schrobenhausen residents. In comparison, 10 of 20 bronchitis and 6 of 20 diabetes patients showed such patterns in study 2. Because of the high interindividual variability of ARIMA-models, patterns of reactivity could not be identified. Differences between sensitive as compared to non- sensitive persons in terms of personality, health or sociodemographic factors failed to reach significance.

To explore to role of appraisal, additional multivariate time- series modelling was carried out. Results indicated immediate effects od SO2 and dust on air quality ratings in polluted areas. For the Ingolstadt residents of study 1 and Nurnberg residents of study 2, increases of SO2 were associated with odour annoyance and ratings of air quality as 'unhealthy". Additionally, for the chronically ill Nurnberg residents of study 2, dust was associated with annoyance due to dirt. In both studies, annoyance due to weather conditions was associated with climatic variables.

Physical state and neuropsychological function

For physical state as assessed daily and analysed with multivariate time series analyses in study 2 volunteers, no effects of pollutants on the bodily complaint checklist total score was apparent. However, SO2 elevations at the previous day led to decreases in physical well-being as documented via a 100 mm visual analogue scale in both patient groups. In addition, in 3 of bronchitis patients with respiratory symptoms during the period of investigation effects of SO2 and dust on self-reported respiratory symptoms (e.g. coughing) weree apparent. In study 1, physical state was assessed weekly in terms of a bodily complaint score and measures of physiological arousal (catecholamines, cortisol, blood pressure). As an example of these measurements, catecholamine concentrations for residents from Ingolstadt and Schrobenhausen over measure points are shown in Figure 2.

Repeated measures ANOVA of physical state variables showed no differences between groups nor over time for physical state variables (with the exception of a fluctuation of catecholamines over time ($F = 3.01$, $DF = 7$, $p = 0.045$). Covariate analyses including pollutant and climatic data at the measure points did not yield significant results – moreover, correlations between physical state and pollutants weree low ($r = .01$ to $r = .16$).

Weekly assessments of neuropsychological function, carried out identically in both studies yielded a complex pattern of results. While healthy residents of study 1 did not differ in concentrational ability and reaction times, chronically ill residents from Nurnberg did so. Bronchitis patients had a higher D2 test score than diabetes controls, but longer reaction

246

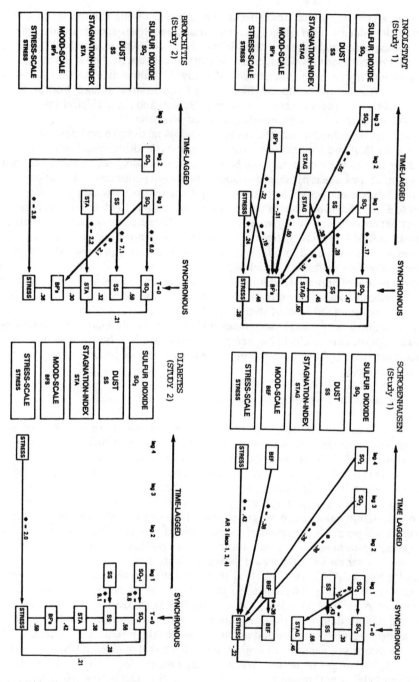

Fig. 1. Multivariate time-series analyses results

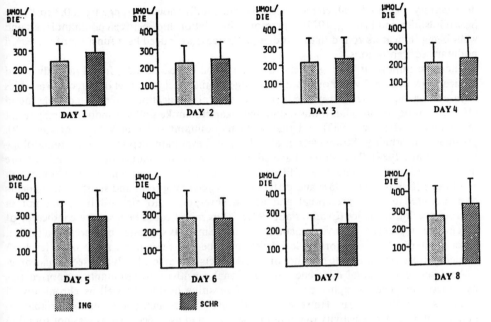

Fig.2. Catecholamine concentrations over 8 weeks from residents of a polluted (ING, n = 11) vs. non-polluted area (SCHR, n = 11)

times to visual and acoustic stimuli. Also, group specific differences over time occured in both studies. These differences over time, however seemed related to pollutant concentrations only in the healthy residents of Ingolstadt in study 1. Here, univariate ANOVA results with pollutants as covariates showed effects especially of SO_2 on concentration (D2-Test) and reaction times especially to choice reaction tasks. No pollutant covariate effects were apparent for the neuropsychological performance of chronically ill subjects from study 2.

Discussion

The investigations presented assessed the emotional, neuropsychological and somatic effects of air pollution with samples of healthy residents of differentially polluted areas (study 1) and differentially chronically ill residents of a polluted area (study 2). For residents of polluted areas in both studies, initial interview results indicated the stressor quality of air pollution. In these areas awareness of pollution and feelings of annoyance were higher than in a non-polluted area. Similar results have been reported in epidemiological studies of environmental perceptions and attitutes (Downing et al., 1971).

The low pollutant concentrations in both studies during the period of investigation reflect a basic problem of quasi- experimental research: even with careful selections of seasons and study sites, pollutant levels may fail to show high concentrations during the period of investigation. In the present studies, however, peak concentrations came close to odour detectability thresholds, which have been experimentally shown to range from 0.3 to 1 ppm SO2 (Golsmith and Friberg, 1977). Therefore, the relationships between pollutant levels and well- being indicators found in the studies presented may indeed be a function of increased pollutant concentrations.

The most important finding is related to the mood effects of SO2, found in study 1 and replicated in study 2. Even though studies using a similar design and set of outcome variables have not yet been described in the literature, several publications point at possible mood effects of air pollution. Thus it was found that cigarette smoke-polluted indoor air may result in aggression (Rotton, 1983) and that ambient pollutant levels may be associated with psychiatric morbidity (Strahilevitz et al., 1975) and especially depressive symptomatology (Jacobs et al., 1984). Two studies have addressed the stressor nature of air pollution more directly and found an interaction of life events and perceived distress under acute ozone peak conditions (Evans et al., 1987) and differences between long-term and short-term residents of a polluted area in preceptual and affective responses to air pollution, but not in neuroendocrine variables (Evans et al., 1982). Thus mood could be a sensitive indicator of psychological sequelae of environmental stressors. Since, as suggested by Burchfield (1979), chronic intermittend stressors at low levels may be as powerful as acute peak levels in inducing stress-reactions, the presence of mood effects also at low levels also seems plausible.

Analyses of individual differences in emotional sensitivity to pollutants supported this hypotheses in yielding higher proportions of sensitive healthy as well as chronically ill residents of polluted areas. However, an influence of personality and coping variables to patterns of affective sensitivity to pollutants as suggested by a recent study was not found in the research presented (Navarro et al., 1987).

Evidence for the role of appraisal in mediating pollutant effects was found for odour annoyance reactions in both studies with annoyance reactions being influenced by previous day SO2 measures. Such correspondence between annoyance and observed pollutats has been found in epidemiological and experimental research, indicating that the experience of impaired air quality is dependent on observed pollutant levels (Koelaga, 1987). In the studies presented, a weak association between pollutants and neuropsychological impairment was only found in healthy but not in chronically ill volunteers. Since ill volunteers had lower performance scores than the healthy volunteers, this finding could reflect a pre-existing impairment of functioning as a result of illness which seems not to be affected by environmental stimuli. The weak relationships between performance and pollutant in healthy volunteers is probably due to the low level of pollution. In experimental studies mainly of CO and O_3, effects on cognitive and psychomotor function occured with markedly higher pollutant concentrations (Breisacher, 1984). However, since neuropsychological testing has not been widely used in field settsing it is not yet clear to what extent functional impairment – as an indicator of stress-reaction – is a probable consequence of air pollution.

With regard to the physical findings in study 2, the epidemiological literature has reported increase of respiratory symptomatology under the presence of low pollutant levels (Whittemore, 1981). Using time series analyses, relationships between mortality (Hechter and Golsmith, 1961), hospital admissions for respiratiry disease (Bates and Szito, 1983) and pulmonary morbidity (Lawther and Waller, 1970; Lan and Sky, 1981) have been found. The finding of only a small proportion of bronchitis-patients having respiratory dysfunction with pollutants parallels findings of interindividual variability in pollution induced respiratory symptomatology (Sky and Muller, 1980). Analysis of physiological arousal in study 1 did not indicate effects of pollutants. Since arousal data might be influenced by a variety of factors in field research, this finding is not suprising. In experimental studies, only stressors with markedly higher potency than the environmental conditions investigated in the present studies induced neuroendocrine and psychophysiological changes (Frankenhauser, 1980). Thus it is unlikely that arousal effects would ocur at the pollutant concentrations measured. The influence of air pollution on physiological precursors od psychosomatic symptoms and disease is yet to be demonstrated (Wortis, 1971).

In summary the results of the investigations suggest a potential stressor nature of air pollution. While evidence for pollutant effects on neuropsychological functioning and physical state still remains questionable, the hypothesis of mood – impairments with air pollution is supported. Since emotional effects seem to be present already at low pollutant concentrations and may be associated with longterm physical dysfunction, mood changes should be taken in to account when assessing environmental health.

References

Bates D.V., Szito, R. (1983): 'Relationships between air pollutant levels and hospital admissions in southern Ontario". *Canadian Journal of Public Health* 74, 117-123.

Baum, A., Singer J.E., Baum C.S. (1981): 'Stress and the environment". *Journal of Social Issues*, 37, 4-35.

Baum A., Deckel A.W., Gatchel, R.J. (1982): 'Environmental stress and health: is there a relationships?' In: Sanders G.S. and Suls I. (eds): Social psychology of health and illness. Hilsdale N.J.: Lawrence Erlbaum As soc. Publ.

Box G.E.P., Jenkins G.M. (1976): 'Time series analysis: forecasting and Control.' San Francisco, Holden Day.

Breisacher P. (1971): 'Neuropsychological effects of air pollution.' *American Behavioral Scientist* 14, 837-864.

Brickenkamp R. (1967): 'Der Aufmerksamkeits-Belastungstest D2.' Stuttgart, Hogrefe.

Bullinger M., Keeser W. (1985): 'Befindlichkeitsverlaufe unter Luft-schadstoffeinfluss in unterschiedlich umweltbelasteten Gebieten – ein zeitreihenanalytischer Ansatz.' In Appelt H., Strauss B. (eds): Er gebnisse einzelfallstatisticher Untersuchungen. Heidelberg: Springer.

Bullinger, M., Schlehaider, G. (1988): 'Psychologische Effekte der Luftverun reinigung. Submitted to *Umweltmedizin*.

Burchfield S.R. (1979): 'The stress response: a new perspective.' *Psychosomatic Medicine*, 41, 661-672.

Campbell J.M. (1983): 'Ambient stressors.' *Environment and Behaviour* 15, 355-380.

Cohen S. (1980): 'Cognitive processes as determinants of environmental stress.' In: Sarason I and Spielberger C. (eds): Stress and Anxiety Vol 7. Washington DC: Hemisphere Press.

Cohen S., Evans G.W., Stokols D., Krantz D.S. (1986): 'Behavior, health and environmental stress.' New York, Plenum Press.

Dixon W.J., Brown M.B. (1983): 'BMDP Statistical Software 1983.' Berkeley CA: University of California Press.

Dowing, P.B. (1971): 'Air pollution and the social sciences.' New York, Praeger Publishers.

Evans G.W., Campbell I.M. (1983): 'Psychological perspectives on air pollution and health.' *Basic and Applied Social Psychology*, 4, 137-171.

Evans G.W., Jacobs S.V. (1981): 'Air pollution and human behavior.' *Journal of Social Issues* 37, 95-125.

Evans G.W., Jacobs S.V., Frager N.B. (1982): 'Human adaption to smog.' *Journal of the Air Pollution Control Association* 32, 1054- 1057.

Evans G.W., Jacobs S.V., Dooley D., Catalano R. (1987): 'The interaction of stressfull life events and chronic strains on community mental health.' *American Journal of Community Psychology* 15:23-34.

Folkman S., Schaefer C., Lazarus R.S. (1979): 'Cognitive Processes as mediators of stress and coping.' In: Hamilton V., Warburton: Human stress and cognition – an information processing approach. London, Wiley.

Frankenhacuser M. (1980): 'Psychoneuroendocrine approaches to the study of stressful person-environment transactions.' In: Selye H. (ed): Selye's Guide to stress Research. New York: Van Nostrand Reinhold.

Goldsmith J.D., Friberg L. (1977): 'Effects of air pollution on human health.' In: Stern A.C. (ed) Air Pollution 3. New York: Academic Press.

Hechter H.H. and Goldsmith J.R. (1961): 'Application of time series analysis to acute air pollution reactions.' *American Journal of Medical Science* 241, 581-588.

Jacobs S.V., Evans G.W., Catalano R., Dooley D. (1984): 'Air pollution and depressive symptomatology: Exploratory analyses of intervening psychosocial factos.' In: Population and Environment 7, 260-272.

Keeser W. (1978): 'Dynamic regression with autocorrelated time series.' Paper presented at10th Meeting of the Psychometric Society, Uppsala.

Koelega H.S. (1987): 'Environmental annoyance: characterization, measurement and control.' Amsterdam, Elsevier Science Publishers.

Lan S.P., Sky C. (1981): 'Effects of air pollution on chronic respiratory disease in the New York City metropolitan area 1972.' *Environmental Health Perspectives* 42: 202-214.

Lawther P.J., Waller R.E., Henderson M. (1970): 'Air pollution and exacerbation of bronchitis.' *Thorax* 25, 525-534.

Lave L., Seskin F. (1977): 'Air pollution and human health.' Baltimore: Johns Hopkins University Press.

Lazarus R.S. and Cohen I. (1978): 'Environmental Stress.' In: Altman I., Wohlwill I. (eds): Human Behavior and Environment 2. New York: Plenum.

Navarro P.L., Simpson-Housley P., de Man A.F. (1987): 'Anxiety, locus of control and appraisal of air pollution.' *Perceptual and Motor Skills* 64, 811-814.

Robinson J.D., Higgins M.D., Bolyard P.K. (1983): 'Assessing environmental impacts on health: a role for behavioral science.' *Environmental Impact Assessment Review* 4, 41-53.

Rotton J. (1983): 'Affective and cognitive consequences of malodorous pollution.' *Basis and Applied Social Psychology* 4, 171-193.

Rotton J., Frey J. (1984): 'Psychological costs of air pollution: Atmospheric conditions, seasonal trends, and psychiatric emergencies.' *Population and Environment: Behavioral and Social Issues* 7, 3-16.

Shy C.M., Muller K.E. (1980): 'Evaluating the effects of air pollution on sensitive subjectivs.' *American Journal of Public Health* 70, 680-681.

Stokols D. (1983): 'Theoretical directions of environment and behavior research: introduction.' *Environment and Behavior* 15, 259-272.

Strahilevitz M., Strahilewitz A., Miller I.E. (1979): 'Air pollutants and the admission rate of psychiatric patients.' *American Journal of Psychiatry* 136, 205-207.

Tiao G.C., Box G.E.P. (1971): 'Modelling multiple time-series with applications.' *Journal of the American Statistical Association* 76, 802-816.

Tiao G.C., Box G.E.P., Grupe M.R., Hudak G.B., Bell W.R., Chang I. (1979): 'The Wisconsin Multiple Time Series Program – a preliminary guide.' Madison: Department of Statistics, University of Wisconsin.

Whittemore A.S. (1981): 'Air pollution and respiratory disease.' Annual Review of Public Health 2, 397-429.

Wortis I. (1971): 'Pollution, somatopsychics and disease.' *Biological Psychiatry* 3, 103-104.

Zerssen D.v. (1976): 'Klinische Selbstbeurteilungsskalen: Die Befindlichkeitsskala (Bf-S) und die Beschwerdeliste (BL).' Weinheim: Belts Verlag.

Index